Forestry Commission Bulletin 65

Advances in Practical Arboriculture

Edited by D. Patch,
Forestry Commission

*Proceedings of a Seminar held at
the University of York, 10–12 April 1985.
Forestry Commission and Arboricultural Association*

LONDON : HER MAJESTY'S STATIONERY OFFICE

First published 1987

ISBN 0 11 710203 2

ODC 270: 946.2

Published in 1987:
European Year of the Environment

Enquiries relating to this publication should be addressed to the Technical Publications Officer, Forestry Commission Research Station, Alice Holt Lodge, Wrecclesham, Farnham, Surrey, GU10 4LH

Contents

Protection

Preface

In the second half of the 19th Century there were two Arboricultural Societies in Great Britain. These eventually became the Royal Forestry Societies (England, Wales and Northern Ireland, and Scotland) of today. However arboriculture did not become acknowledged as a profession until the second half of the 20th Century. In 1964 two organisations the Arboricultural Association and the Association of British Tree Surgeons and Arborists, were formed with the aim of bringing together people of common interests and improving standards in the industry. Later the two organisations amalgamated under the name of the Arboricultural Association.

Since their formation both the Arboricultural Association and the Association of British Tree Surgeons and Arborists stimulated development of the industry, initially by prompting the establishment of educational and training courses. In the 1970s attention turned to the need for research. Professor F.T. Last, chairing an *ad hoc* committee convened after the 1972 National Arboricultural Conference, assessed the state of knowledge and prepared a report titled *Arboriculture: research and advisory needs.* The Department of the Environment accepted some of the recommendations and has been funding arboriculture research since 1974.

In his report, Professor Last gave prominence to the range of relevant research already in hand but about which arboriculturists were unaware. One of the major recommendations of the report was, therefore, that an advisory and information service should be established to communicate results to the industry.

A forum for research results was provided by a seminar organised in 1980 jointly by the Arboricultural Association and the Forestry Commission. This seminar was held at Lancashire College of Agriculture and the papers were published as Forestry Commission Occasional Paper 10 *Research for practical arboriculture* which unfortunately is now out of print. In 1985 the industry was updated on developments in research during another seminar organised jointly by the Arboricultural Association and the Forestry Commission and held at the University of York. About 350 people attended this seminar representing the largest gathering of tree interests ever held in Britain.

Now publication of the papers presented to this second seminar provides arboriculturists and those in allied industries with a useful record of the research results and should assist in applying the results in the field. As research progresses additional information will emerge and new topics will be investigated. Results from this work which will update the following papers will be published as Arboriculture Research Notes* and in the scientific and practical journals.

D. Patch

October 1986

*Available from the Arboricultural Advisory & Information Service, Forest Research Station, Alice Holt Lodge, Wrecclesham, Farnham, Surrey, GU10 4LH.

Introduction

J.C. Peters
Department of the Environment

The Department of the Environment (DOE) has been supporting the Arboriculture Research Programme at the Forestry Commission (which supports the Arboriculture Advisory and Information Service) for 9 years now. There have been some cuts in DOE funding during this period because the programme of research was the first to be the subject of a new system of Chief Scientist reviews within the DOE. These reviews were set up to examine the relevance of DOE participation in any subject receiving DOE funds. This audit system looks at the standard of the research in terms of its value for money and also considers how further improvements can be made either in dissemination of results or the inquiry line of the research itself. The audit also placed arboriculture in the forefront of the minds of high level management in the Department and has resulted in an acknowledgement of the importance of core research at the Forestry Commission and the value of the Arboriculture Advisory and Information Service.

Also clearly indicated by the audit were the many other interested parties in arboriculture, ranging from local authorities through the contracting industry to land-owners and gardeners, and a recommendation was made that DOE should not accept responsibility for more than a core of research of a strategic and applied nature. Our funding of the Forestry Commission has been running at approximately £150 000 annually. This core can be considered to be one on to which other work should be grafted. DOE hope this will be recognised and I, therefore, appeal to all readers to consider not just the research needs and their application but also to think positively about the machinery which could be invoked to ensure that funds are made available in a co-ordinated fashion – the very nature of arboriculture, the long lead times for results – calls for more careful assessment of objectives and linkages at an earlier stage than almost any other line of biological research.

I am mindful of the high standing of both sponsors in this area of work but nevertheless I think they would be the first to agree that we know little about the peculiar conditions in which trees are expected to grow when they are planted for arboricultural rather than forestry reasons. Not only that but the range of species employed, together with the genetic stock from which any one species has been taken, can make for great variation in success which requires careful analysis. There is still a tendency to continue with reliance on trial and error and empirical methods.

The very nature of the papers that are presented in these Proceedings indicates that a wide network of guidance can now begin to be identified for the practitioner on the ground.

The Present Atmosphere

The developing emphasis in rural areas is to move away from high intensity agricultural production of foods that add to European mountains towards many products of which both we in the UK and the European Community as a whole have in short supply. One of these is timber and this is resulting in increasing pressure on farmers to utilise areas of land which can be set aside as being marginal and on which small woods could flourish. Such areas fall between the traditional province of forester and that of the arboriculturist. Although their location may be in marginal areas in terms of normal agricultural production these sites – particularly in the lowlands – may provide very superior growing conditions for trees when compared with traditional forestry locations. It is in such situations that the experience of the arboriculturist can play an important role. In order to play this part the arboriculturist needs a predictive ability in terms of the growth rates that can be expected, the ongoing management which will be required, the timescales involved; as well as the practical advice that needs to be given in choosing the location and the handling of the immature plant. It matters very much whether an individual tree will outgrow another planted only a few yards away under apparently very similar conditions.

Predictive Ability

I should like to dwell for a moment on the real nature of our ability to predict growth potential and the management required to attain it. Many of us have seen the experiments set out at the Forestry Commission's Research Station at Alice Holt Lodge indicating what can be done in terms of improved establishment and growth if a single technique of mulching or other control of competitive species is carried out around trees in a plot. Nevertheless, each plot of 16 or so trees has several doing far less well than the others – even dead. There could be many reasons for such individual failures. We know too little of the variables involved to be certain whether it is a local soil difference or perhaps some physiological variant of the species produced from its genetic origin. The use of statistical methods to compare the effects of management through a well structured experiment such as this allows us to make general recommendations on management techniques which can be employed to allow a high rate of survival and growth in groups of trees, but it does not assist us with individuals where the single tree is really important.

It is often the individual authorised felling of a tree that has been protected by a Tree Preservation Order and the likely success of a replacement planting which can affect the willingness of authorities to replace our ageing trees. In such circumstances we are far away from a scientific solution – we can improve the chances of success but we still rely greatly on the experience of the individual planting the replacement. The experience of such individuals is highly valued. We must ask what it is that they see and feel about a location and how can we better translate this into scientific terms for use elsewhere.

Having said all this many may wonder in this practical world in which we live how, when we have no time or money to monitor individual situations, such improvements can be made in an atmosphere of the need to plant and move on. Perhaps, therefore, we need to examine the success rate as it has been stated.

Background Costs

Plant survival

Actual figures on plant survival are extremely difficult to determine because surveys are not done and practitioners are often loathe to admit their lack of success. However, after the planting drive in 1973 it was estimated that 50 per cent of the trees were dead after the first growing season and that this had risen to probably 70 per cent within 5 years. More recently figures of 30–50 per cent failure are being mooted. For a number of years much of the research funded by DOE and other agencies has been investigating aspects of plant failure.

The reasons for high losses are complex but it is accepted that areas of high risk are:

1. Planting stock quality – plant physiological conditions
   root:shoot ratio
   root desiccation
   plant size
   species selection.
2. Planting site factors – disturbed, denatured soils
   poor water relations of the soil
   competition from herbaceous vegetation
   nutritional status of the soil
   predation

Plant production and quality form the subject of the first five papers in these Proceedings. The problems of site factors are explored in five other papers.

Whenever and wherever trees and shrubs are planted there must be a long term commitment to aftercare and management if there is to be survival to full maturity in the landscape. In the past this has often been lacking in planting schemes because of the method of funding which has often relied on capital funds for the planting, while aftercare is a new item for which no provision was made. This sort of situation is not unique to arboriculture and it is for this reason that the Department of the Environment have funded a research project with the Groundwork Trust in the St Helens area to examine the ongoing costs of various schemes for landscape and nature conservation improvement. This research programme has already produced a large database on the costs of ongoing management.

An illustration of the extent of the management that may be required following a planting scheme is provided by the Department of Transport plantings. These total some one million trees and shrubs per year at an average cost of one million pounds. There is an estimated budget of approximately a further one million for aftercare and maintenance work of plantings during each financial year.

Manpower

On the subject of costs it is worth reminding ourselves that the questionnaire completed by 113 Local Authorities in 1982 for the Arboriculture Advisory Service at Alice Holt Lodge produced returns indicating that approximately 5200 man years were spent on arboriculture – that is at all levels throughout all Local Authorities. It seems reasonable to assume that private sector and other public bodies employ a similar number. However, the amount of time private sector people spend on arboriculture may vary from year to year, especially with the move to increasing privatisation and reduction in the public sector manpower; however, the number of people employed is likely to be fairly constant even though the emphasis of who is the employer may change. Therefore, a total workforce of 10 000 man years for all operations may be regarded as the annual total for the arboricultural industry. If one assumes that the wage earning capacity of this workforce could be based on an average wage of £8000 per annum the resulting work bill is £80 million per annum. To this should be added the hardware items of tools, transport and general equipment which could represent a similar figure.

Value of the product

Nursery production of trees and shrubs is primarily a commercial operation done by specialist growers who sell the produce to the arboriculture industry. The Ministry of Agriculture, Fisheries and Food maintains records which show approximately 3000 hectares of tree and shrub nursery field production in Britain (1984 estimate) and 450 hectares of container growing area in Britain. The estimate of sales production from these nurseries in 1983/84 was in the region of £74 million (Gilbert, 1987).

It is difficult to determine the proportion of this material that is purchased by the public sector. As mentioned earlier the Department of Transport uses approximately 1 million trees and shrubs each year on motorway and trunk road planting schemes. Returns made by the 113 Local Authorities in 1982 indicated that £1.75 million was spent on tree and shrub purchases. If this is extrapolated to represent the Local Authority buying power throughout Britain it suggests a likely demand of £8-£9 million per annum. To this must be added a similar figure for ancillary materials, transport and labour.

Another indication of the size of the tree planting programme is the grant aid paid by the Countryside Commission. This has increased steadily from £25 500 in 1974/75 to £1.5 million in 1983/84, this is 18 per cent of the Countryside Commission's total budget. Care must be taken not to regard this as an addition to the Local Authorities' spending because many Local Authorities act as Agents for the Countryside Commission. Some Authorities then use the grant to purchase plants and do the work for landowners while other Authorities pay owners to buy plants and do the work themselves. Therefore, some of the returns from local Authorities may include some of this Countryside Commission expenditure.

Mature tree problems

Unfortunately the data from the Forestry Commission's Census of Woodlands and Trees does not give information of the relative status of trees in urban and rural situations. As a result it is not possible to assess the need for planting or the extent of the over mature and senile tree population in towns. However, research has shown that in addition to poor returns from investment in tree establishment there are many problems which reduce the useful life of the established tree. Of particular concern are the pests and diseases, the effects and risks of which have been adequately emphasised by Dutch elm disease. It is obvious even in the area of the taxonomy of fungal species that may infect trees that we have a long way to go. There seems to be little possibility that this will be improved overnight because of the difficulties of identifying the material that may be causing the problem.

Man's actions also effectively reduce the useful life of trees and pruning, which is essential for ensuring that man can continue to live comfortably in close proximity to large trees, will continue to be a necessary operation. The exposure of the wood creates points of entry for wound pathogens which can influence safety. We know that research has thrown doubt on the widespread practice of treating wounds with proprietary tree paints; suggested alternative techniques for minimising the effects of pathogens on trees are covered in papers which follow.

In the past man has also failed to appreciate the adverse effects of tree roots on structures. Research is in hand to create an understanding of the relationship between trees, soil and buildings and to formulate reasonable guidelines so that tree planting does not suffer from sterile 'no go' areas in the neighbourhood of buildings.

Tree Distribution

Imperial College Centre for Environmental Technology has indicated the effect of fragmentation of woodlands and the size and frequency of the occurrence of such fragments on the occurrence of species of birds. We know very little about how such patterns may be modified by the presence of single trees or groups in the countryside or even the tree species. Although policy statements can be made in terms of the need to prevent undue loss of broadleaved trees and in particular ancient woodlands in both landscape and nature conservation terms we are far from understanding the best patterns of tree distribution or the species needed to provide for different uses.

Databases

The Department of the Environment have set up a Review Committee under the Chairmanship of Lord Chorley to examine the uses of Geographic Information Systems.

This seems to be an opportunity for arboriculture to indicate a user interest. Linkages between databases ranging from soils and climate to known species distributions could provide one more important tool in determining the type of trees to be planted in any area. Similarly, the Department of the Environment in conjunction with the Countryside Commission has been funding an examination of the distribution in change in landscape features over the past 30 years. This work conducted jointly between Hunting Survey and Consultants and the Forestry Commission is examining linear features such as hedgerows as well as woodlands. The studies are making use of a combination of ground survey, aerial photography and satellite data. In the latter case although the levels of resolution are such that individual trees cannot be seen the new Landsat Thematic Mapper does indicate shelterbelts and lines of trees. The Department are conducting further research with the National Remote Sensing Centre into methods of describing these patterns of distribution in the countryside so that analyses of the type that I have already described in relation to birds could be conducted.

It is exciting that Imperial College claim that their analysis enables them to define with increasing precision the necessary requirements for survival of bird species. It should soon be possible to define in terms of size and configuration patches of habitat of so many hectares which will result in an increased probability of the presence of a particular species. Strategic thinking on such management matters to develop new technologies of joint planting and management could strengthen the understanding of what species should go where for the arboriculturist. This may seem speculative at this time but I consider that this is the type of strategic thinking which is required for the future as our ability to analyse different data sets in association with one another increases in efficiency.

Advice

Finally, we are all individuals in our appreciation of a beautiful tree or landscape and as individuals we dictate what we wish to see. In attaining these requirements we all need advice and that advice should be as uniform and as knowledgeable as possible. Other fields of endeavour such as agriculture are beginning to run into problems of conflicting advice in the present atmosphere of change. This conference should consider how the approach to arboricultural advice can be best maintained and improved upon.

Acknowledgement

I am very grateful to Mr D. Patch of the DOE Arboriculture Advisory and Information Service (Forest Research Station, Alice Holt) for the data he has provided.

Reference

GILBERT, D.G. (1987). Research and development work in ADAS on hardy ornamental nursery stock. In, *Advances in practical arboriculture*, ed. D. Patch. Forestry Commission Bulletin 65, 11–15. HMSO, London.

Research and Development Work in ADAS on Hardy Ornamental Nursery Stock

D.G. Gilbert

Ministry of Agriculture, Fisheries and Food, Cambridge

Summary

Agricultural Development and Advisory Service (ADAS) investigations on hardy ornamental nursery stock are concerned with experimental development work rather than research. Developing ideas and new knowledge or techniques to the stage for commercial adoption is the main aim. Ideas to be developed may originate from research findings, literature, extension staff, growers or elsewhere. The current programme of work is outlined.

Introduction

The Frascati definitions of Research and Development (OECD, 1980) cover three activities.

1. Basic research is experimental or theoretical work undertaken primarily to acquire new knowledge of the underlying foundation of phenomena and observable facts without any particular application or use in view.
2. Applied research is also original investigation undertaken in order to acquire new knowledge. It is, however, directed primarily towards a practical aim or objective.
3. Experimental development is systematic work, drawing on existing knowledge gained from research and/or practical experience, that is directed to producing new materials, products or devices, to installing new processes, systems and services, or to improving substantially those already produced or installed.

Horiculturists in the Agriculture Service of ADAS are involved in R&D at experimental centres and in regions with almost all this work being experimental development activity.

Experimental centres were established as an integral part of the service to provide advisers in the field, and through them the horticultural industry, with new technical, cultural and economic information which had been fully tested under controlled conditions. They enable the investigation of problems encountered in the course of advisory work and provide soundly based technical information leading to the development of new production methods for adoption by the industry.

The overall programme of work undertaken is designed to meet the needs of the horticultural industry for new information on all aspects of production, both to improve its productivity and quality of its produce. The aims are to test, compare and evaluate, under a range of different conditions, research findings and new ideas coming from many sources and to develop from these improved cultural systems and production techniques.

Most of the experimental development work on hardy ornamental nursery stock is carried out at Efford EHS, Lymington, Hampshire and Luddington EHS, Stratford, Warwickshire, but a few topics are investigated in the regions and may be conducted on commercial holdings. The major part of the experimental resources is allocated to a nationally co-ordinated programme to meet identified requirements and priorities. This is complemented by the investigation of local problems, as the need arises and as far as facilities allow. Studies also include the evaluation and development of buildings, machinery and other equipment where these contribute to improved production methods.

Work in progress is continually reviewed and priorities for new lines of work are determined to ensure that the available resources are concentrated on the highest priority requirements.

Regular consultation between ADAS experimental staff, extension workers, science service specialists, research workers and growers is a feature of general organisation.

At the experimental centres the Station Director is assisted by a Station Advisory Committee which includes growers with a wide experience of hardy ornamental nursery stock. They provide guidance on the industry's problems and needs and are active in advising on the promotion of new developments into commercial practice. The success of experimental development programmes could be judged by the products of investigations which are adopted by commercial producers.

Dissemination of Results

The primary channel for communicating experimental results to the industry is through ADAS advisory staff who are in day to day contact with producers in the course of their work. ADAS advisory staff keep in close touch with the work at Efford and Luddington EHSs and have access to the latest information on new developments and techniques.

The stations arrange Open Days to show work in progress and produce an Annual Review in which topics reaching the stage for adoption are reviewed. The horticultural press carries numerous reports and articles summarising ADAS work. In addition, information is disseminated through grower groups, conferences, radio, television and ADAS leaflets.

All experimental development work on hardy ornamental stock is briefly reported, using the headings 'object', 'treatments' and 'results' in the MAFF/ADAS Reference Book 235, ADAS Research and Development Reports, Agriculture Service, *Hardy ornamental nursery stock*. A full report on each experiment is available on request to the station or region named.

At Efford EHS the main emphasis is on container growing of shrubs including the commercially important ornamental conifers, deciduous and evergreen shrubs. Container growing has been the growth sector in the industry during the last 10 years and present output of 76 million plants is valued around £40 million at farm gate prices. Significant completed work at Efford EHS has included composts, nutrition, herbicides, development of Efford sandbeds and energy saving during propagation. More recently, cheap propagation of common shrubs from cuttings using low polythene tunnels (sun frames) and subsequent field growing has been included in the programme.

The main emphasis at Luddington EHS is on ornamental tree production and rose bush growing. Use of understocks, budding and grafting are important

Table 1. Hardy ornamental nursery stock production in England and Wales

(i) Total area; trends 1976–1984 (hectares – field)

	1976	1977	1978	1979	1980	1981	1982	1983	1984
Roses	1071	1032	1006	987	1049	1077	1102	1037	996
Shrubs	1508	1333	1294	1449	1485	1488	1480	1524	1692
Trees	1578	1557	1457	1464	1496	1459	1338	1418	1434
Herbaceous	247	266	256	241	250	258	274	258	268
Others	777	758	800	852	841	830	948	813	859
Total	5181	4946	4813	4993	5121	5162	5142	5050	5249

(ii) Container grown plants

(a) Number of container plants (millions)

1976	1977	1978	1979	1980	1981	1982	1983	1984
26.8	31.0	35.1	46.7	56.4	59.4	59.9	68.7	76.2

(b) Container growing area (hectares)

1976	1977	1978	1979	1980	1981	1982	1983	1984
158	182	206	275	332	349	353	404	448

components in the production cycle for these crops. Collaboration with East Malling Research Station and the Forestry Commission is particularly relevant to this work. Production of ornamental *Prunus* and *Malus* can benefit from the progress made in the fruit sector. The use of chip budding, EMLA quality understocks and scions, hardwood cutting technology and work on growth regulators are examples of research emanating from East Malling Research Station promoted through the experimental programme.

In the late 1960s and early 1970s Luddington EHS co-operated with Dr Blundell at Bangor University to perfect breaking dormancy and achieving high germination of *Rosa dumetorum* 'Laxa' seed for use as rose understocks. At that time around 60 million rose understocks were imported. UK now produces 60 per cent of the requirements for the home industry.

Ornamental tree production, with the exception of *Prunus, Malus* and *Pyrus* relies heavily on the use of seedling understocks. Seedbed technology to produce understocks of, for example, *Sorbus, Betula, Laburnum,* links work at Luddington EHS with investigations of the Forestry Commission. More recently, attempts are being made to examine the scope for developing vegetatively propagated clones of *Sorbus, Fraxinus, Acer, Crataegus, Laburnum* and *Tilia*. Yields and uniformity of ornamental trees may be improved by using clean vegetatively propagated clones rather than seedling rootstocks.

Table 2. Trade in hardy ornamental nursery stock, 1983

IMPORTS

Code No.	Subject	Numbers	Values £
6100	Roses – less than 10 mm – unworked	14 929 915	626 363
6500	Roses – above 10 mm – unworked	2 421 531	504 356
6800	Roses – budded or grafted	2 782 868	545 633
5800	Rhododendrons (Azaleas) – others	1 003 282	1 062 424
7800	Trees, shrubs and bushes – forest trees	22 386 898	2 303 391
8100	Trees, shrubs and bushes – rooted cuttings/young plants	6 435 819	910 632
8300	Trees, shrubs and bushes – other	42 647 882	14 894 172
9200	Perennial plants	12 954 887	1 440 861
9300	Other outdoor plants – NES	6 832 859	735 919
	IMPORTS – TOTAL	112 395 941	£23 023 753

EXPORTS

Code No.	Subject	Numbers	Values £
6100	Roses – less than 10 mm – unworked	789 350	72 131
6500	Roses – above 10 mm – unworked	741 279	172 851
6800	Roses – budded or grafted	61 058	22 024
5800	Rhododendron (Azaleas)	6 964	14 059
7800	Trees, shrubs and bushes – forest trees	1 947 259	196 839
8100	Trees, shrubs and bushes – rooted cuttings/young plants	463 596	205 390
8300	Trees, shrubs and bushes – others	329 773	157 984
9200	Perennial plants	179 107	27 960
9300	Other outdoor plants – NES	14 885	562 443
	EXPORTS – TOTAL	4 533 271	£1 431 681

ADAS
Hardy Ornamental Nursery Stock Experimental Programme

Propagation

Ornamental trees, pre-etiolation of *Crataegus* and *Tilia*
Ornamental trees, propagation from summer cuttings
Ornamental trees from seed, seedbed herbicide trial
Ornamental trees from seed, seedbed treatments to improve establishment
Hardwood cuttings, rooting of rootstock material
Hardwood cuttings, rooting of ornamental tree cultivars
Hardwood cuttings, factors affecting field establishment
Hardwood cuttings, production of clonal alder windbreak trees
Weaning of micropropagules
The effect of etiolation on rooting lilac cuttings
Comparison of mist and contact film for propagation under low polythene tunnels outdoors
Plant densities in low polythene tunnel propagation outdoors
Comparison of slow release fertilisers and cutting densities for summer propagation in low polythene tunnels
Comparison of slow release fertilisers in rooting media
Comparison of hormone treatments for rooting winter-struck cuttings
Comparison of hormone treatment for rooting Rhododendrons
Comparison of sheet and cable heat sources
Comparison of bed heating systems
Comparison of thermal covers under double and single clad tunnels

Open ground nursery stock

Ornamental *Malus* selections, 1979–83
Ornamental *Malus* selections, 1980–83
Ornamental *Malus* selections, 1983
Ornamental trees, comparison of *Crataegus* rootstocks
Ornamental cherries, a comparison of growth on different rootstocks
Rootstocks for ornamental cherries
Comparison of planting methods
Nursery stock, herbicide screening trial
Weed control in field-grown nursery stock
Nursery stock, control of simazine-resistant groundsel
Biological control of *Amillaria mellea*
Horse chestnut, chemical control of leaf blotch
The effect of nematode control treatments on growth of field-grown nursery stock

Container-grown nursery stock

Comparison of slow release fertilisers
Compost media and slow release fertilisers for container-grown heathers
Shrubs, nutrition of container plants
Comparison of proportions of bark, fertiliser and systems of production
Rate and type of nitrogen required with different proportions of Cambark
Rate of lime required with mixes containing Cambark
Use of worm cast organic waste as a supplement to peat
Use of resin-coated straw as a compost supplement
Comparison of wetting agents and superabsorbent polymers in peat-based composts
Effects of using nitric acid to reduce pH of mains water on plant growth
Clonal Selection Scheme, Luddington 1980–83
Clonal Selection Scheme, Efford 1983
Assessment of a woven polypropylene sheet for container standing area
Assessment of specialised tapes for protecting polythene on hoops
Assessment of useful life of netting for protected cropping
Comparison of systems of production for Camellias
Comparison of composts for autumn potting of Camellias
Comparison of potting schedules for Rhododendrons
Comparison of composts for fruit tree production in containers
Container-grown trees, staking methods to counter wind effects
Comparison of rigid pots and polythene bags
Screening of herbicides for weed control in containers outdoors
Screening chemicals for control of moss, liverwort and weeds in containers under protection
Liverwort eradication in outdoor container nursery stock
Control of moss and liverwort in container grown plants
The use of paclobutrazole for growth control of ornamental nursery stock
Hardy nursery stock, fungicide phytotoxicity studies

Roses

Weed control in rose stocks
Roses, herbicide treatments for maiden bush production
Roses, use of growth regulators on micropropagated plants
Rose rust control
Roses, defoliation by chemicals

Land use, markets, outputs and imports for the hardy ornamental nursery stock industry in England and Wales are shown in Appendix 1.

References

MINISTRY OF AGRIGULTURE, FISHERIES AND FOOD (1983). *Hardy ornamental nursery stock*. MAFF Reference Book 235.

MINISTRY OF AGRICULTURE, FISHERIES AND FOOD. *The work of the Experimental Horticulture Stations*. MAFF Booklet UB7.

OECD (1980). *The measurement of scientific and technical activities*. Frascati Manual, Paris.

Discussion

D. PATCH (Forestry Commission Research Station)

It has been emphasised that research and development work is designed primarily to aid the profitability of nursery production: what steps are taken to ensure the plants propagated by ADAS techniques and supplied to customers are suitable for the general markets?

D.G. GILBERT 60 per cent of nursery production goes to the general public, mainly through garden centres, and 40 per cent goes to Local Authorities and contractors. The latter tend to be more 'choosey' in their specifications for the supply of planting stock than is the general public.

T.H.R. HALL (Oxford University Parks)

I am concerned about uniformity in planting schemes and the guidelines given to stock producers which result in the selection of species being governed by the grower. There is insufficient adventure into the use of new species. Is ADAS doing anything about the development of new species?

D.G. GILBERT During the last 5 years nurserymen have lost many thousands of pounds, particularly with ornamental tree production. The mainstay of the industry is the production of the 'bread and butter' species such as *Malus* spp., *Prunus* spp. etc. This does create a problem, and the lover of trees would like a greater variety available. Some specialist nurseries do supply unusual species. However, for wider variety some system of pre-ordering would be needed, and plant production geared to a known demand.

T.H.R. HALL There is an analogy – while there is a mass market for ladies dresses from, for example, Marks and Spencers, there is also a place for the *haute couture*.

D.G. GILBERT This probably applies to nursery stock also. The 10 per cent of specialist trees requested are usually available.

Appendix 1

Major 'markets' for Nursery Stock

PUBLIC SECTOR

(a) National and regional authorities
(b) New towns and development corporations
(c) Local authorities
(d) Landscape contractors
(e) Exports

PRIVATE SECTOR

(a) Garden centres
(b) Mail order
(c) Local nursery sales
(d) Multiple chain store
(e) Sales to nursery trade
(f) Shows, shops, market stalls, etc.
(g) Exports

Number of holdings and size 1983; England and Wales

Up to 5 hectares	2044
5 to 20 hectares	192
Over 20 hectares	60
	2296

(Total area 5454 ha)

Estimate of Hardy Ornamental Nursery Stock Sold 1983/84

	£ million
1. Container plants	41.8
2. Ornamental trees	15.8
3. Field grown shrubs	15.7
4. Field grown roses	11.3
5. Herbaceous plants	2.9
6. Other hardy ornamentals	8.2

Total farm gate value = £95.7 million

Sources of Information

MAFF Agricultural Census and Survey Branch
MAFF Statistics Division
Board of Customs and Excise

Objectives and Opportunities in the Vegetative Propagation of Broadleaved Species

B.H. Howard

East Malling Research Station, Maidstone, Kent.

Summary

Opportunities for overcoming propagation constraints in a wide range of woody broadleaved species are being identified and developed. Innovations such as micropropagation are technically demanding and need careful development to ensure relevant and effective application, while others exploit the principles of long-established techniques such as stooling, applying the inherent benefits of severe pruning and etiolation widely by the pretreatment of cutting hedges. Future progress is likely to depend largely on understanding the underlying processes involved in the enhanced rooting of juvenile material and developing techniques which achieve similar increased sensitivity to auxin and other treatments in adult plants.

Introduction

Deciduous broadleaved species include trees grown for fruit, for their amenity value and for timber. Fruit trees are propagated by vegetative means, amenity trees by both vegetative methods and by seed, and woodland species almost invariably by seed. Examples of research and development opportunities in the vegetative propagation of fruit and amenity trees indicate ways of extending vegetative propagation to woodland species.

Underlying the propagation of most broadleaved species is the need to succeed with plants in the adult phase after their mature characteristics have been assessed, in contrast to the unique opportunity to exploit easy-to-propagate juvenile material by taking cuttings from seedlings raised from genetically improved seed of Sitka spruce (Mason, 1984a) and Hybrid larch (Mason, 1984b).

Objectives of Research

From the standpoint of research on behalf of tree producers detailed objectives include:

1. Rapid introduction into commerce of new cultivars arising from chance selection, traditional breeding and the exploitation of natural and induced variation at cell level in tissue culture. The requirement to develop rapid multiplication systems is given added emphasis by the involvement of development and marketing agencies such as the National Seed Development Organisation and the Agricultural Genetics Company, respectively responsible for the promotion of conventionally bred material and that produced by novel means.

   The need for rapid bulking also arises from the periodic re-introduction of plants with improved health status, such as virus-free material in the EMLA fruit tree scheme, or the introduction of selected ornamental trees following screening of commercial sources for trueness-to-type, as in the Clonal Selection Scheme, both based at East Malling.

2. Methods are required for multiplying difficult-to-propagate plants in the adult phase. Only in fruit breeding has ease of propagation been considered among selection criteria, and even so the propagation techniques used in the initial screening invariably need refining for commercial uptake. In amenity trees visual characteristics are given primary consideration, with propagation achieved by whatever method best lends itself, and often by providing the cultivar with a root system by bud-grafting on to seedling rootstocks.

   Difficulty of propagation is the main factor determining scarcity and hence high cost of many amenity plants. The balance to be sought between locating individual source plants with some capability for vegetative propagation and attempting further improvements in propagation techniques will be a major consideration in future programmes to clone woodland species.

3. It is essential for methods practised by producers in a wide range of situations and management approaches to be reliable and cost effective from year to year. Research workers need to identify factors contributing to propagation success or failure and in particular those for producing, preparing, treating, rooting and establishing cuttings. Timing of these operations is often critical. Nurserymen need to recognise the importance of adhering to requirements identified by critical and systematic research, many of which are interdependent.

4. Trees are required of the highest quality commensurate with their use; in periods of over-supply especially, tree quality is an important sales factor. Quality is variously expressed, ranging from the requirement for branches at the appropriate height above ground in fruit trees to thick trunks to minimise casual vandalism in amenity trees. Woodland species will need to be produced in ways that suppress the laterals that introduce knots into the main bole. An underlying requirement for all types is the ability to transplant successfully, often into disturbed or destructured soil.

5. Flexibility of production is required to meet the varied circumstances operating in commercial nurseries.

Opportunities

Some ways of achieving these objectives can be seen from the matrix used to assist with propagation research planning at East Malling, in which the vegetative propagation techniques available to nurserymen are related to perceived opportunities for improvement (Figure 1). Examples are described below to illustrate each of the five objectives above, indicated in the matrix as 1 to 5 and a suffix letter to distinguish between a number of approaches within the same general objective.

Figure 1.

Research opportunities	Techniques					
	Stooling	Winter leafless cuttings	Summer leafy cuttings	Micro-propagation *in vitro*	Budding, grafting	Tree raising
Genetic selection		4a	4a		4d	
Shoot preconditioning towards rooting	2b	2c	2c	2a		
Seasonal effects		3a				
Techniques		1c 3b 3c 5a	1c	1a	4c	
Environment		3e	3d			4b
Management					1b	4e

1. The need to raise large numbers of plants from limited starting material requires either a system free from seasonal constraints and/or one that exploits all available potential growth centres, such as vegetative buds. Propagation of micropropagules *in vitro* (Figure 1, 1a) lends itself to the rapid production of many small plants throughout the year, with the added advantage that they are likely to be free from systemic diseases, an important attribute early in the life of a clone. Among tree species considerable experience has been gained in the micropropagation of fruit rootstocks and scions (Jones, 1983) but as yet there is relatively little experience of their long-term performance in the field.

   Propagation of scion varieties by conventional methods is limited only by the number of buds available for grafting on to rootstocks, and this principle can be applied to rootstocks by grafting each rootstock bud on to a nurse root system (Figure 1, 1b) to produce a hedge of the rootstock for cutting production.

   Single buds can be exploited also through single-node summer or winter cuttings (Figure 1, 1c) given that the subject roots readily.

2. Methods for propagating difficult subjects from cuttings centre on ways of re-introducing those aspects of juvenility beneficial to adventitious rooting, although the physiological basis of these effects is not understood. Micropropagation *in vitro* often succeeds with recalcitrant subjects (Figure 1, 2a), rooting ability of the propagule increasing with successive subcultures. Evidence is accumulating to suggest that 'juvenility' developed *in vitro* can be maintained in plants restricted in size. Juvenile characteristics of excessive spininess and faster rooting persisted in micropropagated 'Pixy' plum rootstocks when planted into the nursery and maintained as a severely pruned hedge at East Malling. The micropropagated apple cultivar 'Jonathan' failed to flower after 29 months when grown to only 1 m in a 2 litre container, whereas others flowered in the same time when grown 2 m in 50 litre containers (Sriskandarajah, 1984).

   The fact that when propagating a range of species by conventional cuttings some may not require auxin treatment, others respond to treatment, while others cannot respond, illustrates that post-cutting collection treatments are only successful if the plant is in a physiologically receptive condition. It follows that one way to succeed with difficult subjects is to enhance their sensitivity during the shoot growth phase to subsequent cutting treatments. A study of why an apple rootstock rooted in the stoolbed (Figure 1, 2b) but not from hedge-grown cuttings, showed that ultra severe stock-plant pruning and the exclusion of light from the shoot base were central to success (Howard *et al.*, 1985). Many of the field factors governing responses to pruning and etiolation have been identified and consistent rooting responses obtained in fruit and amenity tree cuttings raised on hedges (Report of East Malling Research Station for 1982, 1983). Such shoot preconditioning treatments (Figure 1, 2c) will be important in rooting cuttings of woodland species either applied to regrowth after felling or lopping or to entire plants obtained by grafting, a view supported by studies of stock plant factors influencing rooting of a tropical forest hardwood species (Leaky, 1983).

3. Reliable and cost-effective production requires the essential components of a method to be identified and optimised. These include seasonal trends (Figure 1, 3a) linked to changes in endogenous rooting factors in cuttings (Bassuk and Howard, 1981) and those at bud burst that drain carbohydrate reserves and increase the risk of loss through water stress when planted in the field (Cheffins and Howard, 1982 a & b). Cuttings planted directly into the field without a prior period of basal heat must be propagated in autumn to benefit from residual soil warmth and to allow maximum time for rooting.

   Wounding (Figure 1, 3b) in addition to that incurred when collecting, the cutting may increase the sites available for root initiation (Howard *et al.*, 1984) and the application of both liquid and powder preparations of auxin (Figure 1, 3c) can be standardised to give consistent results (Howard, 1985 a & b). The importance of high incident irradiance as a cause of water stress was described (Grange and Loach, 1983) in a reappraisal of mist and polythene systems for propagating leafy cuttings (Figure 1, 3d), while for leafless winter cuttings the critical environmental requirement (Figure 1, 3e) is to control water availability in the rooting medium and ensure adequate aeration to avoid cutting death (Harrison-Murray, 1982).

4. Quality and cost-effective production are generally more crop specific than other objectives. The opportunity exists to select within seedling populations of *Tilia* spp. for rooting ability from cuttings (Figure 1, 4a) leading to the development of clonal rootstocks and increased uniformity of production (Howard and Shepherd, 1978). Rootstocks can be established effectively in well-drained raised beds of bark:peat:sand:grit compost without planting being hampered by wet weather, and with the opportunity to alter plant density to control plant size (Figure 1, 4b).

Slow or incomplete union formation after budding leads to loss of quality; most rapid and complete union formation results from chip budding (Figure 1, 4c) (Howard *et al.*, 1974) and by selecting superior budding sources of scion wood (Figure 1, 4d) as in the case of *Acer platanoides* 'Crimson King'. Residual herbicide washed into the root zone may cause damage resulting in reduced stem girth increase, the depressed cambial activity resulting in reduced bud-take and production of low quality trees (Figure 1, 4e). Among a group of amenity trees approximately 1 per cent reduction in stem diameter resulted in 1.5 per cent reduction in bud-take, compared with plants not given herbicides (Report for East Malling Research Station for 1982, 1983).

5. Flexibility can be achieved in various ways, the most notable example being production in either the field or in containers to take account of soil-borne diseases and marketing preferences. Currently the opportunity to produce 'instant trees' is being investigated by propagating <2.5 m long cuttings (Figure 1, 5a) to produce self-rooted scions as a possible alternative to budding. In this case a reduction in the time between planting and harvesting the budded crop is traded-off against the need to maintain severely pruned stock hedges. The technique is particularly effective with London plane (*Platanus × hispanica*).

References

BASSUK, NINA L. and HOWARD, B.H. (1981). A positive correlation between endogenous root-inducing co-factor activity in vaccuum-extracted sap and seasonal changes in rooting of M26 winter apple cuttings. *Journal of Horticultural Science* **56**, 301–312.

CHEFFINS, N.J. and HOWARD, B.H. (1982a). Carbohydrate changes in leafless winter apple cuttings. I. The influence of level and duration of bottom heat. *Journal of Horticultural Science* **57**, 1–8.

CHEFFINS, N.J. and HOWARD, B.H. (1982b). Carbohydrate changes in leafless winter apple cuttings. II. Effects of ambient air temperature during rooting. *Journal of Horticultural Science* **57**, 9–15.

GRANGE, R.I. and LOACH, K. (1983). Environmental factors affecting water loss from leafy cuttings in different propagation systems. *Journal of Horticultural Science* **58**, 1–7.

HARRISON-MURRAY, R.S. (1982). Air/water relations of rooting media for leafless winter cuttings. *21st International Horticultural Congress*, Hamburg. Abstract 1787.

HOWARD, B.H. (1985a). The contribution to rooting in leafless winter cuttings of IBA applied to the epidermis. *Journal of Horticultural Science* **60**, 153–159.

HOWARD, B.H. (1985b). Factors affecting the response of leafless winter cuttings to IBA applied in powder formulation. *Journal of Horticultural Science* **60**, 161–168.

HOWARD, B.H. and SHEPHERD, H.R. (1978). Opportunities for the selection of vegetatively propagated clones within ornamental tree species normally propagated by seed. *Acta Horticulturae* **79**, 139–144.

HOWARD, B.H., HARRISON-MURRAY, R.S. and ARJYAL, S.B. (1985). Responses of apple summer cuttings to severity of stockplant pruning and stem blanching. *Journal of Horticultural Science* **60**, 145–152.

HOWARD, B.H., HARRISON-MURRAY, R.S. and MACKENZIE, K.A.D. (1984). Rooting responses to wounding winter cuttings of M26 apple rootstock. *Journal of Horticultural Science* **59**, 131–139.

HOWARD, B.H., SKENE, D.S. and COLES, J.S. (1974). The effects of different grafting methods upon the development of one-year-old nursery apple trees. *Journal of Horticutural Science* **49**, 287–295.

JONES, O.P. (1983). *In vitro* propagation of tree crops. In, *Plant biotechnology*, eds. S.M. Mantell and H. Smith, 139–159. Society for Experimental Biology, Seminar Series 18. Cambridge University Press.

LEAKEY, R.R.B. (1983). Stockplant factors affecting root initiation in cuttings of *Triplochiton scleroxylon* K. Schum., an indigenous hardwood of West Africa. *Journal of Horticultural Science* **58**, 277–290.

MASON, W.L. (1984a). *Vegetative propagation of conifers, using stem cuttings*. I. *Sitka spruce*. Forestry Commission Research Information Note 90/84/SILN. Forestry Commission, Edinburgh.

MASON, W.L. (1984b). *Vegetative propagation of conifers using stem cuttings*. II. *Hybrid larch*. Forestry Commission Research Information Note 91/84/SILN. Forestry Commission, Edinburgh.

Report of East Malling Research Station for 1982 (1983), 59–75.

SRISKANDARAJAH, S. (1984). *Induction of adventitious roots in some scion cultivars of apple (Malus pumila Mill.)*. PhD Thesis, Department of Agriculture and Horticultural Science, University of Sidney, Australia.

Discussion

J.E.G. GOOD (Institute of Terrestrial Ecology, Bangor).

Have you any comments on the use of polythene in propagation, for example, over benches?

B.H. HOWARD It can be used for environmental control, and as a forcing treatment on bushes. Over benches, there is benefit in the use of polythene as it enhances the temperature component. Polythene tents and intimate polythene are good in winter when mist treatment may create conditions that are too wet. In summer intimate polythene is not very good because of lack of ventilation. The temperature may increase to such an extent that the relative humidity falls below 99 per cent. Fogging is probably a better technique in summer.

J. RISHBETH (School of Botany, Cambridge University).

What is the risk of secondary damage to cuttings resulting from the technique of splitting the base to encourage root initiation? Are there any disadvantages, e.g. the possibility of introducing disease?

B.H. HOWARD The major benefits of splitting are the speed of rooting and the increase in the number of roots produced. The principal disadvantage is that it is an extra job to be done at the bench in the preparation of cuttings.

I.R. BROWN (Department of Forestry, University of Aberdeen)

There appear to be very few problems with vegetative propagation in fruit tree species. However, there are problems with forest tree species. Any comments or explanations?

B.H. HOWARD No experience with forest tree species, but East Malling has concentrated on the particular problems that occur with amenity trees. With application many of the individual problems can be overcome.

Tree Improvement by Selection

J.E.G. Good[1] and F.T. Last[2]
[1]Institute of Terrestrial Ecology, Bangor
[2]Institute of Terrestrial Ecology, Midlothian

Summary

Tremendous unexplored potential exists for tree improvement by selection, most species being in their wild state and in possession of the wide range of genetic diversity that this implies. The longevity of trees, the slowness of many species in reaching reproductive maturity, and the fact that many are obligate outbreeders, present problems to tree breeders. Additionally, there is the requirement for large areas of ground, occupied for many years, to accommodate trials testing relatively few provenances, seedlots or clones. Improvements in macro- and micro-propagation techniques, allied with the development of predictive tests enabling various aspects of growth in mature trees to be predicted from the behaviour of immature plant material, are beginning to overcome these problems.

Introduction

Intra-specific variation offers immense possibilities for tree improvement with regard to almost any character that may be of interest to the improver. Significant gain is considered feasible in trees grown primarily for timber (and, one assumes, in those grown for amenity) for resistance to disease (Pawsey, 1960; Butcher *et al.*, 1984), drought (Schmidt-Vogt, 1977; Townsend, 1977), waterlogging (Heybroek, 1982), mineral deficiency (Bell *et al.*, 1979), frost resistance (Wilcox *et al.*, 1980), in addition to fast growth rates (Raulo and Koski, 1977), stem-form (Howard and Shepherd, 1978; Townsend, 1983), wood quality (Heybroek, 1982) and the ease of rooting of cuttings (Komissarov, 1969; Good *et al.*, 1978).

Why has this reservoir of genetic diversity not been more effectively tapped? Part of the answer lies in the unsuitability of the traditional methods used for improvement of agricultural and horticultural crop plants for tree improvement. Whereas most crop plants are annuals or biennials, easily grown in large numbers and thus ideally suited for selection and breeding, trees are large, long-lived and slow to reach reproductive maturity. Thus, in the absence of established relationships between juvenile and mature characteristics, tree improvement has been bedevilled by the need for large areas of land to accommodate a relatively narrow range of genotypes in trials and by the problems, often considerable, of obtaining adequate numbers of suitable plants. Clonal selection has not, until recently, been a practical possibility except in a few genera (e.g. *Populus, Salix*), because of problems associated with vegetative propagation.

All this is now changing. Recent advances have greatly increased understanding of the physiology of both vegetative propagation (Howard, 1987), and of sexual reproductive processes (Longman, 1975; Tompsett and Fletcher, 1977). As a result, tree breeding problems are becoming increasingly predictable with fewer delays, while clonal forestry is now also a serious option (Libby, 1969; Zobel, 1981). It is reasonable to suppose that in the near future it will be possible to produce unlimited numbers of most clones of most species more or less on demand.

Of great importance to both traditional tree breeding programmes and to clonal selection is the development of reliable predictive tests enabling particular charateristics of mature trees (e.g. form, wood quality, disease resistance) to be predicted from the behaviour of immature saplings. Prediction is more likely to be successful for strongly inherited form characteristics than for less strongly inherited quantitative characteristics, e.g. increment. With the development of advanced micro-propagation and tissue culture techniques it may even be possible to develop characterisation procedures whereby physiological and chemical properties of clonal material in aseptic culture may be used to predict morphological, anatomical or physiological traits of the mature whole plant, but such developments are still in their early stages. Should success be achieved, it would be possible to test many more genotypes of a much wider range of species without the need for costly long-term field trials tying up large areas of land.

Work on Tree Improvement in the Institute of Terrestrial Ecology

Over the past decade a number of projects concerned, in one way or another, with tree selection and improvement, have been initiated. These have varied from studies concerned primarily with variation between seedlots of a species, variation in birch (*Betula pendula* Roth. and *B. pubescens* Ehrh.), oak (*Quercus petraea* (Matt.) Lieb. and *Q. robur* L.), Sitka spruce (*Picea sitchensis* (Bong.) Carr.) and Lodgepole pine (*Pinus contorta* Dougl.) having been studied, to the selection of improved clones of obeche (*Triplochiton scleroxylon* K. Schumm.) for use in west African forestry or of birch and willow (*Salix caprea* L. and *S. cinerea* L.) for planting restored coal waste.

In the seedlot studies of birch it has been shown that the latitude of origin of seedlots profoundly influences, among other characters, their height growth (see Figure 1) when grown side by side near Edinburgh. In Sitka spruce, resistance of shoots to cold temperature varied considerably between seedlots with those from more northerly latitudes in Alaska becoming hardened to sub-zero temperatures up to two months sooner in the autumn than those from further south in Oregon. Altitude of origin also has an effect. Last (1975) reported that young trees of *Betula pubescens* from seed collected 960 m above sea level in Norway started seasonal growth earlier and finished earlier than trees grown from seeds collected at the same latitude but within 45 m of sea level in Scotland. During periods of maximum growth, however, the two provenances were equally active. Climatic factors, seemingly related primarily to altitude and shelter, have resulted in inland seedlots of Lodgepole pine having fewer branches per whorl than coastal types. Such variations in form, which are to be found wherever they are sought in wild populations, are of considerable interest to arboriculture, providing the key for the development of tailored trees of known form and growth rate for particular purposes.

Figure 1. Effect of latitude of origin on heights after six seasons of saplings of *Betula pendula* Roth. ■, and *B. pubescens* Ehrh. □, grown at Bush near Edinburgh.

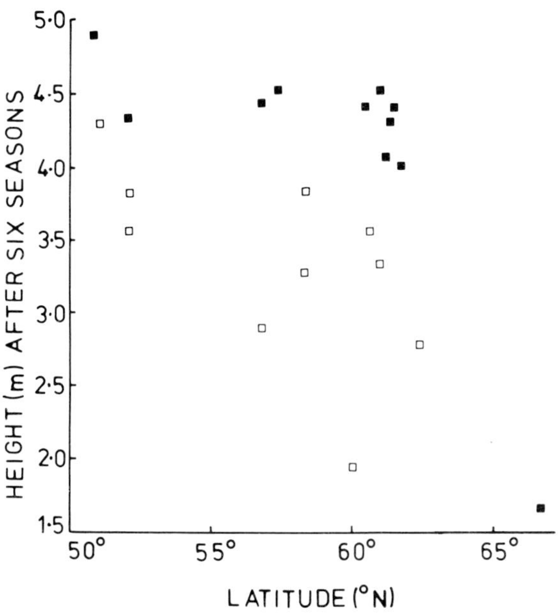

One need not look, however, to differences over large distances, or between markedly different geographical regions, for variation in trees. Seedlots of birch collected from different objectively defined land classes in Scotland varied not only with regard to various growth attributes, but also in the incidence of the rust *Melampsoridium betulinum* (see Table 1) overleaf. Studies of oak seedlots collected on a similar basis in Wales are in their very early stages, but nevertheless it is already clear that the association between land class and seedling characters is much less marked in oak than in birch. This is probably because natural variation in oak is obscured by the many planted trees in the population in all parts of the country, whereas birch is mostly of natural origin. Clearly there is potential for selection on the basis of land class as a means of matching trees to sites in species which have not been extensively planted.

Obeche is a species of major importance in west African forestry (Last *et al.*, 1984). Unfortunately it sets seed erratically, so that there was considerable concern in the mid 1970s over how a consistent re-afforestation programme could be sustained with this species. Clonal propagation of the best trees from the wild was proposed and a system was developed which enabled most clones to be propagated from softwood cuttings with a reliable high success rate (Howland, 1975; Leakey *et al.*, 1982). It is one thing to get cuttings to root, however, quite another to obtain a consistent supply of uniform, easily-rooted cuttings: stockplant management is of fundamental importance (Last *et al.*, 1984).

While experimenting to try and improve the number and quality of obeche cuttings Leakey (1983) noticed that some clones produced many more lateral branches when decapitated than others. Studies of the same clones

Table 1. Growth attributed and incidence of rust *Melampsoridium betulinum* at the end of the second season after germinating and growing saplings of *Betula pubescens* originating from different UK land classes, together, near Edinburgh

Land classes*	Stem diameter (mm)	Leaf length (mm)	Proportion (%) of saplings infected with rust
27	13.32	40.60	11
29	12.60	34.00	14
31	11.89	29.00	24

All differences between land classes 27 and 31 are statistically significant ($p<0.05$)

*The land class descriptions taken from Bunce *et al.* (1981) are as follows:

Land class 27 – Occurring in central, east and north-east Scotland, mainly at low or medium altitudes, and comprising well-fenced (mainly arable) agricultural land, often with mixed woodland. Soils are brown earths or gleys.

Land class 29 – Occurring in west Scotland, at low or medium altitudes, and comprising uneven topography with mainly open range grazing. Soils are mainly acid peats but there are also rankers and brown earths.

Land class 31 – Occurring in north Scotland and the Isles, mainly at low or medium altitudes, and characterized by severe wind exposure. Land use mainly rough grazing. Soils are brown earths, acid peats or podzols.

growing in field trials in Nigeria revealed a strong positive correlation ($r=0.93$) between: (a) total numbers of branches per tree, counted 5 years after establishment of field trials of rooted cuttings, and (b) the percentage of axillary buds developing after decapitating stockplants (see Figure 2). Thus growth form in the field could be predicted from the behaviour of stockplants in the glasshouse. We recommend the widespread search for similar predictive tests in other tree selection programmes as a means of determining the extent of variation of particular characters in certain species and as an aid to the selection of the most appropriate seedlots or clones for specified purposes.

A further finding to emerge from the work with obeche was the importance of observing individual trees within populations when seeking variation, rather than restricting observations to seedlots. A similar conclusion was reached by Kleinschmidt and Sauer (1976) when analysing variation in Norway spruce (*Picea abies* Karst.). Theory suggests that by restricting plantings of obeche to the 33 per cent of clones with form scores and stem volumes both above average, if there are no interactions with silvicultural practices or severe within and between clone competition, that stem volumes would be increased by 30 per cent. The use of only the best 10 per cent of clones might increase volumes by 81 per cent (Last *et al.*, 1984). Furthermore, an economic crop of obeche is likely in 35 years, compared with 150+ in nature. Is it not possible that the rapid capture of genetic gain, facilitated by the development of predictive tests, might enable comparable rewards to be reaped in temperate hardwoods such as oak and beech?

In our work on selection of tolerant clones of birch and willow for use in revegetation of restored opencast coal sites, we have shown that some propagated clones developed from saplings growing on colliery spoil heaps have significantly higher mean survival than unselected controls when planted on a range of opencast sites

Figure 2. Correlation between branching in *Triplochiton scleroxylon* K. Schum. clones in the field, 5 years after planting at initial 2.5 m spacing, and in the 'Predictive Test' in ITE glasshouses

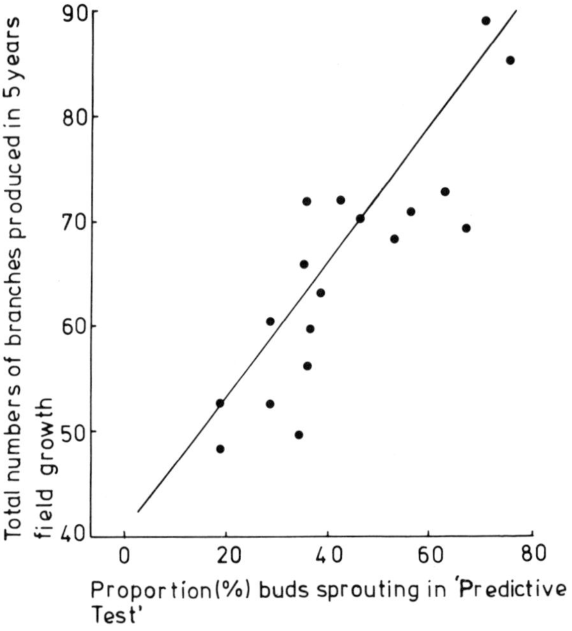

throughout Britain (see Table 2). The greatest gains in survival occurred on the most difficult sites, but in all cases there was considerable variation between clones. Contrary to expectations, the growth rates of unselected birch and willow equalled or exceeded those of the selected clones at most sites. In retrospect this was not really surprising, however, since several studies have shown that ecotypes of a range of species obtained from low-nutrient sites grow more slowly than high-nutrient adapted ecotypes even when nutrient supply is limited (Goodman, 1969; Chapin, 1983; Clark, 1983). In this way the low-nutrient adapted types are able to conserve nutrients, maintaining a slow rate of growth throughout periods of shortage. High-nutrient ecotypes, on the other hand, are liable to gradually run out of nutrients during periods of shortage, death resulting if low levels continue indefinitely (Clarkson, 1967; Grime, 1979; Rorison, 1968).

Table 2. Mean survival (%) of selected clones and unselected controls of birch and willow when planted on a range of restored coal sites

Species and selected clones planted		*Mean survival (%)* Selected clone		Unselected control
Birch (*Betula*)	CLONE			
B. pubescens	7	95	++	52
	47	94	++	70
B. pendula	28	72		86
	34	85	++	60
	64	36		22
	73	66		63
	94	89	+	70
Mean		77	+	60
Willow (*Salix*)	CLONE			
S. cinerea	12	86	+	74
	90	93	++	61
S. caprea	16	86	+	56
	29	95		88
	50	68		73
	51	46	++	83
	76	85		82
Mean		76		76

\+ Means for selected and unselected significantly different ($p<0.05$).

++ Means for selected and unselected significantly different ($p<0.01$).

This work has indicated that selection of tolerant clones of woody plants from among the very variable seedlings of open-pollinated species occurring on derelict and degraded soils is a feasible approach to improvement of planting stock for these other difficult sites. It is very likely that similar success could be achieved in selecting woody plants for other desirable characters, but our results indicate that rigorous clone testing is necessary – not all clones from nutrient-deficient sites are tolerant of similar deficiencies under all circumstances.

Conclusion

In this paper we have attempted to demonstrate the immense potential for tree improvement by selection and some of the approaches we have adopted for exploiting it. The concept of predictive testing as a means of simplifying selection procedures and enabling many more individual trees to be tested over a given period of time is, in our view, an essential element in the advancement of tree improvement procedures. It is dependent for its success as a technique on improvements in both macro- and micro-propagation methods, a field in which progress of late has been rapid and effective. We shall soon reach the stage where clonal propagation presents few problems. Before that stage is reached decisions should be made as to what the priorities for improvement should be. Our feeling is that there is an urgent need for trees suited to particular soils and situations and that, initially at least, such selection should be restricted in the main to those species most widely used in arboriculture. We already have clones of birch and willow able to tolerate specific soil problems, but these represent only a tiny fraction of the useful clones available. Clones of these and other species could be selected which would provide a wide range of form and growth rate as well as improvements in such individual features as bark colour (*Betula* spp., *Prunus* spp.), flowers (*Crataegus* spp., *Malus* spp., *Prunus* spp.), fruits (*Cotoneaster* spp., *Malus* spp., *Prunus* spp., *Sorbus* spp.), foliage (*Acer* spp., *Cotoneaster* spp., *Prunus* spp., *Sorbus* spp., the ornamental conifers). In making selections it is very important that they be tested under a wide range of planting conditions before being propagated *en masse*.

References

BELL, H.E., STETLER, R.F. and STONECYPHER, R.W. (1979). Family × fertilizer interaction in one-year-old Douglas fir. *Silvae Genetica* **28**, 1–5.

BUNCE, R.G.H, BARR, C.J. and WHITTAKER, H.A. (1981). *Land classes in Great Britain: preliminary description for users of the method of land classification.* Merlewood Research and Development Paper 86. Insititute of Terrestrial Ecology, Grange-over-Sands.

BUTCHER, T.B., STUKLEY, M.J.C. and CHESTER, G.W. (1984). Genetic variation in resistance of *Pinus radiata* to *Phytophthora cinnamomi*. *Forest Ecology and Management* **8**, 197–220.

CHAPIN, F.S. (1983). Adaptation of selected trees and grasses to low availability of phosphorus. *Plant and soil* **72**, 283–287.

CLARK, R.B. (1983). Plant genotype differences in the uptake, translocation, accumulation and use of mineral elements required for plant growth. *Plant and Soil* **72**, 175–196.

CLARKSON, D.T. (1967). Phosphorus supply and growth rate in species of *Agrostis* L. *Journal of Ecology* **55**, 111–118.

GOOD, J.E.G., BELLIS, J. and MUNRO, R.C. (1978). Clonal variation in rooting of softwood cuttings of woody perennials occurring naturally on derelict land. *International Plant Propagators' Society Combined Proceedings* **28**, 192–201.

GOODMAN, P.J. (1969). Intraspecific variation in mineral nutrition of plants from different habitats. In, *Ecological aspects of the mineral nutrition of plants*, ed. I.H. Rorison, 237–253. Blackwell, Oxford.

GRIME, J.P. (1979). *Plant strategies and vegetation processes* (222pp.) Wiley, New York.

HEYBROEK, H.M. (1982). The right tree in the right place. *Unasylva* **34**, 15–19.

HOWARD, B.H. (1987). Objectives and opportunities in the vegetative propagation of broadleaved species. In, *Advances in practical arboriculture*, ed. D. Patch. Forestry Commission Bulletin 65, 17–21. HMSO, London.

HOWARD, B.H. and SHEPHERD, H.R. (1978). Opportunities for the selection of vegetatively propagated clones within ornamental tree species normally propagated by seed. *Acta Horticulturae* **79**, 139–141.

HOWLAND, P (1975). Variations in rooting of stem cuttings of *Triplochiton scleroxylon* K. Schum. In, *Proceedings of the symposium on variation and breeding systems of Triplochiton scleroxylon (K. Schum.).*, Ibadan, Nigeria, 110–114.

KLEINSCHMIDT, J. and SAUER A. (1976). Variation in morphology, phenology and nutrient content among clones and provenances and its implications for tree improvement. In, *Tree physiology and yield improvements*, eds. M.G.R. Cannell and F.T. Last, 508–517. Academic Press, London.

KOMISSAROV, D.A. (1969). *Biological basis for the propagation of woody plants by cuttings.* Israel Program of Scientific Translation, Jerusalem. (Translated by Z. Shapiro, ed. by M. Kohn from original Russian text published Moscow, 1964).

LAST, F.T. (1975). Some aspects of the genecology of trees. *Report of the East Malling Research Station for 1974*, 25–40.

LAST, F.T., LEAKEY, R.R.B and LADIPO, D.O. (1984). Safeguarding the resources of indigenous West African trees: an international venture exploiting physiological principles. In, *Technology transfer in forestry*, eds. G.H. Moeller and D.T. Seal. Forestry Commission Bulletin 61, 61–68. HMSO, London.

LEAKEY, R.R.B. (1983). Stockplant factors affecting root initiation in cuttings of *Triplochiton scleroxylon* K. Schum., an indigenous hardwood of West Africa. *Journal of Horticultural Science* **58**, 277–290.

LEAKEY, R.R.B., CHAPMAN, V.R. and LONGMAN, K.A. (1982). Physiological studies for tropical tree improvement and conservation. Factors affecting root initiation in cuttings of *Triplochiton scleroxylon* (K. Schum.). *Forest Ecology and Management* **4**, 53–56.

LIBBY, W.J. (1969). Some possibilities of the clone in forest genetics research. *Genetics lectures*, ed. R. Bogart, **1**, 121–136. Oregon State University Press.

LONGMAN, K.A. (1975). Tree biology research and plant propagation. *International Plant Propagators' Society Combined Proceedings* **25**, 219–236.

PAWSEY, C.K. (1960). Heredity in relation to some disorders and defects of *Pinus radiata* (D. Don) in South Australia. *Australian Forestry* **24**, 4–7.

RAULO, Y. AND KOSKI, V. (1977). Growth of *Betula pendula* Roth. progenies in southern and central Finland. *Communicationes Instituti Forestalis Fenniae* **90**, 1–38.

RORISON, I.H. (1968). The response to phosphorus of some ecologically distinct plant species. 1. Growth rates and phosphorus absorption. *New Phytologist* **67**, 913–923.

SCHMIDT-VOGT, H. (1977). Investigations on the drought resistance of conifer provenances. In, *Proceedings of the Third World Consultation on Tree Breeding, Canberra, 1977, Section 2: Advances in species and provenance selection.*

TOMPSETT, P.B. and FLETCHER, A.M. (1977). Increased flowering of Sitka spruce (*Picea sitchensis* (Bong.) Carr.) in a polythene house. *Silvae Genetica* **26**, 84–86.

TOWNSEND, A.M. (1977). Characteristics of red maple progenies from different geographic areas. *Journal of the American Society of Horticultural Science* **102**, 461–466.

TOWNSEND, A.M. (1983). Selection and breeding of urban trees. *Arboricultural Journal* **7**, 87–92.

WILCOX, M.D., FAULDS, T., VINCENT, T.G. and POOLE, B.R. (1980). Genetic variation in frost tolerance among open-pollinated families of *Eucalyptus regnans*. *Australian Journal of Forest Research* **10**, 169–184.

ZOBEL, B. (1981). Vegetative propagation in forest management research. *Proceedings of the 6th Southern Forest Tree Improvement Conference, Blacksburg, Virginia, USA*.

Discussion

B.H. HOWARD (East Malling Research Station, Kent)

There is a demand for dwarfing forms of trees, are there any easy methods of achieving this? It has been noted that slow growth types are prone to early loss of leaves.

J.E.G. GOOD No special study has been made with a view to selecting dwarfing types. Of course this could be done if finance was available. Slow growth types of birch shed leaves too early, as suggested. A good type of dwarfing Lawson cypress has been identified.

C. YARROW (Chris Yarrow and Associates)

Has there been selection for tree shape, timber properties, timber shake, fibre length – that is, examination within the tree?

J.E.G. GOOD Melvin Cannell (ITE, The Bush) has worked on wood density.

H.G. MILLER (Department of Forestry, Aberdeen University).

Comment on work being done on shake by Dr Denne at University College of North Wales, Bangor.

I.R. BROWN (Department of Forestry, Aberdeen University)

There appeared to be a great variation between performance of birches depending on site, some only performed well on particular sites and others could be planted anywhere. It is not always clear therefore whether a particular provenance or genotype of birch would suit a particular site, and there appeared a need to select birch plants according to land classes through identifying the tree's physiological adaptability to the particular site types, and that probably one needed to select clones which would perform well on a broad range of sites.

J.E.G. GOOD This comment emphasises the need to do more work on selection of birch and other species.

Dormant Tree Seeds can Exhibit Similar Properties to Seed of Low Vigour

P.G. Gosling

Forestry Commission Research Station,

Alice Holt Lodge, Wrecclesham, Farnham, Surrey GU10 4LH

Summary

Several types of laboratory seed tests are available to seed analysts. The actual tests applied to a particular seed lot, the method of interpreting the results, how the results are extrapolated to the nursery and the field conditions after sowing are all shown to influence the accuracy of predicting field emergence. Recent evidence suggests that in the case of many conifer and some broadleaved species, computer analysed comparisons between germination data obtained ± pretreatment may provide a measure of the vague and illusive property known as seed vigour and thereby increase forecasting accuracy.

Introduction

The ultimate aim of the seed test is to provide standard results which will give potential purchasers or nurserymen a guide to the overall quality of seed and hence enable comparisons to be made between one seed lot and another. It is extremely important to appreciate that tests on tree seeds are not specifically designed to predict field emergence; they are laboratory tests carried out under standard, near optimum conditions and although results are indicative of whether one seed lot has the potential to produce more plants in the nursery than another, as every nurserymen knows laboratory results are no guarantee of field performance.

This paper focuses on the relatively unpredictable emergence of tree seeds in the nursery and explores reasons for this behaviour with particular reference to conifer species. It will go on to discuss how it is hoped that increased forecasting accuracy may be obtained from the interpretation of computer analysed laboratory germination data.

Types of Laboratory Seed Test

The maximum number of potential plants in any given quantity of 'seed' will always depend on three things:

1. the proportion of pure seed:impurity in the bulk;
2. the average weight of individual pure seeds;
3. the proportion of pure seeds which are able to germinate.

During a test at an Official Seed Testing Station, seed of every species is tested in the same way for the physical attributes described in 1 and 2. However, the method of assessing 3 can depend upon the physiological properties of the species under test and the urgency with which the seed test results are required. These considerations make it necessary to decide between a germination test or a viability test. Precise details of the conditions used for seed testing are contained in *International Rules for Seed Testing* (Anon., 1976).

As the name implies a germination test is used to assess the percentage of seeds which are capable of germinating under standard laboratory conditions. A viability test on the other hand measures the number of seeds which are 'alive' (= viable) according to different criteria dependent upon the type of viability test employed.

Viability tests have two advantages over a germination test. They can be applied to virtually every species; and they can be conducted comparatively rapidly, making the results available much sooner than from a germination test. However, viability tests also have two distinct disadvantages. They do not provide a direct measure of a seed's ability to germinate, only an estimate of how many seeds are 'alive' at the time of testing; and they cannot be used to identify the presence or absence of seed dormancy.

In practice germination tests are preferred for those tree species capable of being induced to germinate within *c.* 8 weeks (including pretreatment) and viability tests tend to be reserved for those seed lots either where more rapid results are required, or where seeds take longer than 8 weeks to germinate, or when germination tests need to be preceded by a lengthy (> 4 weeks) pretreatment.

The results from germination or viability tests are normally reported as the maximum percentage of germinable or viable seeds, but in addition to this they can be combined with the results of the physical tests 1 and 2 to give a measure of the overall quality of the seed lot in terms of the 'number of germinable (or viable) seeds/kg'.

The remainder of this paper will only consider the results of laboratory germination tests.

Extrapolating Laboratory Germination Results to the Field

Since the conditions used in testing the germination of tree seed are designed to be optimal, it is not surprising to find that a seed lot capable of, for example, 95 per cent germination in an official seed test often achieves less than 95 per cent emergence in the nursery. Although growers seek to provide optimal seed bed and growing conditions the weather usually makes this impossible and some depression of germination is to be expected.

To date, the only method for overcoming this limitation of laboratory germination results has been to combine official seed test results with empirically obtained 'field survival factors' (fsf). This is the ratio of seedlings surviving at the end of the year to the number of germinable seeds sown (Aldhous, 1975) and can be calculated for each site/species. But this is still inconsistent. The modified germination percentage only provides a good approximation of a seed lot's performance when the species shows a fairly constant relationship between laboratory germination capacity and field emergence, e.g. Scots pine (*Pinus sylvestris*); but in many cases, different seed lots of the same species do not show a constant relationship between laboratory germination and field emergence, e.g. Sitka spruce (*Picea sitchensis*). When this is so the application of a correction factor in the form of a fsf is less useful.

Seed Vigour

Focusing attention on seeds with the above type of unpredictable emergence leads to discussion of seed vigour. This is a property which is easy to appreciate when observing the performance of seeds in the field, but very difficult to define in terms that are easy to measure in the laboratory.

To illustrate this, where field stress is low and conditions are least hostile the emergence of any seed lot is likely to approach laboratory germination capacity. However, as field stresses increase emergence levels decline in some seed lots more than others. Seed lots most resistant to stressful field conditions are described as high vigour; seed lots suffering proportionally greater decreases in emergence are then low vigour. In practical terms, vigour embraces those seed qualities which result in disproportionate differences in seed performance as field stresses increase (Figure 1).

Figure 1.

(a) To illustrate how the same seed lot can give rise to significantly different field performances

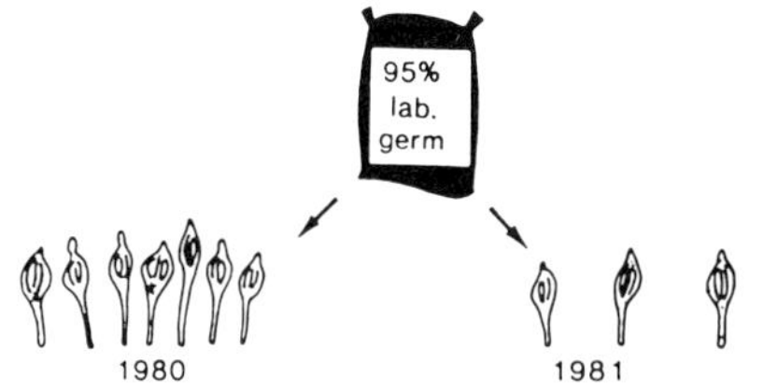

Same seed lot, same nursery, different years

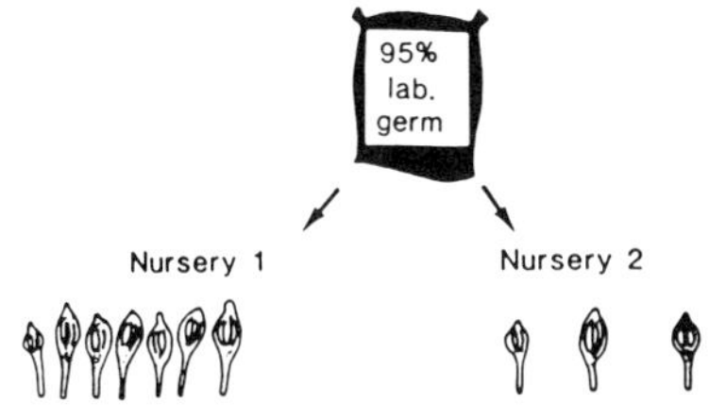

Same seed lot, different nurseries, (same or different years)

(b) To illustrate how different seed lots (of equivalent lab quality) can give rise to significantly different field performances

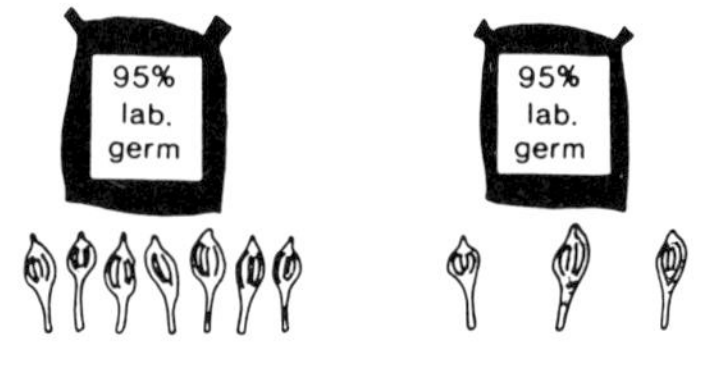

Different seed lots, (same lab germination), same nursery, (same or different sowing times)

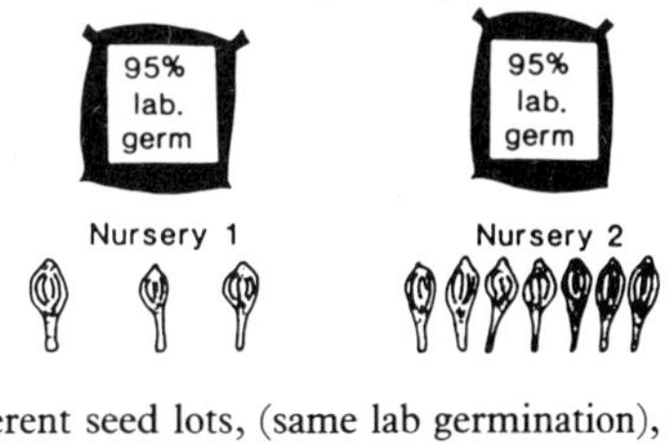

Different seed lots, (same lab germination), different nurseries, (same or different sowing times)

Unfortunately, despite extensive work in agricultural and horicultural seed testing it has so far proved impossible to identify a readily measurable parameter from any laboratory germination tests which would reliably reflect seed vigour in the field for all species. However, the laboratory testing of most of these species only involves observations on the rate and maximum percentage germination under one set of conditions. Many temperate conifer species require an additional germination test after being exposed to a moist chilling period. This pretreatment or prechill is designed to measure the presence or absence of dormancy, but recent close analysis has revealed that it may also have additional uses.

In the past, the identification of conifer seed dormancy at the Forestry Commission Official Seed Testing Station has relied upon the following criteria. When the maximum percentage germination after pretreatment was at least 5 per cent more than untreated seed, the lot was termed 'dormant'. When the maximum percentage germination of pretreated seed did not differ by more than ± 4 per cent from untreated seed, the lot was called 'non-dormant'. This method of distinguishing between 'dormant' and 'non-dormant' was neither statistically correct nor has it provided an adequate reflection of how seeds perform in the less optimum conditions of the nursery. Recent work has, therefore, refined the analysis.

The terms 'dormant' and 'non-dormant' are still retained but it will be shown below that the new type of analysis may reveal properties tantamount to seed vigour as well as providing a means of enhancing seed vigour prior to sowing, even when seeds appear to be 'non-dormant'.

Figure 2. Interpretation of routine germination results

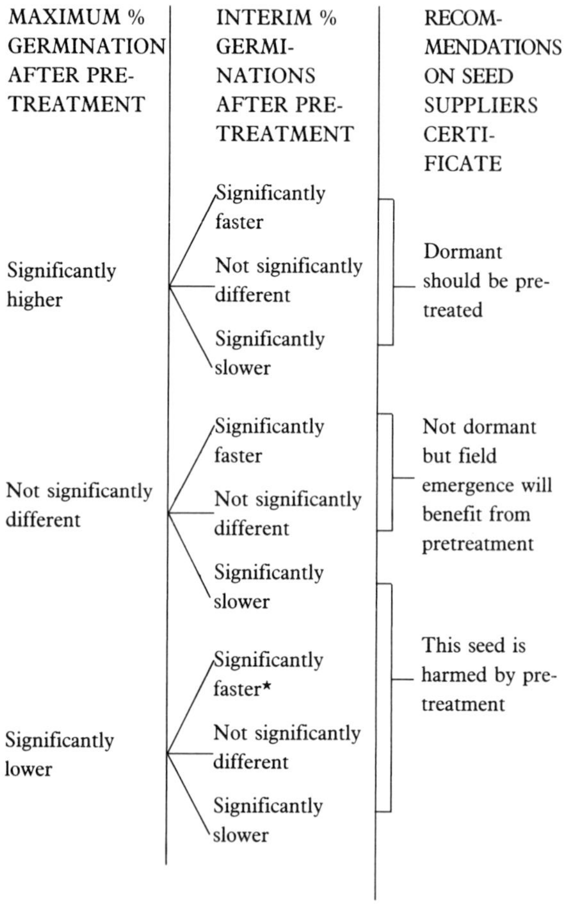

*Although lab germination is reduced by pretreatment the effects on field emergence are unknown

Interpretation of Laboratory Germination Data for Conifers

During the course of germination tests ± pretreatment, interim observations on the number of germinating seeds are made to confirm when the maximum percentage germination has been reached. It is therefore possible to compare not only the maximum percentage germination obtained by prechilled and unchilled seeds but also the respective rates at which these are achieved. Figure 2 shows how comparisons between the germination of prechilled and unchilled seeds can (in most cases) be divided into nine separate categories.

Until recently seed test data has been processed manually and the number of calculations incorporating the necessary statistical analyses made the above type of objective analysis impossible. However, recently data processing has been computerised and the necessary comparisons are now readily calculable. The precise methods used to identify whether percentages are statistically significantly different need not concern us here, except to say that an analysis is employed taking into consideration, 1) the total number of seeds in the two tests, 2) the highest percentage germination obtained, 3) the lowest percentage germination obtained, and 4) the mean of the two percentages.

When pretreatment results in a significant improvement to the maximum percentage germination in the laboratory there can be little doubt that it will benefit field emergence. However, recently the response of apparently non-dormant seed lots to rechilling has been investigated over a range of germination temperatures. This has shown that even for 'non-dormant' seed lots, when either the germination rate does not show a significant benefit for prechilling, but especially when it does, the range of

Figure 3. Maximum percentage germination of 'non-dormant' Sitka spruce seeds at various temperatures without pretreatment (● – ●) and with 3 weeks moist prechill (■ – ■). (Bars signify 97.5% confidence limits).

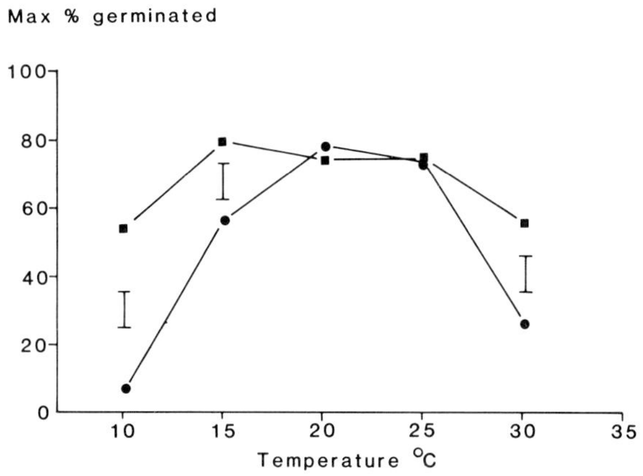

temperatures over which the seed is subsequently able to germinate is widened (Figure 3). In other words the seeds become more resistant to extreme or previously stressful conditions.

It is clear from this that the characteristics normally associated with 'seed vigour' can be mimicked by an extension of those properties more commonly taken to reflect 'seed dormancy'.

Conclusion

It is hoped that the method of germination data analysis described above will do three things:

1. increase the objectivity with which recommendations can be made on the benefit or injuriousness of prechilling;
2. lead to the possibility of predicting which seed lots are inherently more likely to be sensitive to field stresses, by providing an objective means of measuring the vague and illusive property known as seed vigour;
3. provide a means of increasing seed vigour prior to sowing.

In addition, since the seeds of deeply dormant species, e.g. cherry (*Prunus* spp.), ash (*Fraxinus* spp.), lime (*Tilia* spp.) and rowan (*Sorbus aucuparia*) lose their dormancy gradually during pretreatment and pass these stages with germination characteristics similar to those illustrated for conifers in Figure 3 (Gosling, unpublished data), a thorough understanding of the characteristics of this type of dormancy/vigour will lead to more effective propagation of these species from their seed.

Acknowledgements

I would like to thank Mr. D.C. Wakeman for supervising the seed testing; Miss K. Spriggs, Miss A. Pocock and Mrs Y. Samuel for conducting the seed tests; Mr. A. Peace for computer programming and statistical analysis; and Mr. J. Williams for drawing the Figures.

References

ALDHOUS, J.R. (1975). *Nursery practice.* Forestry Commission Bulletin 43. HMSO, London.

ANON. (1976). International rules for seed testing. *Seed Science and Technology* **4** (I) & (II).

Discussion

T.H.R. HALL (Oxford University Parks).

Is there any available information on the relationship between the age of seed and the rate and amount of germination?

P.G. GOSLING No work has been done specifically on this subject although it has been observed that age effects are more pronounced at high levels of moisture and high temperature storage. There is a general decrease in germination at the extremes of the ranges of both moisture and temperature.

J.G. GRACE (Royal Borough of Kingston).

What is the effect of light, particularly red and far red, and its relationship to seed depth in the soil?

P.G. GOSLING Unaware of information on that subject. In the laboratory the work is always done with the safest of safe lights.

The Growth of the Nursery Tree Root System and its Influence on Tree Performance after Transplanting

D. Atkinson[1] and T.E. Ofori-Asamoah[2]
East Malling Research Station, Maidstone, Kent

Summary

The basic processes involved in the growth of the root system of a deciduous woody plant are reviewed. The production of new roots and the survival of some of these as woody roots are both important. The growth of the root system in the nursery can be affected by treatment and both direct effects and those on the tree's ability to build up a reserve of nutrients are important. In the nursery and at transplanting the balance between shoot mass and root length is as important as the development of the root system alone. The success of transplanting is affected by its timing, at least partly due to interactions with natural root growth cycles, the quality of the nursery trees and the degree of root pruning at transplanting. Infection of the root system with vesicular-arbuscular mycorrhizas can be important.

Introduction

All fruit trees, most trees used in urban or amenity horticultural situations and many forest trees are raised in nurseries then transplanted to their final, or a second intermediate, location. The nursery stage nurtures the trees during the initial phase of growth, when mortality can be high, and so improves the probability of survival and the vigour of subsequent growth. It also makes effective use of space. The performance of the tree post-transplanting is a function of both the environment to which it is transplanted and the 'quality' of the tree. The root system of the tree at transplanting, its capacity to produce new roots and the balance between the root and shoot systems are all important.

Post-transplanting, the root system must provide the tree with water and mineral nutrients (and hormones?), so that early growth is not limited. It must also grow so as to supply a shoot system which is rapidly increasing in size. The 'reserves' of nutrients and assimilates built up during the nursery phase contribute to this. Most plants growing in field situations are infected, at least to some extent, with mycorrhizas, either vesicular arbuscular (endomycorrhizas) or sheathing (ectomycorrhizas), and so the degree of infection of the tree, at the time it leaves the nursery, and the potential of the root system developed in the field for infection are also important, perhaps especially where trees are being transplanted to urban or amenity sites of low fertility status.

This paper examines the characteristics of the woody root system, its development in the nursery and its impact on tree performance post-transplanting.

Root System Development

Woody plant root systems have a number of features in common. Some basic characteristics and attributes are discussed here as a basis for latter sections of a more applied nature.

All roots in the root system of a woody plant begin as white roots of primary structure. Roots survive in this condition for a period which can be as short as 2 weeks or as long as several months (over winter) (Atkinson, 1985a). Eventually all roots turn brown and lose all the tissues external to the pericycle. Some roots develop secondary tissues, secondary xylem (wood) and phellogen (bark), others remain in the soil as isolated steles or decay as a

---

1. Current address: Department of Soil Fertility, Macaulay Institute for Soil Research, Craigiebuckler, Aberdeen, U.K.
2. Current address: Oil Palm Research Institute, Kade, Ghana.

result of animal and microbial activity (Atkinson, 1980, 1983).

The proportion of new growth surviving varies with tree age. For apple trees on M9 rootstock Atkinson (1980, 1985) estimated survival as 57–63 per cent of the root length produced by one year trees but 21–25 per cent when averaged over the first 5 years. This decrease in survival is related to the type of root produced. New white roots divide into a) extension roots; of large diameter (usually 0.8–5.0 mm) with unlimited growth potential, b) lateral roots; of smaller diameter (usually 0.3–1 mm) with very limited growth potential (usually 10–20 mm). In newly transplanted trees (Atkinson, 1980) the ratio of lateral to extension roots is low, 0.2–0.9 but higher, 3–7 (Atkinson, 1983) subsequently.

All tree root systems, in both the nursery and in the field post-transplanting, except for those in early nursery stages comprise a mixture of new white roots, woody roots of variable diameter, and smaller diameter roots with either limited secondary development or with none of the primary tissues external to the pericycle remaining (brown roots). The balance of new white root length to woody plus brown roots varies during a season. Atkinson and Thomas (1985) found for 3-year apple trees on M9 rootstock that the proportion of total root length contributed by white root increased from 1–6 per cent in June to 23–28 per cent in July.

In nursery trees the proportion contributed by white roots will be high, especially in the year following the first planting of a cutting. Following planting in the nursery, either as a rooted seedling or a rootstock cutting, initial root growth will be either by the growth of a tap root system or, for vegetatively propagated plants, the development of adventitious roots from the stem. In subsequent years, root development can be:

a) new adventitious roots from the stem;
b) new extension or lateral roots from the larger diameter roots formed in a previous season;
c) new growth from brown roots.

To produce nursery plants with root systems with good regenerative capacity the relative importance of these sources of new growth should be understood.

Asamoah and Atkinson (1985) investigated the effects of root pruning and the growth regulator paclobutrazol (PP333) on the growth of Colt cherry roots (Table 1). Root pruning greatly reduced subsequent new root growth in control plants. When the plants were treated with paclobutrazol, which increases the production of new adventitious roots from the stem (Atkinson and Crisp, 1982), the effects of root pruning were much less marked, suggesting that initiation of new roots from the stem is important. Abod (personal communication) found production of new adventitious roots from the stem was the dominant factor in the growth of the root systems of both M9 and MM106 apple rootstocks. Asamoah and Atkinson (1985) found the improved root production of root pruned trees treated with paclobutrazol was associated with a smaller relative decrease in stem growth (Table 1) due to root pruning. The importance of new adventitious root growth from the stem to the mature tree can be seen by comparing the number of major roots originating from the rootstock of a tree at transplanting (usually 3–6) with the number on an established tree, for M9 up to 20 (Atkinson *et al.*, 1976).

The root system of a tree is important not just for nutrient uptake. The woody tissues, including the root system, of a tree can act as a nutrient reservoir for new growth (Figure 1). Early in the season the roots and woody

Table 1. The effect of root pruning and a soil application of paclobutrazol at 4 kg ha^{-1} on the root and stem growth of Colt cherry rootstocks (g)

	Control		Paclobutrazol-treated	
	Intact roots	Half roots	Intact roots	Half roots
Root	1.0	0.1	1.5	0.6
Stem	4.0	2.4	2.5	1.8

After Asamoah and Atkinson (1985).

Figure 1. Variation in the total amount (g $plant^{-1}$) of magnesium present in root, rootstock, scion wood and scion leaves of trees of Cox/MMlll budded the previous year. Early season reductions in the amounts in rootstock tissues indicate the potential value of nutrient reserves to early growth.

Redrawn from Asamoah (1984).

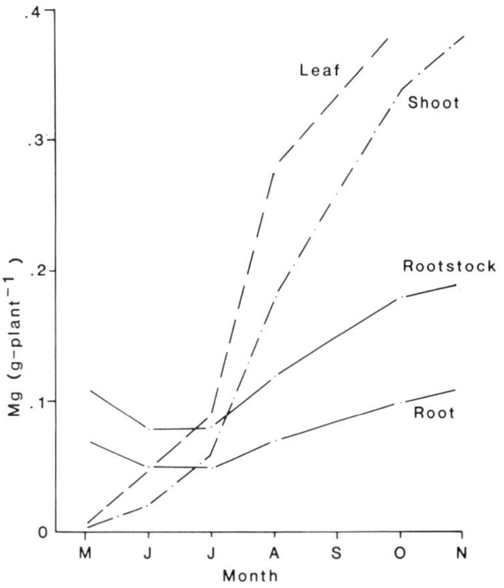

tissues of a rootstock, budded with a scion variety the previous summer, decreased in their nutrient content. The decrease was similar in amount to that needed for the early leaf and stem growth of the scion variety. Atkinson (1985b) calculated that reserves in woody tissues could supply a significant proportion of the nutrients needed by transplanted trees in the immediate post-transplanting phase.

Root System Growth in the Nursery

In a fruit nursery a rootstock is lined out as a single stem with a small number of roots budded in mid summer with a scion variety. It is grown to produce a maiden tree during the subsequent year then transplanted to the field. During this two year period the root system to support rapid post-transplanting growth is developed. There is less information on growth during this phase than by established trees. Asamoah (1984) found (Figure 2) that early in the season total root length decreased to reach a minimum in July after which it increased to a maximum in November. Studies of the periodicity of new growth showed peaks in June and in the autumn, August or October. This is similar to the pattern described for established trees by Atkinson (1980). Even during a period of net total root decrease, as in spring (Figure 2), new root production can still occur. During the period August to November the contribution to total root weight by woody roots increased from 58 per cent to 76 per cent. However even at this time white + brown roots made up 84 per cent of total root length.

Figure 2. The leaf area (dm^2 $plant^{-1}$) and root length (m $plant^{-1}$; white + brown + woody root) for trees Cox/MMlll budded the previous year.

Redrawn from Asamoah (1984).

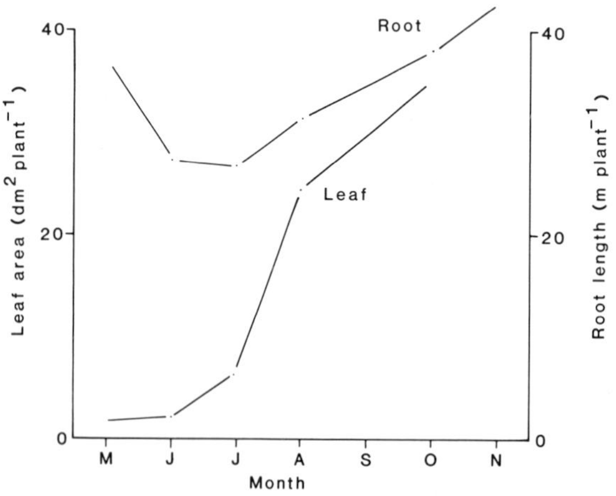

Table 2. The effect of irrigation, a polythene mulch and soil-applied fertiliser on white root length (cm per tube) at depth for trees of Cox/M26 in the year following budding

Depth (cm)	Control	Irrigation + fertiliser	Polythene + fertiliser	Control + fertiliser
0–10	1	13	6	1
11–50	13	7	12	8

After Asamoah (1984)

There have been many studies of the effects of nutrient applications on growth in the nursery. Graca and Hamilton (1981, 1982) and Gillam *et al.* (1980) found that root growth in cotoneaster was unaffected by nutrient additions although shoot growth was usually increased. Under deficient conditions Hu (1975) obtained improved root and shoot growth in *Paulownia fortunei* seedlings. In apple Asamoah (1984) found nutrient additions did not increase growth although mulching improved shoot but not root growth and irrigation altered root distribution but not total root length (Table 2). Irrigation and mulching both increased the proportion of the root system in the surface soil. Watson and Himelick (1982) found, for a range of ornamental species, that in the nursery most 'fine' roots were in the surface 12 cm of soil. Many of the treatments applied in the nursery affect root and shoot growth to different extents and during the course of a single season (Figure 2) the relationship between leaf area and root length can change greatly. This can have large effects post-transplanting.

The Effect of Transplanting on the Root System

At the time of transplanting trees are usually under cut and then removed in a 'bare root' condition. The subsequent effectiveness of the root system will be affected by the extent and timing of this 'root pruning'. Fuchigami and Moeller (1979) investigated the effect of time of transplanting for a range of ornamental species and found that, except for *Prunus laurocerasus*, all species showed a definite periodicity. Most species rooted well after July and December transplanting but relatively poorly if moved between August and November. Swanson (1977) compared the success of transplanting a range of species in October or May and found substantial damage with the autumn date. Whitcomb *et al.* (1978) found that *Platanus occidentalis*, *Pistacia chinensis*, *Nyssa sylvatica*, *Cornus florida* and *Betula maximowicziana* all grew better if transplanted in July than August. Lee and Hackett (1976) found that root regeneration was greatest at the time of terminal bud formation and lowest during dormancy for *Pistacia chinensis*. In disbudded plants root growth was highest at bud break and at terminal bud formation. There is substantial variation in response to transplanting time but the period when autumn root growth has either just begun, or during this peak, seems relatively poor, perhaps because transplanting damages new growth giving an imbalance with shoot growth.

The effects of root pruning have been recently reviewed by Geisler and Ferree (1984). Root pruning has been deliberately used to give a dense compact root system although generally it results in substantially reduced growth with the reduction persisting for substantial periods. After root pruning root activity and root growth rate can be increased (Richards and Rowe, 1977). Plant species seem to have characteristic functional relationships between root and shoot. The consequence of root pruning is therefore increased root growth to restore the balance. Atkinson and Thomas (1985) found that the ratio root length to leaf area was unaffected by tree planting density and for established apple trees relatively constant during a season; an increase from 3.5 to 4.1 cm cm^{-2}.

The Effect of Nursery Treatment on Post-Transplanting Performance

Links between the quality of nursery stock and subsequent performance are well established. Preston (1967) found that even after 3 years in the field trees planted with well developed lateral shoots (well-feathered-maidens) made more growth than trees planted without laterals. The relationship between the root system at planting and subsequent root and shoot growth is more equivocal. For a range of forest trees Sutton (1983) found that post-transplanting root growth was inconsistently related to root growth capacity (RGC). In Jack pine (*Pinus banksiana*) and spruce (*Picea* spp.) increased root growth was sometimes related to subsequent shoot growth. In contrast Burdett (1983) found that the survival and growth of first year Lodgepole pine (*Pinus contorta*) and White spruce (*Picea alba*) increased with increasing RGC. Trees produced with an increased number of lateral roots, induced by chemical root pruning, had improved growth and stability. Nambiar (1983) found that the vigour and number of first order laterals was important to root system regeneration and improved water and nutrient supply.

Asamoah (1984) found that nursery treatment, including treatments which had no effect in the nursery, could affect post-transplanting performance. As a result of increased shoot growth in the absence of any change in post-transplanting root volume a polythene mulch treatment increased the shoot length:root volume ratio from 6.6 to 8.9. During the year following transplanting, compared to an unmulched control, shoot growth decreased but root growth increased. In the second year after transplanting the mulched trees had more fruit buds. Trees which had received foliar nutrient additions in the nursery showed increased growth.

The Effect of Vesicular Arbuscular Mycorrhizas (VAM)

In many trees the degree of infection of the root system with VAM is important to the supply of mineral nutrients, i.e. phosphorus. A survey of papers dealing with the effects of VAM (Table 3) on a range of species indicated

Table 3. The effects of vesicular-arbuscular mycorrhizas on woody species

Species	VAM	Effect	Reference
Apple	Mixed	Improved growth on high arsenic soils	Trappe *et al.* (1973)
	Glomus mossae or *G. fasciculatus*	Increased shoot growth	Benson and Covey (1976)
	Mixed	Increased weight, leaf area and root volume	Plenchette *et al.* (1981)
	G. mossae	Increased growth, smaller P fertiliser need	Koch *et al.* (1982)
Avocado	*G. fasciculatus*	Increased growth	Menge *et al.* (1978)
Ornamental species	*G. fasciculatus* *G. mossae*, *G. etunicatus* or Mixed *Glomus* + *Gigaspora*	Increased shoot and root weight	Kormanik *et al.* (1982)
Peach	Mixed	Increased Zn and P uptake	Larue *et al.* (1975)
	Mixed	Increased growth	Lambert *et al.* (1979)

that infection can improve growth and mineral nutrition, even with relatively high levels of soil phosphorus. It can also give increased resistance to toxic elements, e.g. arsenic. In some situations soil fumigation may reduce growth by reducing the level of infection with VAM (Lambert *et al.*, 1979).

In many nurseries the use of chloropicrin or formaldehyde as fumigants can lead to the production of VAM-free plants. If these are transplanted to sites of marginal fertility, not uncommon in amenity sites, it may well result in retarded growth.

Conclusions

The growth of the tree root system in the nursery follows a similar pattern to that of trees post-transplanting, although the relative importance of factors may vary. In the nursery and after transplanting the balance between leaf area and root length may be as important as the size of the root system itself. Treatments in the nursery affecting tree nutrient status may be important in the post-transplanting stage because of the build up of nutrient reserves. A whole range of factors can influence the performance of the transplanted tree, but where a tree has an effective root system in the nursery then nutrient reserves should be available, if the tree is able to use them, for the rapid production of new roots after transplanting.

References

ASAMOAH, T.E.O. (1984). *Fruit tree root systems: effects of nursery and orchard management and some consequences for growth nutrient and water uptake.* PhD Thesis, Wye College, University of London.

ASAMOAH, T.E.O. and ATKINSON, D. (1985). The effect of (2RS, 3RS)-1-(4-chlorophenyl)-4, 4-dimethyl-2-2(1H-1,2,4 triazol-1-yl) pentan-3-ol (paclobutrazol:P-P333) and root pruning on the growth, water use and response to drought of Colt cherry rootstocks. *Plant Growth Regulation* **3** (1), 37–45.

ATKINSON, D. (1980). The growth and activity of fruit tree root systems under simulated orchard conditions. In, *Environment and root behaviour,* ed. D.N. Sen. Giobios, Jodhpur, India, 171–185.

ATKINSON, D. (1983). The growth, activity and distribution of the fruit tree root system. *Plant and Soil* **71**, 23–35.

ATKINSON, D. (1985a). Spatial and temporal aspects of root distribution as indicated by the use of a root observation laboratory. In, *Ecological interactions in the soil : plants, microbes and animals,* eds. A.H. Fitter, D. Atkinson, D.J. Read and M.B. Usher, 43–65. Blackwell, Oxford.

ATKINSON, D. (1985b). The nutrient requirements of fruit trees: some current considerations. In, *Advances in Plant Nutrition* **2.**

ATKINSON, D., NAYLOR, D. and COLDRICK, G.A. (1976). The effect of tree spacing on the apple root system. *Horticultural Research* **16**, 89–105.

ATKINSON, D. and CRISP, C.M. (1982). Prospects for manipulating tree root systems using plant growth regulators : some preliminary results. *Proceedings 1982 British Crop Protection Conference – Weeds,* 593–599.

ATKINSON, D. and THOMAS C.M.S. (1985). The influence of cultural methods on the water relations of fruit trees. *Acta Horiculturae* 171, 371–382.

BENSON, N.R. and COVEY, R.P. (1976). Response of apple seedlings to zinc fertilization and mycorrhizal inoculation. *Hortscience* **11**, 252–253.

BURDETT, A.N., SIMPSON, D.G. and THOMPSON, C.F. (1983). Root development and plantation establishment success. *Plant and Soil* **71**, 103–110.

FUCHIGAMI, L.H. and MOELLER, F.W. (1978). Root regeneration of evergreen plants. *Proceedings of International Plant Propagators' Society* **28**, 39–49.

GEISLER, D. and FERREE, D.C. (1984). Response of plants to root pruning. *Horticultural Reviews* **6**, 155–187.

GILLIAM, C.H., FRETZ, T.A. and SHEPPARD, W.J. (1980). Effect of nitrogen form and rate on elemental content and growth of pyracantha, cotoneaster and weigela. *Scientia Horiculturae* **13**, 173–179.

GRACA, M.E.C. and HAMILTON, D.F. (1981). Effects of nitrogen and phosphorus on root and shoot growth of *Cotoneaster divaricata* Rehd. and Wils. *Scientia Horticulturae* **15**, 77–85.

GRACA, M.E.C. and HAMILTON, D.F. (1982). How cotoneaster cuttings respond to nitrogen and potassium treatments. *American Nurseryman* **155**, 46–49.

HU, H.T. (1975). Influence of mineral nutrition on the development of *Paulownia fortunei* seedlings. I. Leaf symptoms, root development and height growth. *Memoirs, College of Agriculture, Taiwan University* **16**, 33–44.

KOCH, B.L., COVEY, R.P. and LARSEN, H.J. (1982). Response of apple seedlings in fumigated soil to phosphorus and vesicular-arbuscular mycorrhiza. *Hortscience* **17**, 232–233.

KORMANIK, P.P., SCHULTZ, R.C. and BRYAN, W.C. (1982). The influence of vesicular-arbuscular mycorrhizae on the growth and development of eight hardwood tree species. *Forest Science* **28**, 531–539.

LAMBERT, D.H., STOUFFER, R.F. and COLE, H. (1979). Stunting of peach seedlings following soil fumigation. *Journal American Society for Horticultural Science* **104**, 433–435.

LARUE, H.J., McCLELLAN, W.D. and PEACOCK, W.L. (1975). Mycorrhizal fungi and peach nursery nutrition. *California Agriculture* **29**, 7.

LEE, C.I. and HACKETT, W.P. (1976). Root regeneration of transplanted *Pistacia chinensis* Bunge seedlings at different growth stages. *Journal American Society for Horticultural Science* **101**, 236–240.

MENGE, J.A., DAVIES, R.M., JOHNSON, E.L.V. and ZENTMYER, G.A. (1978). Mycorrhizal fungi increase growth and reduce transplant injury in avocado. *California Agriculture* **32**, 6–7.

NAMBIAR, E.K.S. (1983). Root development and configuration in intensively managed Radiata pine plantations. *Plant and Soil* **71**, 37–47.

PLENCHETTE, C., FURLAN, V. and FORTIN, J.A. (1981). Growth stimulation of apple trees in unsterilized soil under field conditions with VA mycorrhizal inoculation. *Canadian Journal of Botany* **59**, 2003–2008.

PRESTON, A.P. (1967). Using feathers as primary branches on apple trees. *Report of East Malling Research Station for 1966*, 211.

RICHARDS, D. and ROWE, R.N. (1977). Effects of root restriction, root pruning and 6-benzylaminopurine on the growth of peach seedlings. *Annals of Botany* **41**, 729–740.

SUTTON, R.F. (1983). Root growth capacity : relationship with field root growth and performance in outplanted Jack pine and black spruce. *Plant and Soil* **71**, 111–112.

SWANSON, B.T. (1977). Transplanting woody plants effectively. *American Nurseryman* **146**, 7–8.

TRAPPE, J.M., STABLY, E.A., BENSON, N.R. and DUFF, D.M. (1973). Mycorrhizal deficiency of apple trees in high arsenic soils. *Hortscience* **8**, 52–53.

WHITCOMBE, C.E., STORJOHANN, A. and GIBSON, J. (1978). Effect of time of transplanting container grown tree seedlings on subsequent growth and development. *Research Report, Agricultural Experiment Station, Oklahoma State University* 777, 37–39.

Discussion

A.D. BRADSHAW (Department of Botany, Liverpool University).

Could you comment further on the 'loading' of plants with nutrients while in the nursery.

D. ATKINSON Nutrients can be stored in the rootstock wood. If the volume of the roots is known then the reserves can be calculated and these are often substantial. Tissues give up water and nutrients in response to demand elsewhere in the plant. This movement can be traced by N_{14} labelling. Uptake of excess nutrients in the nursery stage will assist healthy growth subsequently.

J.E.G. GOOD (Institute of Terrestrial Ecology, Bangor).

On planting out, some clones always die back more than others but may ultimately be the best survivors. Could this be the result of re-establishing a good root:shoot ratio?

D. ATKINSON I have found no effect later of N application in the first year but P given at this time shows subsequent benefit.

A.J. GRAYSON (Forestry Commission Research Station).

If looked at over a period of time the story seems diverse with lines on a graph crossing once or perhaps twice. The results are rarely unequivocal.

A.D. BRADSHAW (Department of Botany, Liverpool University).

Different results in different experiments may both have applicability.

D. ATKINSON It should also be borne in mind that what holds for one species may not apply to another.

Reclamation of Mineral Workings to Forestry

K. Wilson
*Forestry Commission**

Summary

Recent legislation has placed powers on the Local Planning Authorities to ensure that sites damaged by mineral extraction are reclaimed to an economic use. To be successful, reclamation must be planned well in advance of the operation and provision made to ensure the soil conditions are appropriate for healthy tree growth. The main problems are reviewed and solutions suggested.

Introduction

Throughout Britain there are 121 × 10^3ha of derelict and despoiled land and the largest single cause of this is the extractive industries (Department of the Environment, 1984). Prior to the enactment of the Town and Country Planning (Minerals) Act 1981 there was no requirement for reclamation of despoiled land. Local Authorities backed by Department of the Environment grants have accepted responsibility for these historic sites. Now, under the Act, the extractor is responsible for and required to undertake reclamation once extraction is completed.

A literature survey supported by field observations and discussions with 40 organisations was started in 1983 with a brief to review the problems and practices of reclaiming surface workings. Deep-mined colliery spoil, which is covered by Jobling and Stevens (1980) and Jobling (1981 and 1987), and open cast coal sites were excluded from the review. Nevertheless experience from open cast sites is applicable to other mineral sites.

Most worked sites will be restored to agriculture, but many areas are more suitable for tree planting and these include plateau gravels of south and south-east England, sites with thin or non-existent topsoils, steeply sloping sites and sites in the uplands. Also sites adjacent to houses may be developed for open space use which will involve tree planting.

Soil Compaction and Impeded Drainage

Mineral extraction and reclamation involves heavy machinery such as box scrapers and bulldozers which move overburden and topsoil to create desirable land forms. These machines "exert a pressure on the soil of 2–3 kg/cm^2" and compaction of the order of 15–20 per cent has been reported (Downing, 1971). This compaction closes the large pores in the soil and increases the proportion of small pores. As a result the rate of water and gas movement into and through the profile is reduced and water stored in the soil is held more tightly because of the higher capillary forces present in the narrower pores. Less water is therefore available to the plants which are only able to exert a finite suction before wilting occurs. Furthermore, if the bulk density of the material is increased beyond a certain point roots cannot penetrate the soil (Rimmer, 1979).

Soils are much more prone to compaction when worked wet and all earthmoving operations, including ripping and discing, should therefore only be carried out during dry conditions which usually occur in the June to August period. Discs have been used to draw up small ridges to provide an improved planting position, but this technique did not relieve the compaction or improve the site drainage. Complete ripping to 50–75 cm (a depth

*Formerly at the Forest Research Station, now Forest District Manager, Northants Forest District.

calculated to hold sufficient water to sustain the growing trees during normal dry periods – and about the maximum practicable depth) resulted in a suitable planting medium. It soon became obvious that where the site was relatively flat with poor drainage, the soil structure, which had been created by the ripping, rapidly slaked down under the influence of the winter's rainfall with predictable results. Russell (1961) in his standard text *Soil conditions and plant growth* said "If, but only if, the drainage system allows surplus soil water to be removed rapidly its efficiency can often be increased by deep tillage, either ploughing or subsoiling, . . . if the drainage of the subsoil is very impeded, deep ploughing merely increases the volume of the largest pores that can become waterlogged in wet weather and helps to turn the soil into a marsh".

Adequate movement of water can be produced in a wide range of soil textures by slopes of 1:10 (about 6°). Thus the recommendation for treatment for restoring compacted areas is a combination of slopes and deep ripping. Compacted sites are most effectively loosened by ripping down-slope with three winged tines mounted in a parallelogram linkage on a Caterpillar D8 or equivalent 300 HP machine (Binns and Fourt, 1981 and 1984a).

Where natural slopes are absent they are best created as a series of ridges, 30 m and 40 m wide and 1.5 to 2 m high in the centre, giving 1:10 or 6° slopes. On permeable sites ripping along the furrow bottom helps through-percolation. On impermeable sites open drains must be dug in the furrows and the ridges themselves should slope at between 1° and 2°. Ridges of sandy materials should be ripped to a depth of 75 cm, clayey materials to 50 cm. Discing downslope to raise small planting ridges is beneficial in the heavier spoils. Sites with natural slopes of between 6° and 20° should be ripped to 75 cm downhill, and provided with open contour drains, or benches on steeper slopes. Slopes steeper than 20° should be regraded to produce shallower slopes to reduce the risk of erosion.

It is easier and cheaper to incorporate the required slopes into the site during the course of restoration rather than having to construct them afterwards – indeed regrading may be quite impracticable.

Some Problems of Tipped Sites

If, as often happens, mineral extraction from a site is followed by tipping of domestic refuse, specific problems will have to be catered for. The combination of gases and heat from decomposition may cause root death and uneven settlement of the landfill can lead to localised waterlogged hollows. Good slopes and a thick final cover of rootable material are both essential, as are measures to prevent contamination of the rooting zone by gaseous or liquid pollutants. These can require construction of gas vents and leachate sumps. Treatment of waste disposal sites for tree planting is, however, less well researched than for normal mineral sites. Development of amelioration techniques is hampered because these problems may not appear till several years after reinstatement and planting. Recommendations are given by Wilson (1985) for modifications to these standard treatments, or for additional measures which are considered necessary for the restoration of specialised sites, including: very rocky sites, phytotoxic spoils, washings or tailings, lagoons and 'hill and dale'.

Loose tipping of material on sites designated for woodland produces uncompacted soils which are freely drained. Trees planted on these sites are showing good initial growth without any expensive earth moving or cultivation.

Species Selection and Planting

Having created physical conditions in the soil suitable for planting and sustained tree growth, care must be taken to ensure that the trees themselves are appropriate species, of good quality, well planted and given adequate after-care.

On all sites the conditions are likely to be harsh. The climate of reclamation sites is likely to be extreme. Pioneer tree species such as Corsican pine (*Pinus nigra*), larch (*Larix* spp.) and alder (*Alnus* spp.) must be the first choice. Climax species such as oak (*Quercus* spp.) and beech (*Fagus sylvatica*) should be kept for the second rotation by which time a suitable soil structure should have developed.

Care must be taken to ensure that apparently good plants are not damaged by desiccation between the nursery and final planting (Insley, 1979). Bare-root transplants are probably most suitable for reclamation sites and they must be sealed in plastic bags, shaded from sunlight and planted as soon as possible after lifting from the nursery or else stored in cold stores.

Nutritional Status of Reclamation Sites

Most spoils contain sufficient mineral nutrients for tree growth, but sites with little or no topsoil will be nitrogen deficient. Nitrogen from artificial fertilisers is rapidly leached from most spoils and should not be applied until trees have developed a new root system. Nitrogen (and phosphorus) is also supplied in animal slurries and sewage sludges; these materials may however be unpleasant to handle, difficult to apply, and may also contain weed seeds or toxic levels of heavy metals. There is little practical experience of their use on reclamation sites and the results of the few trials show that application should be delayed for 3–5 years, until a spread of roots is present to take up the nutrients released. Nitrogen-fixing plants, such as legumes and alders, can be used as a source of nitrogen (Binns and Fourt, 1984b). Alders can be used in mixture with other tree species while shrubby or herbaceous legumes can be used as ground cover plants.

It has been found that grass or clover swards, including those recommended in the Department of Transport mixture, which contains *Lolium perenne* and *Festuca rubra*, compete strongly with planted trees for water and nutrients (Insley, 1981); deep rooting herbaceous legumes appear to be less competitive. Swards must be controlled around trees (Davies, 1983 and 1985). Any suggestion of nutrient deficiency should be checked by analysis of foliage (Binns *et al.*, 1983).

Treatment of Planted Reclamation Sites

Some sites which have been reclaimed and planted with trees in the past are showing only moderate or poor tree growth, often because of untreated compaction. It may be practicable to rip a sloping site if the tines and tractor tracks can be positioned between the rows of trees; otherwise it will be necessary to determine the best time to clear all the area, re-form the surface topography if necessary and rip to disrupt the compaction.

After-care of Planting

The after-care of the plantations established on restored sites should be monitored by the Mineral Planning Authority and the operator to detect problems in their early stages when remedial action may well be simpler and cheaper than later on.

Conclusion

Successful restoration can most easily be accomplished if it is planned for from the earliest stages in the consideration of any mineral extraction operation. Such planning should make provision for contouring, cultivation, the use of appropriate species and sizes of plants which are of good quality and well handled. Once planted on the site all trees should be monitored so that necessary after-care can be implemented at an early stage.

Some sites will have specific problems and these must be recognised and specialist guidance sought before reclamation commences.

Acknowledgements

I am very grateful to those colleagues in the Forestry Commission, local authorities and industry who gave valuable assistance at all stages of the preparation of the report. The work was made possible through a Department of the Environment contract with the Forestry Commission.

References

BINNS, W.O. and FOURT, D.F. (1981). Surface workings and trees. In, *Research for practical arboriculture*. Forestry Commission Occasional Paper 10, 60–75. Forestry Commission, Edinburgh.

BINNS, W.O. and FOURT, D.F. (1984a). *Reclamation of surface workings for trees. I. Landforms and cultivation.* Arboriculture Research Note 27/84/SSS. DOE Arboricultural Advisory and Information Service, Forestry Commission.

BINNS, W.O. and FOURT, D.F. (1984b). *Reclamation of surface workings for trees. II. Nitrogen nutrition.* Arboriculture Research Note 28/84/SSS. DOE Arboricultural Advisory and Information Service, Forestry Commission.

BINNS, W.O., INSLEY, H. and GARDINER, J.B.H. (1983). *Nutrition of broadleaved amenity trees. I. Foliage sampling and analysis for determining nutrient status.* Arboriculture Research Note 50/83/SSS. DOE Arboricultural Advisory and Information Service, Forestry Commission.

DAVIES, R.J. (1983). Transplant stress. In, *Tree establishment*, Proceedings of a Symposium at Bath University.

DAVIES, R.J. (1985). The importance of weed control and the use of tree shelters for establishing broadleaved trees on grass-dominated sites in England. *Forestry* **58** (2), 167–180.

DEPARTMENT OF THE ENVIRONMENT (1984). *Results of the survey of land for mineral working in England 1982.* Department of the Environment, London.

DOWNING, M.F. (1971). Earth works: outline of methods used within the project. In, *Landscape reclamation, a report on research into problems of reclaiming derelict land*, **1**, 43–52. University of Newcastle-upon-Tyne.

INSLEY, H. (1979). *Damage to broadleaved seedlings by desiccation.* Arboriculture Research Note 8/79/ARB. DOE Arboricultural Advisory and Information Service, Forestry Commission.

INSLEY, H. (1981). Roadside and open space trees. In, *Research for practical arboriculture*. Forestry Commission Occasional Paper 10, 84–92. Forestry Commission, Edinburgh.

JOBLING, J. (1981). Reworked spoil and trees. In, *Research for practical arboriculture*. Forestry Commission Occasional Paper 10, 76–83. Forestry Commission, Edinburgh.

JOBLING, J. (1987). Reclamation: colliery spoil. In, *Advances in practical arboriculture*. Forestry Commission Bulletin 65, 42–51. HMSO, London.

JOBLING, J. and STEVENS, F.R.W. (1980). *Establishment of trees on regraded colliery spoil heaps*. Forestry Commission Occasional Paper 7. Forestry Commission, Edinburgh.

RIMMER, D.L. (1979). Effects of increasing compaction on grass growth on colliery spoil. *Journal of Sports Turf Research Institute* **5**.

RUSSELL, E.W. (1961). *Soil conditions and plant growth* (10th ed.). Longmans, London.

WILSON, K. (1985). *A guide to reclamation of mineral workings for forestry*. Forestry Commission Research and Development Paper 141. Forestry Commission, Edinburgh.

Discussion

T. RENDELL (Gwynedd County Council)

What are the costs of implementing the various recommendations for reclamation?

K. WILSON As costs tend to be very variable it is not possible to give meaningful figures, but it is important to do the land formation process to the specification when the material is being replaced on site. This should minimise the additional cost of the very expensive earth moving operation.

Reclamation: Colliery Spoil

J. Jobling
Forestry Commission

Summary

A programme of research carried out by the Forestry Commission Research Division between 1980 and 1985 is briefly reviewed. The research, carried out under contract to the Department of the Environment, was undertaken to determine the most effective engineering and cultural practices to obtain high survival and growth rates of trees on reclaimed colliery spoil heaps. Some 30 experiments were established and assessed during the contract, mostly on reclaimed sites in the north of England. A large amount of non-experimental work was also conducted on reclaimed sites and, in addition, supporting research was carried out at Alice Holt Lodge. Cultivation practices, planting methods and root growth in compacted spoil, type and size of planting stock, choice of species and choice of ground cover receive special attention in this paper.

Introduction

The chemical and physical properties of colliery spoil and some spoil heap reclamation practices may seriously hinder tree planting and adversely affect tree establishment and vigour (Jobling and Stevens, 1980). Between 1980 and 1985 a programme of research was carried out by the Forestry Commission's Research Division, under contract to the Department of the Environment, to determine the most effective engineering and cultural practices to obtain consistently high survival and growth rates of trees on reclaimed colliery spoil heaps. Thirty experiments were started and regularly assessed during the contract, most of them on regraded heaps in the north of England, to observe the responses of trees over several seasons to different experimental treatments. A large amount of supporting work was undertaken both in the field and at the Forest Research Station, Alice Holt Lodge.

The research has proved to be informative and a comprehensive review, appraising the accumulated evidence and drawing conclusions helpful to practitioners, has been undertaken. Whenever possible the report includes guidance on the best available practices. The research carried out on cultivation practices, planting methods and root growth in compacted spoil, and on type and size of planting stock, choice of species and choice of ground cover, is discussed in this paper.

Cultivation to Relieve Compaction

Regraded colliery spoil is severely compacted and highly resistant to penetration. Newly regraded heaps are intensively ripped to a depth of 45 to 60 cm on completion of major earth-works – the operation is known as 'rooting' and is carried out to expose large, solid objects in the spoil which are taken from the site – and are then subsoiled to a depth of at least 45 cm to incorporate limestone and fertilisers prior to sowing grass seeds mixtures. However, the spoil is usually seriously re-compacted due to settlement and the movement of tractors and other agricultural equipment over the site when tree planting follows 6 to 18 months later. As a consequence, planting is physically difficult to carry out.

Many local authorities relieve compaction prior to tree planting by ripping to a depth of 60 cm along planting lines during dry weather in late summer-early autumn. Experiments on reclaimed heaps in Northumberland, County Durham and West Yorkshire have revealed that planting in ripped spoil, some 10 to 15 cm away from the slot, leads to several important benefits. In the first place the rate and quality of planting are improved, and large increases in stem and root biomass and in the depth and lateral extent of rooting can be readily detected over several seasons. In the case of Scots pine (*Pinus sylvestris*) needle size and colour may be improved by planting in ripped spoil and the risk of socketing and basal bowing of the stem much reduced. More important, however, trees planted in ripped spoil have markedly better survival and growth rates than trees planted direct into unripped spoil.

A more intensive ripping system has been developed by reclamation officers in West Yorkshire. Ripping to a depth of 60 cm is carried out first along planting lines and then a second ripping is undertaken at right angles at the same spacing. Trees are planted about 15 cm away from the intersection of the slots. The experiments disclosed that trees in cross-ripped spoil tended to have higher survival and growth rates than trees in spoil ripped only along planting lines though none of the increases was found to be statistically significant. The improvements in tree performance were not sufficiently large to suggest that cross-ripping should be generally adopted, but the practice may be justified where it facilitates planting and its supervision. Figure 1 shows the changes in the rate of survival and Figure 2 the changes in the mean total height of Scots pine in ripped and unripped spoil at North Beechburn, County Durham.

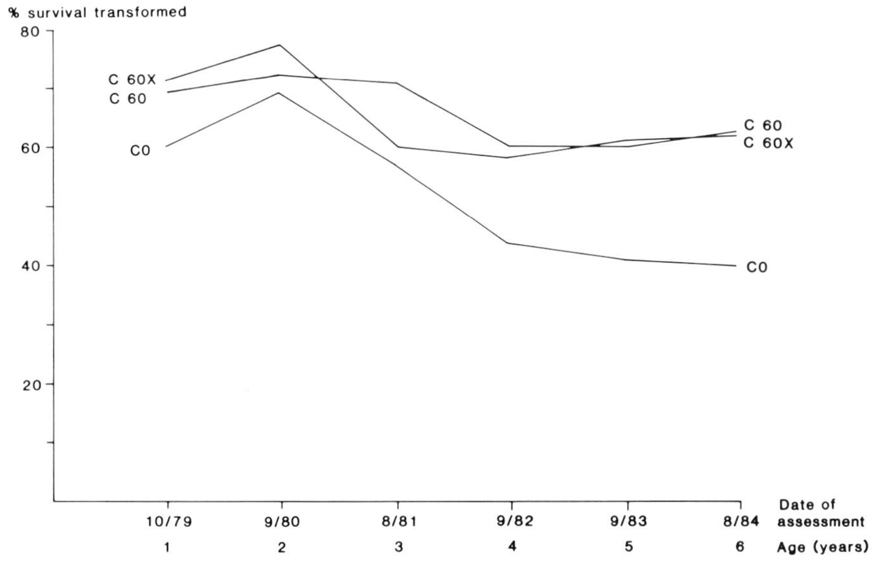

Figure 1. Mean survival rate over six seasons of Scots pine (*Pinus sylvestris*) in ripped and unripped spoil. North Beechburn, County Durham.

CO = no ripping, C60 = ripping to 60 cm along planting lines, C60x = ripping to 60 cm along planting lines and at right angles.

Increases in survival rate are due to changes in assessment and data analysis procedures.

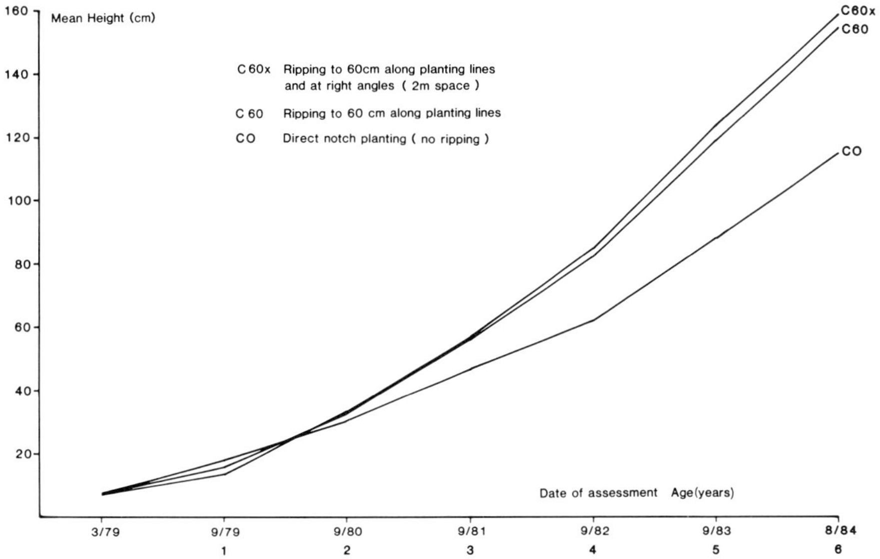

Figure 2. Mean total height at planting and over six seasons of Scots pine (*Pinus sylvestris*) in ripped and unripped spoil. North Beechburn, County Durham.

CO = no ripping, C60 = ripping to 60 cm along planting lines, C60x = ripping to 60 cm along planting lines and at right angles.

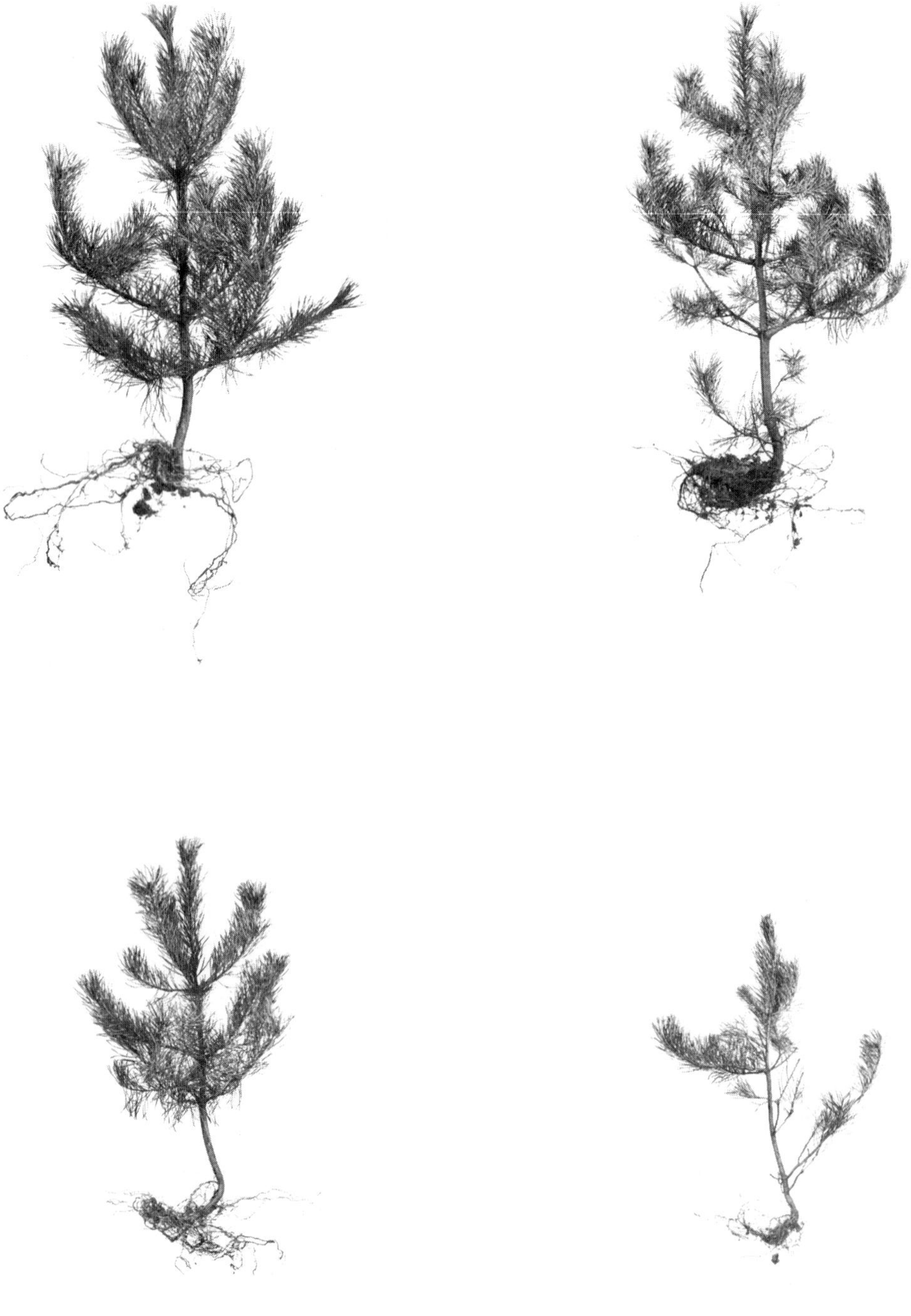

Plate 1. Three-year-old Scots pine from ripped spoil to 60 cm (top) and unripped spoil (bottom). Ashington/Woodhorn, Northumberland. Scale 1 : 10

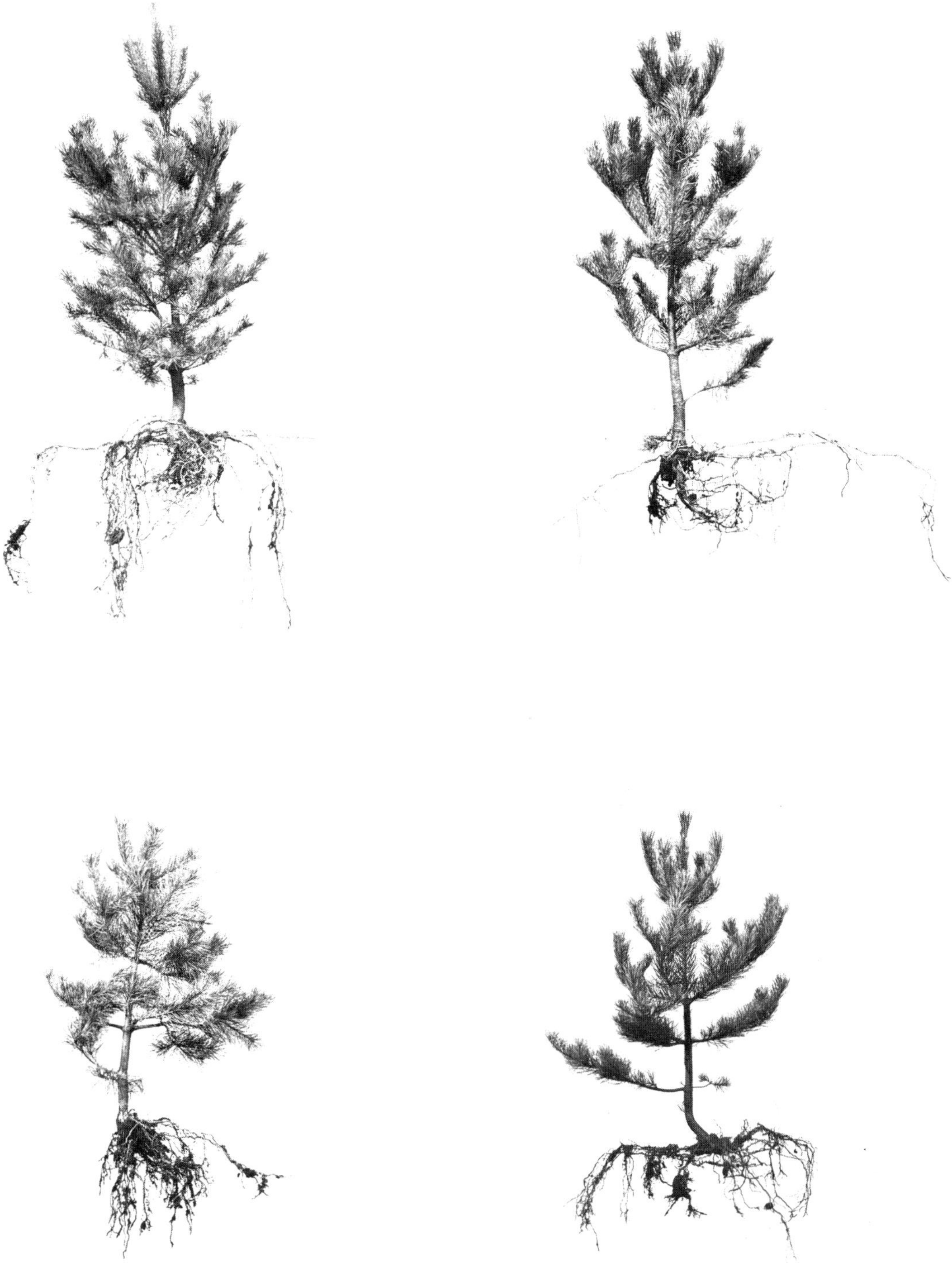

Plate 2. Four-year-old Scots pine from ripped spoil to 60 cm (top) and unripped spoil (bottom). North Beechburn, County Durham.

In spoil ripped only to a depth of 30 cm, tree behaviour over five growing seasons was remarkably similar to that of trees in spoil ripped to 60 cm. However, the long term performance of trees in shallow-ripped spoil cannot at this stage be fairly predicted and shallow ripping cannot be recommended. Tree planting and growth in compacted spoil, and the benefits of pre-planting ripping have been reviewed by Jobling and Carnell (1985).

Root Growth

Excavation of Scots pine roots in cultivation experiments in Northumberland and County Durham revealed that trees in spoil ripped to 60 cm along planting lines had a much larger root biomass and considerably longer roots, both laterally and downwards, than trees in unripped spoil. There were indications that early root extension had taken place primarily along the line of the rip. However, most roots had subsequently grown at a wide angle to the rip and by the fourth and fifth seasons root development appeared to be unrelated to the direction of ripping. In cross-ripped spoil the influence of directions of cultivation on initial root growth was just as marked – early root extension was found to be primarily in four directions – but it was also apparent that subsequent root development had occurred independently of direction of ripping. In cross-ripped spoil rooting was slightly deeper than in spoil ripped only along planting lines, but the difference was not considered important. Perhaps the most encouraging aspect of the work was the detection of roots extending downwards into compacted spoil below the prescribed depth of ripping. Strong tap root development was also found in spoil ripped to 60 cm and a large proportion of the root system had extended downwards throughout the whole of the cultivated profile (Plates 1 and 2) see previous page spread.

In spoil ripped only to 30 cm, though a few roots were found well below this depth, there was no tap root development and a large proportion of the root system was comparatively close to the surface. This finding, in particular, casts doubt on the merits of shallow pre-planting ripping.

Root systems of other conifers and of several broadleaved species were also examined. Most showed widespread lateral development and downward growth into compacted, uncultivated spoil. Common alder (*Alnus glutinosa*), Grey alder (*A. incana*), Locust tree (*Robinia pseudoacacia*) and European larch (*Larix decidua*) had the most vigorous and spreading roots. Nodules were regularly found on the roots of the two alders and Locust tree. Silver birch (*Betula pendula*) had the most surface roots of any tree. None of the birch specimens examined showed downward root growth into uncultivated spoil but, as if to compensate, the lateral roots usually extended over greater distances than the roots of trees of other species. Corsican pine (*P. nigra* var. *maritima*) had the most surface roots of any conifer but their lateral roots were exceptionally long. As with Scots pine, direction of rooting after the first year or two appeared to be independent of previous cultivation. On sites which had not been deeply cultivated some windblow – mainly of pines – was occurring d'ue to restricted root development near the spoil surface.

Methods of Planting

Trees in woodland areas are notch planted. Pit planting is limited to container trees and to well grown broadleaved transplants possessing large root systems. In compacted spoil planting is difficult and time consuming. Roots may be seriously bent, curled or broken during planting, and many are placed just below the spoil surface and are liable to dry out quickly. Shallow planting increases the risk of socketing and windblow. Standards of planting are much improved by planting in ripped spoil (ripping is usually done in dry weather at the end of the summer). Roots are then better spread, they are less contorted and damaged, and they can be placed more at their proper depth with less risk of drying out or socketing. Planting is always easier in wet spoil than in dry spoil.

On slopes that are too steep for safe tractor movement and cannot be ripped, or where erosion might follow ripping, soil augers or crumblers are used to cultivate spoil locally at the planting position. The behaviour of trees in this locally cultivated spoil was compared with that of trees in ripped spoil in two experiments. The holes in both were made by a Stihl soil crumbler. Trees in locally cultivated spoil generally had lower survival (Figure 3) and growth rates (Figure 4) than trees in ripped spoil but their performance was better than trees in compacted, uncultivated spoil.

Because of fears that serious root deformation or root spiralling had occurred in holes prepared by soil crumbler, root systems of Scots pine were excavated at the end of the fifth season. Only one of the trees had a tap root, which extended only to a depth of 40 cm, and all had wide-spreading lateral roots just a few centimetres below the spoil surface. These lateral roots were by far the thinnest of any encountered and they were growing in all directions.

While planting into locally cultivated spoil is preferable to notch planting into compacted, uncultivated spoil, it is a practice which should be confined to parts of sites which cannot be ripped.

The benefits of using plastic shelters to establish broadleaved trees was also examined experimentally.

Figure 3. Mean survival rate after five seasons of Scots pine (*Pinus sylvestris*), Common oak (*Quercus robur*), Common ash (*Fraxinus excelsior*) and sycamore (*Acer pseudoplatanus*) in ripped, locally cultivated and uncultivated spoil. Water Haigh/Oulton Brickworks, West Yorkshire.

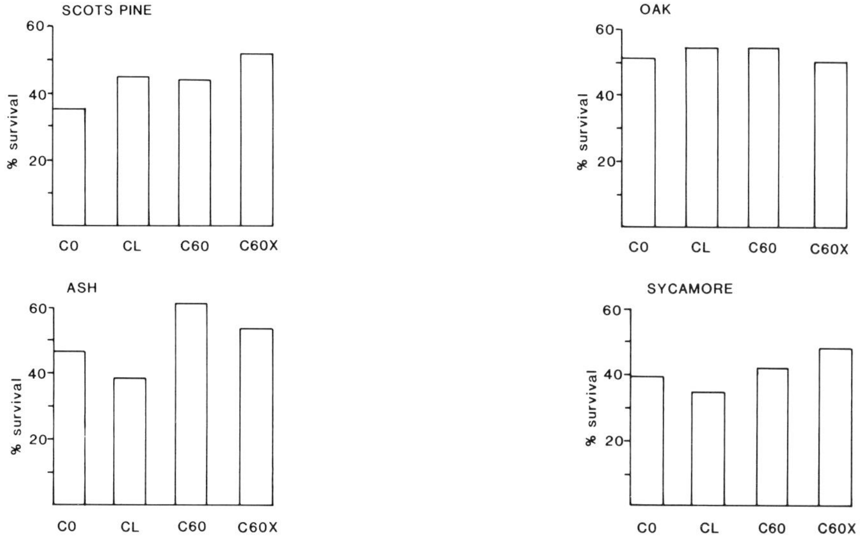

CO = no cultivation, CL = local cultivation by soil crumbler, C60 = ripping to 60 cm along planting lines, C60x = ripping to 60 cm along planting lines and at right angles.

Figure 4. Mean total height after five seasons of Scots pine (*Pinus sylvestris*), Common oak (*Quercus robur*), Common ash (*Fraxinus excelsior*) and sycamore (*Acer pseudoplatanus*) in ripped, locally cultivated and uncultivated spoil. Water Haigh/Oulton Brickworks, West Yorkshire.

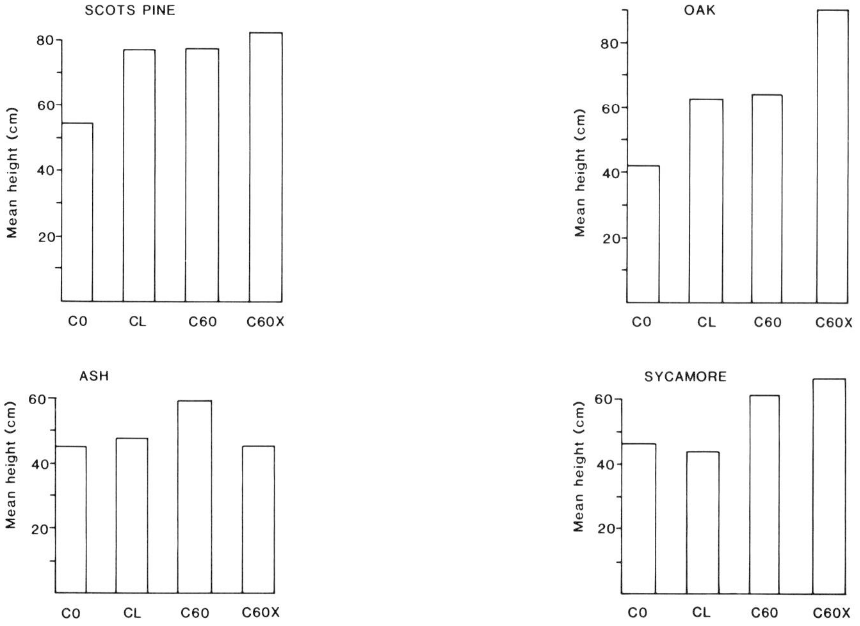

CO = no cultivation, CL = local cultivation by soil crumbler, C60 = ripping to 60 cm along planting lines, C60x = ripping to 60 cm along planting lines and at right angles.

Figure 5. Survival of three types of planting stock of Corsican pine (*Pinus nigra* var. *maritima*), sycamore (*Acer pseudoplatanus*) and birch (*Betula pendula*) after three seasons. Ashington, Northumberland, Mainsforth and Trimdon Grange, County Durham.

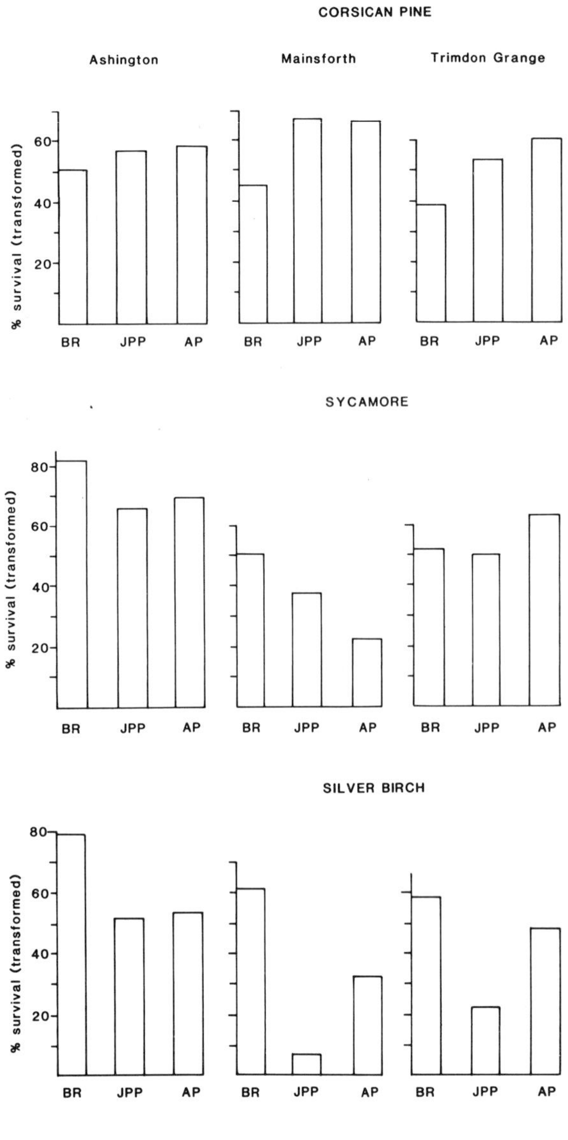

BR = bare-root trees, JPP = trees in Japanese paperpots, AP = trees in amenity pots.

Unfortunately, stakes were not easily driven into the spoil vertically or firmly, because of the high stone content, and many broke, and much socketing and damage to shelters occurred. But by using pointed stakes free of knots and shorter shelters (80 cm) than normal, reliable evidence was obtained at one site. After two seasons, sycamore (*Acer pseudoplatanus*), Silver birch, beech (*Fagus sylvatica*,) Common oak (*Quercus robur*) and rowan (*Sorbus aucuparia*) had a significantly greater height increase and were significantly taller when grown in shelters. At all sites, trees in shelters were adequately protected against injury by sheep, rabbits and hares while trees without shelters suffered serious damage or were killed.

Type of Planting Stock

The behaviour of 1-year-old seedlings of Silver birch, sycamore and Corsican pine raised in Japanese paperpots or in amenity pots was compared at four sites with that of 2-year-old bare-root transplants. Only in the case of Corsican pine were the survival rates (Figure 5) of the container trees consistently higher than those of the bare-root trees (three out of four sites). Even then, in three of the experiments bare-root trees had the same increase in height over three seasons as trees in amenity pots, and both grew more than trees in Japanese paperpots. In the case of sycamore, trees in amenity pots had a higher survival rate than bare-root trees at only one of the four sites, while the height growth of the two types of tree showed little difference overall. The Japanese paperpot sycamore had the poorest survival rate of the three types. Bare-root birch had consistently higher survival and growth rates in all four experiments than birch in either type of container. Survival rates of all three species are shown in Figure 5.

Hardly any of the experimental evidence suggested that 1-year-old seedlings raised in Japanese paperpots or in amenity pots should be preferred to 2-year-old bare-root transplants.

Size of Planting Stock

The behaviour of four grades of tree of six species was compared in four experiments. Details of the grades are shown in Table 1. The specifications were laid down by the Standing Local Authority Officers' Panel on Land Reclamation Vegetation and After Management Subgroups, and show a greater stem diameter for a given height than the minimum standards published by the British Standards Institution (BS 3936, Part 4).

Corsican pine was the only species of the six which did not show a marked improvement in survival rate due to the use of trees with larger stem diameter than the minimum

Table 1. Sizes of planting stock of six species in experiments at Ashington/Woodhorn, Northumberland, Mainsforth and Trimdon Grange, County Durham and Water Haigh/Oulton Brickworks, West Yorkshire.

	Height (cm)	Diameter (mm)	Grade
Sycamore *(Acer pseudoplatanus)*	45–60	<9.0	G1
	45–60	>9.0	G2
	60+	<12.0	G3
	60+	>12.0	G4
Common alder *(Alnus glutinosa)*	45–60	<10.5	G1
Grey alder *(A. incana)*	45–60	>10.5	G2
	60+	<14.0	G3
	60+	>14.0	G4
Silver birch *(Betula pendula)*	45–60	<7.0	G1
Hairy birch *(B. pubescens)*	45–60	>7.0	G2
	60+	<9.0	G3
	60+	>9.0	G4
Corsican pine *(Pinus nigra* var.	20–30	<5.0	G1
maritima)	20–30	>5.0	G2
	30+	<7.0	G3
	30+	>7.0	G4

specification. The largest improvement in survival rate due to the use of sturdy trees was shown by sycamore.

The differences in survival rate between trees in the shorter height class and those in the taller height class were small overall for sycamore, Common and Grey alder and Corsican pine. In the case of the two birch species, the shorter trees at planting survived much better than the taller trees.

It may be concluded that while there may be a strong landscaping argument for planting broadleaved trees taller than 60 cm, trees that exceed this height may be more expensive to buy and, as shown by the experimental evidence, no cultural advantages are likely to be gained by using them. If trees taller than 60 cm have to be planted, sturdy stocks should be chosen because of their potentially higher survival rate. However, in the case of the birches, trees shorter than 60 cm should certainly be preferred. With all species, stocks which meet published minimum root collar diameter specifications should be planted so far as is possible.

Choice of Species

In the late 1960s and early 1970s local authority reclamation officers placed a good deal of faith in so-called pioneer species which, numerically at least, tended to dominate plantings. With a few exceptions they had high survival rates and several increased in height rapidly in the early years after planting.

Now, only 10 to 15 years later, many have been overtaken in total height and current height increment by species which were comparatively slow growing to begin with or which, in one or two cases, even had low survival rates.

These slower developing, though subsequently more vigorous and perhaps more tolerant species, now appear capable of dominating existing crops, of forming the over-storey in crops and of continuing to grow well enough to produce acceptable semi-mature to mature woodland. Six species of outstanding potential were assessed at different sites during the contract; a forecast of their likely height growth, based on the assessments was made to 50 years of age (Figure 6). The species, Corsican and Lodgepole pine (*P. contorta*), Japanese larch (*L. kaempferi*), Silver birch and Common oak, are ordinarily tolerant of a wide range of climates and forest soils, and their outstanding behaviour on reclaimed colliery spoil is

Figure 6. Forecast of height increase of six major species. Sites in Northumberland, County Durham and West Yorkshire. Estimated Yield Classes are shown in brackets.

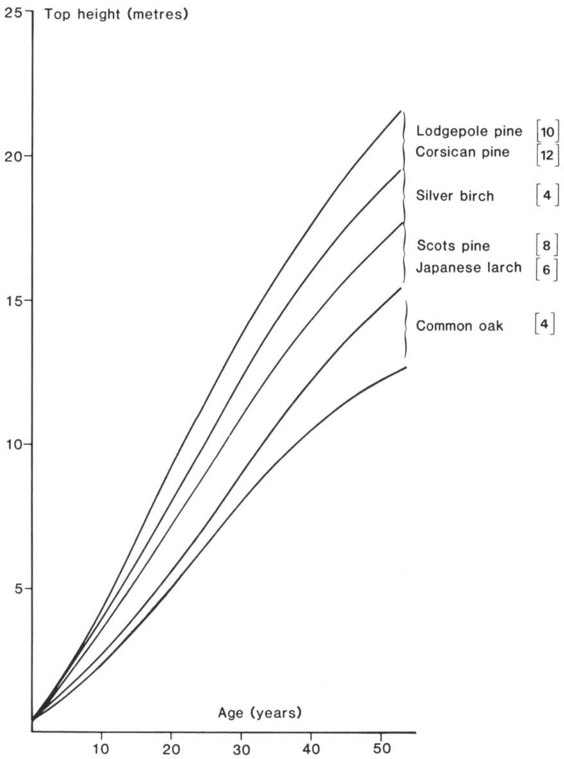

Table 2. Mean survival rate after one season of Common ash *Fraxinus excelsior* and Common oak *Quercus robur* in different ground covers. Pegswood, Northumberland.

Survival (transformed)	Ground cover								Level of significance
	Bare spoil	Grass spot weeded	Grass not spot weeded	Wild white clover	Birdsfoot trefoil	Common vetch	Lucerne	Sweet clover	
Common ash	90.0 a	90.0 a	81.7 a	70.4 ab	83.4 a	90.0 a	54.2 b	90.0 a	*
Common oak	60.3 a	38.9 bc	20.5 d	35.5 cd	37.3 bcd	54.6 ab	0.0 e	38.4 bc	***

Key: Wild white clover = *Trifolium repens*
Birdsfoot trefoil = *Lotus corniculatus*
Crown vetch = *Coronilla varia*
Lucerne = *Medicago sativa*
Sweet clover = *Melilotus alba*

Values sharing a common letter do not differ significantly.
Survival values are shown after angular transformation.
Differences are statistically significant at $p < 0.05$ (*) or $p < 0.001$ (***) levels.

perhaps not surprising. Successfully afforested spoil heaps are increasingly likely to resemble established woodland on light, acid and somewhat infertile heathland soils of southern England.

Choice of Ground Cover

Areas on reclaimed spoil heaps earmarked for tree planting are sown with grass seed mixtures as soon as possible after the completion of the physical and chemical amendment operations that follow regrading. The mixtures usually contain a high proportion of fescue (*Festuca* spp.) and Common bent grasses (*Agrostis tenuis*) and a small (10 per cent) clover (*Trifolium* spp.) component. A vigorous grass sward may have developed by the time tree planting takes place. Colliery spoil has very poor moisture retention properties and trees planted in a grass sward can be killed unless weeding is thorough. Davies (1985) has reviewed the importance of weed competition on grass dominated woodland sites.

Two experiments compared tree survival and growth in bare spoil, in a grass-clover sward spot weeded and not spot weeded and in five pure legume ground covers. The legumes were chosen for their potential soil-forming and nitrogen fixation properties, because of their ability to over-winter, so maintaining the green appearance of a site and limiting the risk of surface erosion, and for their visual attractiveness. They were also selected on account of the ready availability of tested seed of known germination capacity from agricultural seed merchants.

Two important results emerged in the experiment at Pegswood, Northumberland. By far the best survival of trees (Common oak and Common ash (*Fraxinus excelsior*)) occurred in bare spoil, in Crown vetch (*Coronilla varia*) which had not established well so that a large proportion of the surface spoil was bare and weed free, and in the grass sward spot weeded around the trees (Table 2). There was also good tree survival in Sweet clover (*Melilotus alba*) which, like Crown vetch, had established poorly leaving much of the surface of the plots bare.

The worst tree survival occurred in the grass sward which had not been spot weeded and in Lucerne (*Medicago sativa*) which was extremely dense and vigorous. There was also poor survival, of ash, in Wild white clover (*T. repens*) which, like Lucerne, was also dense and vigorous.

The results demonstrated that the rates of tree survival and growth are improved by weed free conditions and the rates are reduced by the presence of herbaceous weed growth. Further, the results show that legumes sown at high rates or that germinate well and then develop vigorously may compete more for soil moisture and nutrients than a grass-clover sward. Conversely, a low stocking of feebly growing legumes may not be seriously competitive. It is evident there is no proven advantage in planting into a legume ground cover.

Conclusions

The research has shown that tree crops can be satisfactorily established on reclaimed colliery spoil heaps with expectations of thriving woodland being produced. A programme of experiments has confirmed that by planting in recently cultivated spoil the speed and standards of work are improved, and the rates of survival, height growth, downward and lateral root extension and stem and root biomass production are increased. The evidence suggests that ripping along intended planting lines to a depth of 60 cm is the optimum method of pre-planting cultivation. Pit planting in mechanically prepared holes is not a satisfactory substitute for notch planting into ripped spoil.

With the exception of Corsican pine, prospects of crop establishment are not improved by planting 1-year-old seedlings raised in Japanese paperpots or amenity pots instead of bare-root transplants. Even in the case of Corsican pine, potential improvements in survival and growth rates have to be weighed against the greater cost of producing, transporting and planting Japanese paperpot or amenity pot trees. Moreover, there are no cultural advantages in planting broadleaved trees taller than 60 cm but sturdy trees which meet published minimum root collar diameter specifications should be preferred to thinner trees on account of their better survival rate. The research also confirmed that several tree species on reclaimed spoil have growth rates after the establishment period that compare favourably with those of trees on some forest sites.

References

BRITISH STANDARDS INSTITUTION (1984). *BS 3936: Nursery stock. Part 4. Specification for forest trees.* British Standards Institution, London.

DAVIES, R.J. (1985). The importance of weed control and the use of tree shelters for establishing broadleaved trees on grass dominated sites in England. *Forestry* **58** (2), 167–180.

JOBLING, J. (1981). Reworked spoil and trees. In, *Research for practical arboriculture*, Proceedings of the Forestry Commission/Arboricultural Association Seminar, Preston 1980. Forestry Commission Occasional Paper 10, 76–83. Forestry Commission, Edinburgh.

JOBLING, J. and CARNELL, R. (1985). *Tree planting in colliery spoil.* Forestry Commission Research and Development Paper 136. Forestry Commission, Edinburgh.

JOBLING, J. and STEVENS, F.R.W. (1980). *Establishment of trees on regraded colliery spoil heaps. A review of problems and practices.* Forestry Commission Occasional Paper 7. Forestry Commission, Edinburgh.

Acknowledgements

I am extremely grateful to the large number of people, both in the Forestry Commission and in local government, who contributed to the programme of research conducted by the Forestry Commission on reclaimed colliery spoil. I am especially appreciative of the help provided by Wendy Varcoe, who was co-author of the Commission's report to the Department of the Environment upon which this paper is based.

Discussion

T.W. HARFORD (Amey Roadstone Corporation Limited).

Invited comments on the use of large birch on reclamation sites; such trees had to be crown pruned but nevertheless he had found good results.

J. JOBLING The purpose of planting large trees was to produce a visual effect. If one then heavily prunes the crown it seems that one would be destroying their appearance. Overall, not in favour of using large trees, but no comparative trials have been done by Forestry Commission.

M.J. HAXELTINE (Domestic Recreation Services).

Does the tree's search for 'food' make the tree more root firm, i.e. can useful growth be stimulated by provision of nutrients at strategic points to encourage rooting in that direction?

J. JOBLING Trees do not 'anticipate' presence of nutrients and proceed towards them, but if a nutrient is located root proliferation in that direction could be expected to be encouraged. This aspect has not been studied in the series of experiments reported.

Tree Establishment: Soil Amelioration, Plant Handling and Shoot Pruning

R.J. Davies
Forestry Commission

Summary

Addition of peat to planting pits improved survival and growth in one experiment, reduced growth in another, and had no effect in a third; these differing responses may have been caused by varying degrees of waterlogging and soil anaerobism. Pre-planting site preparation must relieve soil compaction and allow excess water to drain from the soil and off the site.

Root pruning, to simulate root loss during nursery lifting, had no effect on survival and growth in two experiments; it is concluded that desiccation during plant handling is usually more important.

In one experiment shoot-pruned trees regained their lost height but did not overtake control trees; while in another, shoot pruning increased height growth and reduced dieback; shoot pruning appeared to depress diameter growth in both. Shoot pruning reduces root growth and may thus increase trees' moisture stress.

Introduction

Water is vital to trees, plant cells being killed or injured by desiccation. Drought stress is a common cause of poor growth and death of newly planted trees. Desiccation of bare-rooted trees between lifting and replanting is particularly serious since roots cannot take up water to replace losses. After planting most moisture is lost by transpiration through the leaves. This is unavoidable as the stomata which facilitate photosynthetic gaseous exchange also allow water vapour to escape. A healthy root system permeating a large volume of soil with good moisture storage enables the tree to replace these losses.

This paper examines three factors which influence the moisture status of newly planted trees: soil amelioration to increase its moisture storage; the handling of bare-rooted stock between nursery lifting and replanting; and shoot pruning at planting to reduce trees' leaf areas and consequent transpirational water requirements.

Soil Amelioration

Soil compaction and lack of structure, common features of British landscape tree planting sites, impede water movement through the soil; even sandy soils become impermeable. Thus when precipitation exceeds evapotranspiration the site becomes waterlogged while in dry summer weather soil moisture tension rises rapidly. Compaction and waterlogging restrict rooting depth further increasing plants' drought susceptibility.

Pits are usually dug to accommodate the trees' root systems. Where topsoil has been lost or degraded during site works, imported topsoil may replace the excavated material in the pits. Peat or other bulky organic material is mixed into the pits to improve moisture storage. Occasionally trees may be irrigated during droughts.

Three soil amelioration experiments

An experiment planted on a cutting adjacent to the M5 at the Michael Wood Services compared the effects of zero, 20, 50 and 80 per cent (by volume, not weight) sedge peat mixed into the planting pit backfill on newly planted hawthorn (*Crataegus monogyna*) and birch (*Betula pendula*). It also compared bare-rooted transplants with seedlings raised in Japanese paperpots. Each pit was approximately 21 litres. Table 1 summarises survival, height and diameter growth after three growing seasons. The bare-rooted stock performed better than the

Table 1. The effect of backfill mix, B (zero, 20, 50 and 80 per cent peat) and stock type, S (BR = bare-rooted transplants; JPP = seedlings raised in Japanese paperpots) on hawthorn and birch survival and growth after three seasons. Michael Wood.

Backfill treatment:		0%	20%	50%	80%	Significance [1] B	S	B × S Interaction
				Hawthorn				
Height growth (cm)	BR	29.2	48.9	45.3	59.3	***	NS	NS
	JPP	27.8	36.9	49.4	46.3			
Diameter growth (mm)	BR	6.3	10.3	10.0	11.8	***	***	NS
	JPP	5.4	6.6	9.1	8.3			
Survival [2] (%)	BR	96	100	100	100	**	***	NS
	JPP	76	93	96	88			
				Birch				
Height growth (cm)	BR	36.8	41.0	48.5	44.9	NS	***	NS
	JPP	9.0	14.5	16.8	8.5			
Diameter growth (mm)	BR	8.0	6.6	6.9	7.6	NS	**	NS
	JPP	4.9	5.3	6.2	6.1			
Survival [2] (%)	BR	19	33	57	53	*	NS	NS
	JPP	8	40	60	46			

[1] NS = effect not significant at 5% level
* = effect significant at 5% level
** = effect significant at 1% level
*** = effect significant at 0.1% level

[2] Survival transformed to angles for statistical analysis.

container-grown stock. Increasing the proportion of peat in the backfill improved hawthorn performance. Erratic survival obscures the birch results, but up to 50 per cent peat appears to have improved their survival and growth, although 80 per cent peat was slightly detrimental.

A similar experiment at Milton Keynes compared the effects of zero, 20 and 50 per cent sedge peat mixed into 12 litre planting pits on ash (*Fraxinus excelsior*), Norway maple (*Acer platanoides*) and Western red cedar (*Thuja plicata*) transplants. Peat depressed ash diameter growth in the first year, but unfortunately the experiment was destroyed before any treatment effects emerged with the other species.

The Michael Wood and Milton Keynes experiments were on grassy sites. A third experiment at Islington in Greater London used standard trees in a paved street area. Rowan (*Sorbus aucuparia* 'Fastigiata'), 3 m tall, were planted into 72 litre planting pits with zero, 10, 20 or 40 per cent sedge peat. These treatments had no significant effect on survival or growth. In the first 2 years nearly 40 per cent of the trees died – waterlogging and vandalism were the major causes. Investigation of soil oxygen availability using steel rods (Carnell and Anderson, 1986) showed that peat did not improve the anaerobic soil conditions.

Discussion

Addition of peat was beneficial at Michael Wood, detrimental at Milton Keynes and had no significant effect on tree establishment at Islington. Clearly, it is not a panacea for difficult sites. It increases soil moisture storage but is harmful unless excess water can drain away. All three sites had fine textured soils prone to waterlogging. It may be significant that Michael Wood was the only site with sufficient slope to prevent excessive waterlogging. Waterlogging restricts gaseous diffusion through soil and decomposition of organic matter uses soil oxygen; both factors increase soil anaerobism, restricting root growth and thus rendering trees drought susceptible.

Pre-planting site preparation must therefore ensure that compaction and waterlogging do not restrict root growth. Excess water must be able to drain from the soil and off the site. Binns and Fourt (1984) recommend the creation of slopes, and ripping to a depth of at least 0.5 m, to achieve these objectives on reclamation sites; these techniques could be adapted for use on many other sites. Where these techniques are impractical other methods to achieve the same objectives should be designed.

With such site preparation soils of most textures will store sufficient moisture to supply newly planted trees through British summers, provided weeds are prevented from competing for moisture. The addition of organic matter and irrigation then become luxuries for tree establishment. Such site preparation should also benefit long-term tree stability and drought resistance.

In addition to improving tree establishment the extensive site preparation advocated by Binns and Fourt (1984) is not necessarily more expensive. It may be possible to notch-plant transplants into ground which because of its compaction would otherwise require pit planting. Pit preparation is expensive.

Organic matter influences nutrition as well as soil gas and water regimes. Micro-organisms which decompose the organic matter compete with plants for available nitrogen. This is particularly serious with materials such as straw, wood and bark having a high carbon:nitrogen ratio. Schulte and Whitcomb (1975) found that neither bark, peat moss, sand or vermiculite mixed into planting pits improved the early growth of Silver maple (*Acer saccharinum*). Bark reduced nitrogen availability, and tree growth, although fertilising compensated for this to some extent.

Conclusion

Pre-planting site preparation must relieve soil compaction and ensure that excess water can drain from the soil and off the site; addition of organic matter is no substitute.

Plant Handling

When plants are lifted from a nursery bed, transported and replanted, they are inevitably damaged. Firstly, part of the root system is severed and left in the nursery soil. Secondly, plants dry out unless kept in a saturated atmosphere for the whole time that their roots are unable to extract soil moisture.

Desiccation during plant handling

Insley (1980) intercepted plants being supplied by commercial nurseries for roadside planting in 1978 and 1979 and found that 10 and 3.5 per cent respectively had lost more than half the water they contained at lifting. As most plants were well packaged the damage probably occurred in the nursery between lifting and packaging. He also demonstrated experimentally that desiccation caused by exposure reduced the survival of broadleaved seedlings. Subsequently growth of the survivors of some species was also reduced.

Trees with finely structured root systems, such as *Betula* species, have a large surface-area:volume ratio; they generally dry out faster and are damaged more readily than, for example, *Tilia* species, which have coarser roots. Roots must be protected from desiccation while out of the ground. Shoots, even when dormant and leafless, also lose moisture, albeit at a slower rate, so they too should be protected as much as possible; this is especially important for seedlings and transplants which have a larger surface-area:volume ratio than bigger trees.

Root loss during lifting

In an experiment at Tern Hill, Shropshire, London plane (*Platanus x hispanica*) and Small-leaved lime (*Tilia cordata*) had their roots pruned at planting by half their length to simulate careless nursery lifting. As much of the root system had been lost during lifting, removing half of what remained was a severe treatment. However, survival was excellent and root pruning had no significant effect on height and diameter growth over the following three seasons. (A shoot pruning treatment, discussed below, was also included in this experiment; results are given in Table 2.)

In a second experiment 3.2–4.5 m tall ash, sycamore (*Acer pseudoplatanus*) and Box elder (*Acer negundo*) had all their fine roots shaved off prior to planting. Again, survival was excellent and root-shaved trees grew as well as untreated controls.

Discussion

These experiments indicate that severance of part of the root system between lifting and replanting is not necessarily detrimental. By contrast desiccation during plant handling reduced both survival and growth.

It appears that physiological status, particularly moisture content, has a greater influence on transplanting success than the size of the root system; quality rather than quantity is required. But in practice the purchaser of bare-rooted trees assesses their quality largely by the size and form of their root systems, if the roots are examined at all.

Physiological quality is largely overlooked because it is not easily assessed. Plants may arrive well packaged and appear moist, but they may have dried out once and then

Table 2. The effect of root (R) and shoot (S) pruning on height and diameter (measured 0.7 m above ground) of London plane and Small-leaved lime after three seasons. Tern Hill.

Pruning treatment:	Control	R	S	R+S	Significance [1] R	S	R × S interaction
			London plane				
Initial height (m)	1.99	1.86	1.62	1.59	NS	***	NS
Initial diameter (mm)	18.1	17.2	17.6	17.5	NS	NS	NS
Height (m) after three seasons	3.21	3.00	3.17	3.16	NS	NS	NS
Diameter (mm) after three seasons	46.8	44.2	46.9	46.0	NS	NS	NS
			Small-leaved lime				
Initial height (m)	2.44	2.39	1.90	2.08	NS	***	NS
Initial diameter (mm)	23.4	24.0	22.3	24.1	NS	NS	NS
Height (m) after three seasons	2.84	2.74	2.69	2.76	NS	NS	NS
Diameter (mm) after three seasons	46.6	47.2	43.7	45.0	NS	*	NS

[1] See note 1 below Table 1.

been soaked for a period – soaking will not resuscitate dead wood. Some months later when the tree is found to be dead it may be impossible to ascertain the cause.

In order to present a balanced picture it is necessary to add that root loss is often harmful. Many researchers (e.g. Beckjord and Cech, 1980) have found this. I am not advocating small root systems, merely suggesting that other frequently ignored factors are usually more important.

Fine roots dry out rapidly (Coutts, 1981) and will die after only brief exposure. This may explain the lack of response to root-shaving in the second experiment, only dead material being excised. In that experiment the trees were carefully handled in the short period between lifting and replanting. This suggests that bulk of thick roots from which new fine roots can regenerate is more important than masses of fine roots.

Conclusion

Trees must be protected from desiccation at all times between lifting and replanting.

Shoot Pruning at Planting

Shoot pruning is often advocated (e.g. Kozlowski and Davies, 1975) to restore the root:shoot ratio which is depressed through root loss during nursery lifting. Dormant pruning is expected to delay or reduce leaf area development, transpiration rates and consequently the demands made on the root system.

Two shoot pruning experiments

In the Tern Hill experiment root pruning (described above) and shoot pruning treatments were applied factorially giving four experimental treatments. Shoot-pruned trees had all branches pruned to the bud nearest the mid point of their length; leading shoots were cut back to a bud mid-way between the tip and first main branch. Thus shoot pruning reduced tree heights. Shoot-pruned trees regained their lost height in the first season, thereafter growing at the same rate as the unpruned trees. Heights and diameters after three seasons are given in Table 2. After three seasons it appears that shoot pruning has slightly depressed lime diameter growth.

The second shoot pruning experiment, planted with 2 m tall sycamore and birch at Barton Stacey, Hampshire, used four experimental treatments: (A) control, with no pruning, (B) half of the branches removed, (C) all branches pruned to half length, and (D) all branches removed. Survival was excellent but growth poor. Many trees, particularly sycamore, died back severely. This was probably caused by summer drought and nutritional problems. Shoot pruning reduced the incidence of die-back (only significantly for birch) and increased height growth (Table 3).

Table 3. The effect of four pruning treatments (A = no pruning, B = half of the branches removed, C = all branches pruned to half length, D = all branches removed) on dieback percentage, height and diameter (measured 1 m above ground) growth of birch and sycamore two seasons after planting. Barton Stacey.

Pruning treatment:	A	B	C	D	Significance[(1)]
			Birch		
Dieback[(2)] percentage (%)	7.5	5.0	0	0	**
Height[(3)] growth (cm)	13.2	24.4	20.9	43.0	***
Diameter[(3)] growth (mm)	1.9	1.7	1.5	0.9	NS
			Sycamore		
Dieback[(2)] percentage (%)	26.7	10.0	7.1	13.3	NS
Height[(3)] growth (cm)	6.4	8.4	8.4	14.9	**
Diameter[(3)] growth (mm)	1.4	0.8	1.1	0.8	NS

(1) See Note (1) below Table 1.

(2) Dieback figures are based on the number of trees which were taller at planting than after two seasons.

(3) All trees with height dieback excluded from these analyses.

In neither experiment did sufficient moisture stress develop to kill trees and test the effect of pruning on survival. Pruning increased height growth and reduced the incidence of dieback. There were indications that pruning may have depressed diameter growth.

Discussion

Pruning does not necessarily reduce the tree's leaf area and transpiration. Shoup *et al.* (1981) reported that various pruning treatments had no effect on total first year leaf production of Kieffer pear (*Pyrus communis* 'Kieffer') and dwarf Alberta peach (*Prunus persica* 'Elberta'). Pruned Stuart pecan (*Carya illoensis* 'Stuart') broke bud slightly earlier than unpruned controls. Pruning method can influence the results; Evans and Klett (1984) found that branch thinning reduced leaf production of Sargent crab apple (*Malus sargentii*) but heading the branches back did not.

Shoot pruning affects root growth. Head (1967) observed a reduction in root growth associated with vigorous shoot growth which was stimulated by pruning of Victoria plum on Myrobalan B and Worcester Pearmain on MM104 rootstocks. Richardson (1957) using Silver maple seedlings found that root formation required the presence of an active shoot apex, and that the leaves supplied a second growth factor which controlled root elongation.

Therefore, even when dormant pruning does delay or reduce leaf area development, it may still increase the moisture stress suffered by the tree – the reduction in transpirational water requirements being outweighed by the reduction in root growth. Pruning may have other adverse effects; if it reduces leaf area, nett photosynthesis will be curtailed. Pruning removed stored carbohydrates from the tree. Species, previous nursery treatment, type and severity of pruning, soil, climate and other factors may determine whether pruning helps or hinders tree establishment.

Root:shoot ratios are often calculated using total root and shoot dry weights. But tree growth at Tern Hill was little affected by the root and shoot pruning treatments which must have produced trees with widely differing ratios. Fine roots have both a greater surface area per unit weight and will absorb more water per unit surface area than coarse roots. However, as discussed above under Plant Handling, these fine roots may not survive transplanting. Thus the water balance of the newly transplanted tree largely depends on the regenerative capacity of the surviving roots. Leaf:new root ratio is a better, but still imprecise, measure of the tree's water balance (Evans and Klett, 1984).

Conclusion

It is naive to assume that shoot pruning necessarily reduces moisture stress. Shoot pruning reduces root growth and may thus increase the stress suffered by the tree.

Acknowledgments

The Department of Transport, Milton Keynes Development Corporation, Islington Borough Council and the Property Services Agency provided sites for experiments. J.B.H. Gardiner, T.J. Davies and R.E. Warn did much of the experimental work, R.C. Boswell conducted statistical analysis and Dr. W.O. Binns made helpful criticism on the text of this paper. The work was initiated by Dr. H. Insley and funded by the Department of the Environment.

References

BECKJORD, P.R. and CECH, F.C. (1980). Effects of various methods of root pruning on the establishment of transplanted Red oak seedlings. USDA Forest Service, *Tree Planters Notes* **31** (4), 10–11.

BINNS, W.O. and FOURT, D.F. (1984). *Reclamation of surface workings for trees. 1. Landforms and cultivation.* Arboriculture Research Note 37/84/SSS. DOE Arboricultural Advisory and Information Service, Forestry Commission.

CARNELL, R. and ANDERSON, M.A. (1986). A technique for extensive field measurement of soil anaerobism by rusting of steel rods. *Forestry* **59** (2), 129–140.

COUTTS, M.P. (1981). Effects of root or shoot exposure before planting on the water relations, growth and survival of Sitka spruce. *Canadian Journal of Forest Research* **11** (3), 703–709.

EVANS, P.S. and KLETT, J.E. (1984). The effects of dormant pruning treatments on leaf, shoot and root production from bare-root *Malus sargentii. Journal of Arboriculture* **10** (11), 298–302.

HEAD, G.C. (1967). Effects of seasonal changes in shoot growth on the amount of unsuberized root on apple and plum trees. *Journal of Horticultural Science* **42** (2), 169–180.

INSLEY, H. (1980). Wasting trees? – The effects of handling and post planting maintenance on the survival and growth of amenity trees. *Arboricultural Journal* **4** (1), 65–73.

KOZLOWSKI, T.T. and DAVIES, W.J. (1975). control of water balance in transplanted trees. *Journal of Arboriculture* **1** (1), 1–10.

RICHARDSON, S.D. (1957). Studies of root growth of *Acer saccharinum* L. VI. Further effects of the shoot system on root growth. *Kon. Ned. Akad. Wetensch* C.**60**, 624–629.

SCHULTE, J.R. and WHITCOMB, C.E. (1975). Effects of soil amendments and fertilizer levels on the establishment of Silver maple. *Journal of Arboriculture* **1** (10), 192–195.

SHOUP, S., REAVIS, R. and WHITCOMB. C.E. (1981). Effects of pruning and fertilizers on establishment of bare-root deciduous trees. *Journal of Arboriculture* **7** (6), 155–157.

Discussion

R. WEBB (An Foras Forbartha, Dublin).

(i) Have any soil ameliorants other than peat been tested?

(ii) Should roots be pruned at planting?

R.J. DAVIES

(i) Experiments have been established in the 1984/5 planting season to examine proprietary soil ameliorants. The main problems are poor drainage and overall compaction of the site which cannot be relieved by small quantities of material in a small planting pit.

(ii) Root pruning is not advocated though it seems less harmful than might be expected. Pruning makes trees easy to handle and plant which is probably why people do it.

J. GRACE (Department of Forestry and Natural Resources, Edinburgh University).

Suggested greater use of instrumentation should allow causal factors to be determined while in all establishment research water stress should be monitored.

R.J. DAVIES Accepted the points and described current work with instruments.

R. SKERRATT (Arboricultural Consultant, Cleveland).

Stressed that it is important to state the type of peat used because quality is very variable.

R.J. DAVIES	Sedge peat used in the experiments described.
H.G. MILLER	(Department of Forestry, Aberdeen University). Urged caution because there is a risk of soil deoxygenation with fine textured peat.
R.N. HUMPHRIES	(Midland Research Project). Expressed a belief that the papers had not adequately emphasised the difficulties caused by inadequate drainage on destructured soils. He suggested that rigg and furr plus ripping was not sufficient to get water off sites and that attention should be given to agricultural pipe drainage systems.
D. KENNEDY	(Department of Forestry, Aberdeen University). Birch has established well when shoots were pruned *before* planting.
D. ATKINSON	(East Malling Research Station). Were there any reasons for one stock type responding to peat on one site and other species on other sites not responding?
R.J. DAVIES	No explanations found.

Root Growth and the Problems of Trees in Urban and Industrial Areas

P. Gilbertson, A.D. Kendle and A.D. Bradshaw
Department of Botany, University of Liverpool

Summary

The availability of water and nutrients to transplanted trees is assessed, and ways of improving the uptake of these essential plant resources are outlined. Particular importance is placed on the effects of early planting, pruning, watering, fertilising and weed control, in increasing the contribution of the root system to the growth of the tree in its new environment.

Urban Tree Performance

In any study of tree growth on landscape sites it soon becomes evident that, despite our efforts, performance often falls sadly below that desired, to the cost of the planting agency and the detriment of the client. Results obtained from a recent survey of the annual extension growth of 1000 newly planted standard trees in some towns in the north of England (Gilbertson and Bradshaw, 1985) have shown that a large percentage of trees measured were in fact putting on no extension growth or were even 'growing backwards' (Figure 1). At the same time the mean growth is far below the level that can be achieved with good practice. Such performance may be attributed to the poor made-up soils so commonly found in urban areas, low in nitrogen and organic matter, and which forms the background to our considerations.

Figure 1. The distribution of mean annual shoot increments of 1000 urban trees in 1983 (all species together).

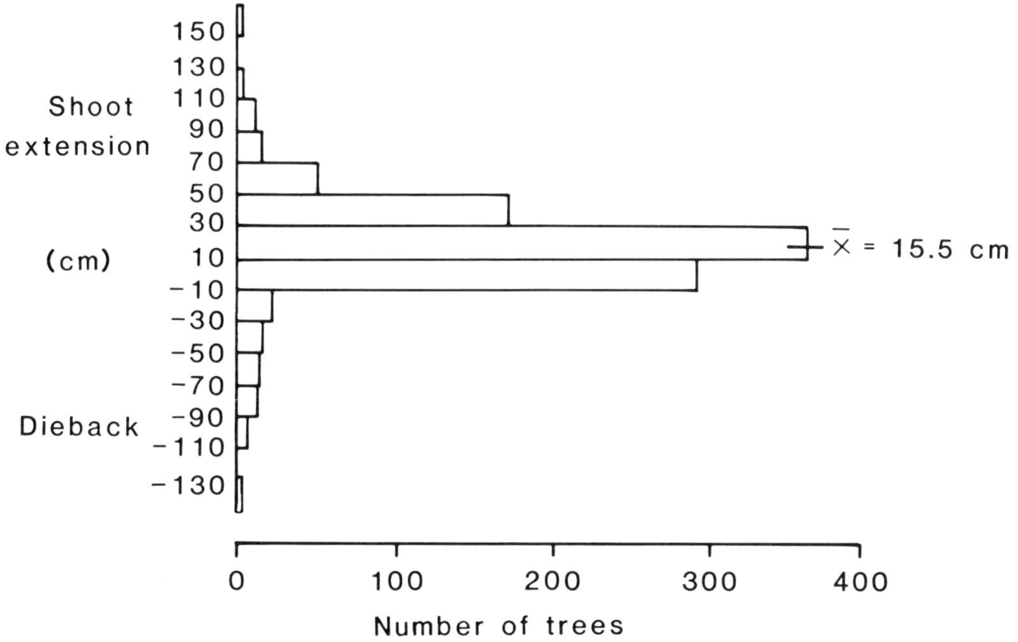

Tree Requirements and Root Growth

When trying to determine why trees are unable to grow on such sites it often pays to return to basic principles and consider what resources a plant needs to enable this growth to occur. Essentially a tree requires light and carbon dioxide absorbed from the air by its leaves, and water and nutrients extracted from the soil by its roots. Of these, it is usually only the latter two which are likely to be seriously limiting in open landscapes. To overcome this limitation adequate root extension is obviously important. However, the ability of the plant to exploit aerial resources may be restricted if it fails to develop adequate leaves, or they fail to function, because of severe soil induced stresses. Usually a drought, which physically inhibits growth, leads to stomatal closure, and inevitably reduced photosynthesis (Levitt, 1972). This serves to underline the importance of the roots and the volume of the soil they exploit, in determining potential growth.

This inter-relationship can be explored by some simple calculations. Taking nitrogen in particular, a soil typical of any derelict urban area into which the majority of new trees are planted will have a total nitrogen supply of 500 ppm (Dutton and Bradshaw, 1981). If we assume a typical mineralisation rate for this nitrogen of 1/16, the annual release of mineral nitrogen for a given volume exploited by roots can be used to determine the maximum potential biomass increase. We have assumed that N forms 1.5 per cent of dry weight increment and that 100 per cent of the mineralised N is captured by the roots. Lateral movement of nutrients does not occur over great distances in the soil, so that there will be a direct relationship between rooting volume and nutrient supply (Figure 2). We can go on to determine what nitrogen is required by (a) a standard tree and (b) a 50 cm transplant, and what volume of soil each will be able to exploit immediately following planting. On such degraded soils the remarkable conclusion is that trees planted in pits of appropriate size for their roots will have barely enough nitrogen to ensure full leafing out, which typically represents about a third of the tree's annual biomass increments, let alone to ensure any shoot extension or girth increment. If, as is likely, at least 25 per cent of the N is not recaptured or there is a competing grass sward, the situation is even more serious.

Similarly, using figures for the amount of water transpired by a lime (*Tilia* sp.), which we have taken to be 2 litres per day (Bradshaw, 1985), we can determine, assuming 100 per cent uptake, the potential number of days support that a given volume of soil can give in the absence of water input in relation to the water capacity of the soil (Figure 3). Again, the rooting volume directly affects the supply of resources, and typically newly

Figure 2. The potential dry matter production for a given rooting volume on urban wasteland.

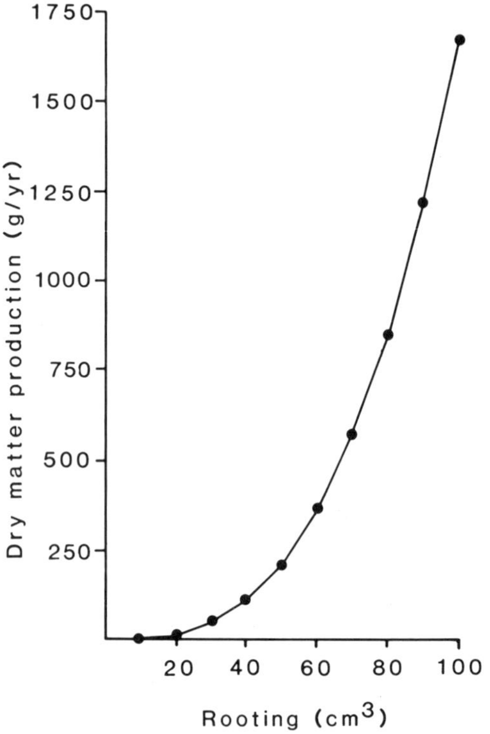

Figure 3. Water supply of a newly transplanted lime tree (assuming a water loss of 2 litres/day).

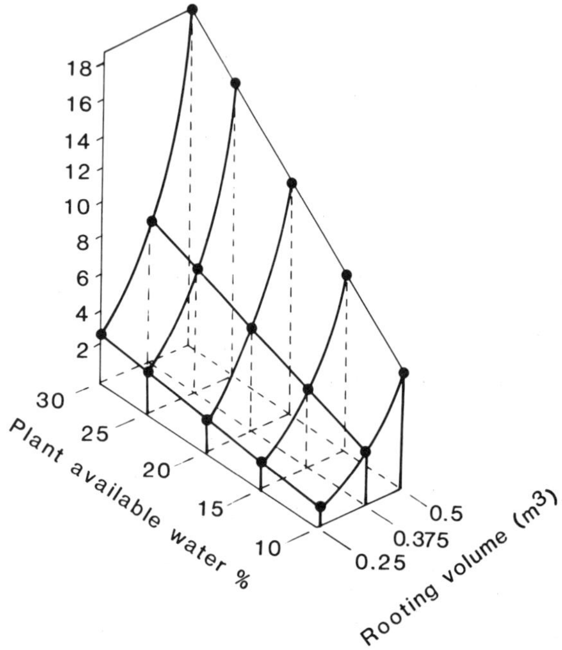

planted trees, such as the normal standard in Figure 2, with a rooting volume of 0.1m^3, have only about 8 days of supply available to them in the absence of rain or irrigation before they begin to experience deficits on the poor soils with low available capacities (15 per cent) typical of urban areas. The situation is not very much better on good soils, 15 days with water capacities of 30 per cent, and will be far worse where there is a competing grass sward. Any periods of drought will not only cause the tree to become water deficient but also prevent the uptake of mineral nutrients.

It is inevitable that the combination of practices in the nursery, whether containerisation of the roots or frequent undercutting and unavoidable root losses at lifting, produce a plant that has an artificially low root to shoot ratio for its size and age. A properly established 5-year-old apple tree (*Malus* sp.), i.e. of comparable age to a typical standard tree, will, by contrast, exploit a soil volume of 7m^3 (Atkinson and Wilson, 1980). A well grown transplant can usually survive on its internal nutrient and food stores for the first few months after planting – sufficient to allow some new roots to grow and aid exploitation of the surrounding soil and minerals if present, although the extent to which a tree can continue to rely on such reserves and remain independent of its soil environment has yet to be determined. It is possible that some species planted into nutrient poor sites may exhibit no apparent deficiency in the first year, but the process of producing and storing metabolites with which to fuel the following year's growth will be severely curtailed and subsequent growth will suffer.

However, even on nutrient rich soils water relations in the spring following planting will be the most critical factor. This is not to say that the tree will die within the 8 to 15 days mentioned. The tree will take 'defensive measures' such as stomatal closure and reduced leaf growth. But this will cause a restriction of photosynthesis that will limit food supplies and the plant's ability to grow subsequently. There will also be reduction of root growth, so that the tree will remain vulnerable to further drought for much longer. The decline to final death will take much longer.

Improving the Contribution from the Root

A newly planted tree in a typically poor urban soil is at considerable risk. Whilst the soil can be improved somewhat, part of the problem remains due to the restricted root system. Fortunately there are management practices that can be employed to improve the situation for the tree.

Early planting

Firstly, try to improve the amount of rooting that has taken place by the time the tree leafs out and experiences full transpiration demand. This can be achieved partly by ensuring that the plant is healthy and vigorous when planted. It has been suggested that some form of 'acclimatisation' of plant material can be obtained by growing it prior to final planting out, in nursery areas that in some way reflect the difficult conditions to be encountered. We know of no evidence to support this idea. Indeed the key to good tree growth is to remove limiting conditions. To deliberately subject a plant to pre-chosen difficulties will almost certainly have a greater effect in terms of depleting reserves than any benefit that may be gained by 'toughening it up'.

What is much clearer is that, wherever practical, planting early in the dormant season should be employed. In an experiment carried out over winter 1983/4 it was shown that trees planted in late November may have twice as much root by full leafing out compared with those planted in late March (Figure 4), although the actual amount of root growth will be dependent on soil temperatures. It is to be regretted that the strictures of the financial year often act to encourage local authorities to plant nearer April than November.

Figure 4. The influence of planting time on root growth. Vertical bars = 99% confidence limits.

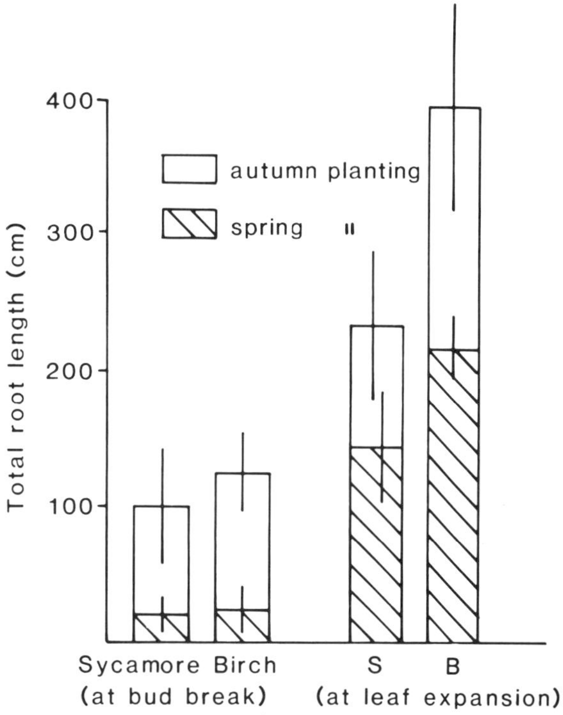

Physical improvement of the substrate

In some situations, especially after earthmoving, soils can be so consolidated that they form a genuine barrier to root growth (Russell, 1977), quite apart from restriction of aeration. In such sites ripping may be essential (Binns and Fourt, 1981). On urban sites large loosely filled planting pits have to be relied on.

Reduction of canopy

Reduction of the total leaf area, and thereby the transpiration demand the shoots put on the truncated root system of newly planted trees, can be achieved by pruning. It has been suggested that shoot pruning, by restricting carbohydrate and growth regulator flow down the stem, can actually restrict the ability of the tree to produce roots (Fayle, 1968). This will be tested experimentally during the 1985 growing season, but preliminary results (Figure 5) demonstrate that shoot pruning is beneficial. As well as an overall increase in shoot extension, there was a depression in water deficits. This will lead directly to increased photosynthesis and the production of more carbohydrates for root growth. There was also a parallel increase in individual leaf areas, and an improvement in aesthetic value of the trees. Even when the initial pruning gives the tree a somewhat stark appearance, regrowth over the first season makes pruned trees more attractive than where stress induced dieback is evident in the canopy. Whilst it is true that dieback can be looked on as the tree's own attempt at shoot pruning, by taking these steps at the planting stage, not only can we constructively influence which shoots are to be retained but also ensure that the tree avoids the full severity of internal water deficits.

Water supply

The supply of water in the soil can be increased by restricting unnecessary evapotranspiration losses from the soil by using weed control and mulch techniques and by irrigation. Many authorities regard watering as an impossible expense but this is false economy if the tree dies and has to be replaced. Even if a commitment to a full season's maintenance is truly impossible, there is no reason to dismiss the idea completely. An experiment at the 1984 International Garden Festival, Liverpool, showed that after a single watering, stomatal resistances were significantly reduced, by the order of 50 per cent, in both feathered transplants and standard alders (*Alnus* sp.). Such dramatic responses were measurable as early as 24 hours after watering and persisted for many days, despite the severity and length of drought that the trees had been subjected to (Figure 6). Perhaps more importantly measurable growth increases were evident at the end of the season (Figure 7).

Figure 5. Effect of two levels of pruning on the shoot extension of five species of amenity tree. Vertical bars = 95% confidence limits.

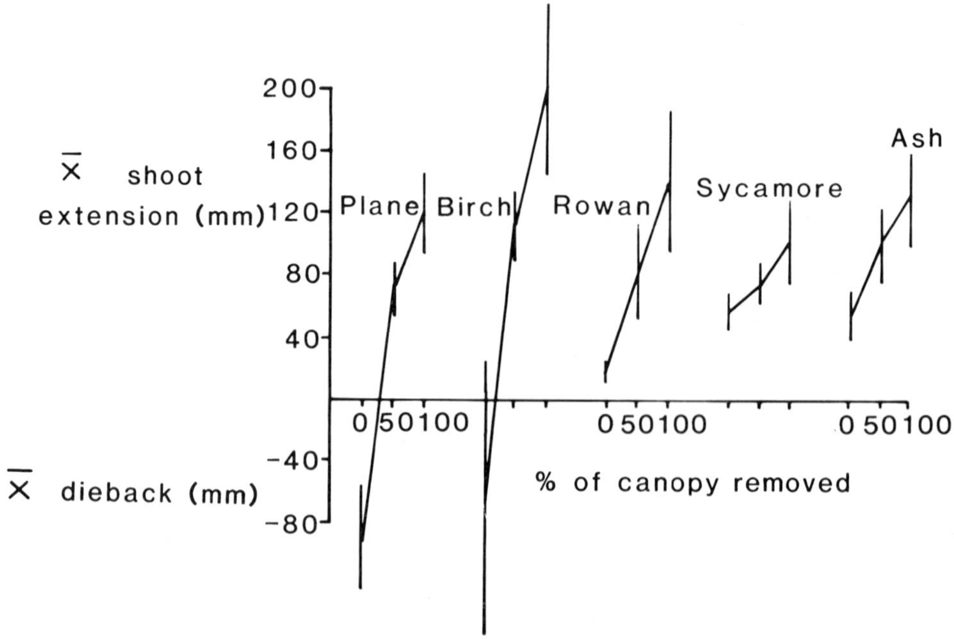

Figure 6. a. Effect of watering on the stomatal resistances of standard trees at the IGF.
b. Effect of watering on stomatal resistances of transplants at the IGF.

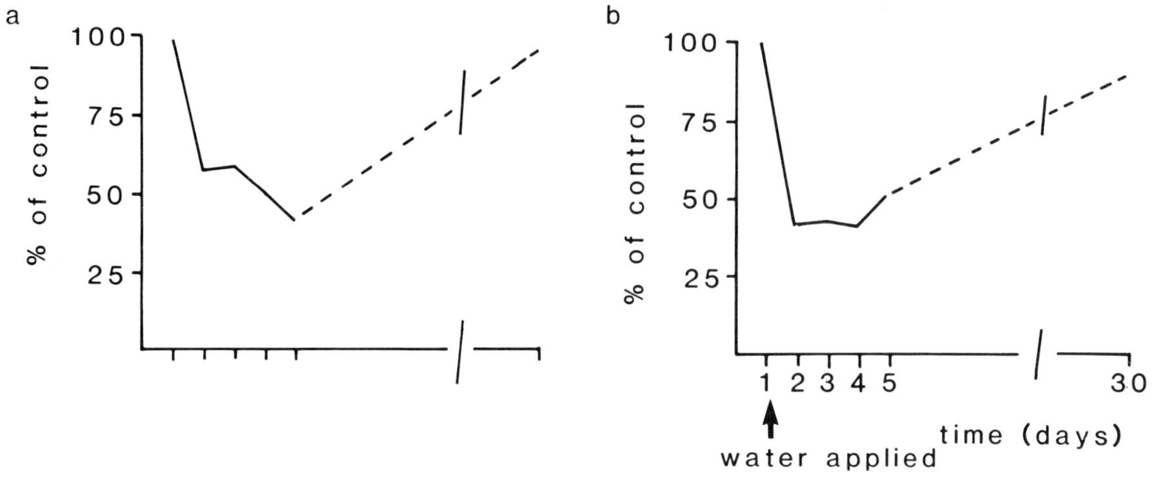

Figure 7. The effect of watering on subsequent shoot growth of alders at the IGF.
Vertical bars = 95% confidence limits.

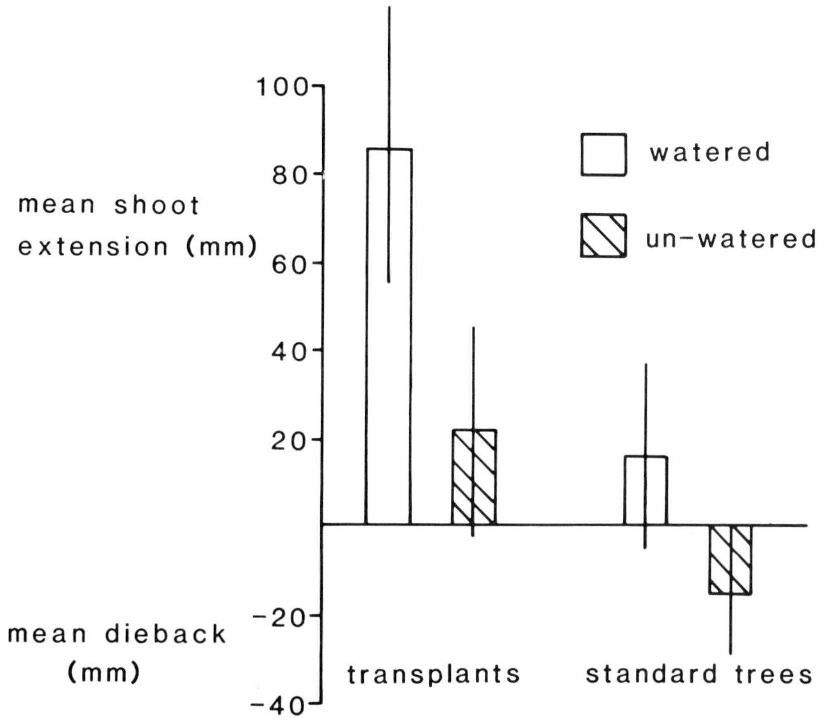

Nutrient supply

Apart from ensuring that root extension is encouraged as far as possible by early planting the most obvious course to take is to increase the nutrient status of the soil. This can be done using inorganic fertilisers, organic fertilisers or by encouraging biological nitrogen fixation, the best approach being determined by considerations of soil type, location and costs. There has been a lot of debate about the effects of different nutrients on morphology and plant growth. Any nutrient that is deficient will have a limiting effect on both root and shoot growth. Although even acute nutrient shortage will not directly kill a tree it will restrict the tree's ability to grow and establish properly and so survive other stresses such as drought or pathogen attack.

This can be illustrated by the results of an experiment performed on severely nitrogen deficient china clay wastes. On these sites it is common for trees to develop severe chlorosis of the leaves due to nutrient shortage in the first year. Excavation of such trees illustrates that almost no root growth has taken place since planting. In trials, three levels of N were applied to 1+1 birch (*Betula* sp.) and sycamore (*Acer pseudoplatanus*) transplants which had already been given a base dressing of phosphate and potassium. The levels chosen were 50, 100, and 150 kgN/ha. Although on the waste leaching is severe and in any one year it is almost impossible to gauge how much of the application reaches the plant, the results show that addition of N had a beneficial effect on the growth of all plant organs including the roots (Figure 8). Indeed, root extension of birch in the high N treatment allowed the trees to exploit a soil volume of approximately ten times that of the control trees and brought this volume up to a level comparable with that quoted for a standard tree of several times the size.

Figure 8. Effect of increasing nitrogen on the volumetric (ml) root and shoot increments after one year of newly planted transplants on china clay wastes.

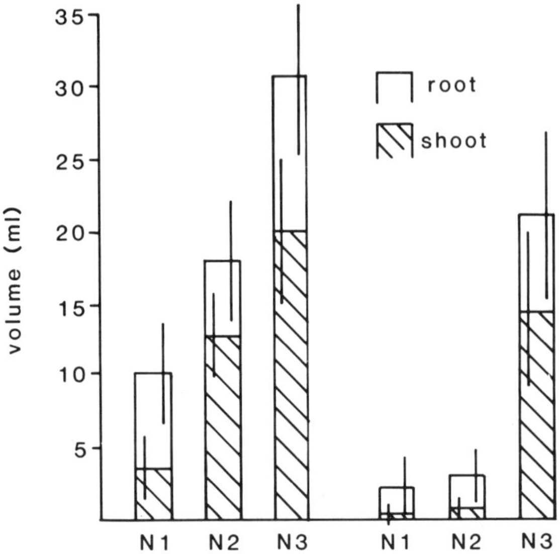

Vertical bars = 95% confidence limits.

There seems little evidence for the common belief that N promotes shoot growth whilst P is for roots. Increasing N on this site slightly increased root:shoot ratio, which perhaps may be attributed to the excessively truncated roots of the transplants, being preferentially regrown compared to shoots (Figure 9). Even so, the principal conclusion is that N does not inhibit recovery, and that the recovery cannot take place when N is limiting. As a direct result of severely limited root extension, as in the N1 treatment, long term survival must be thrown into question.

Although industrial wasteland sites such as china clay may seem extreme in their properties, experience has shown that the magnitude of the problem, at least with regard to nitrogen is not much greater than can be encountered in urban areas where planting is taking place into brick waste, subsoil or 'topsoil' of extremely poor quality. The widespread importance of nitrogen to general growth on these sites was reported previously (Binns and Fourt, 1981).

Interactions between Water and Nutrients

It is rare that the effects of nutrient deficiency or drought can be separated. Nutrient deficiencies will limit root growth and make plants more susceptible to drought. Similarly moisture deficits will restrict nutrient uptake in the soil. The situation can be further complicated when the drought is caused primarily by weed competition. Davies (1987) has shown that adding nutrients in such a situation can actually increase the vigour of the weeds and cause more harm to the trees. This was confirmed by an experiment on a site where soil nutrient status was low enough for weeds to respond to fertiliser but high enough to support tree growth without fertiliser, although even very fertile agricultural soils will also show increased grass growth following nutrient addition. There was a marked negative effect of fertiliser in the presence of weeds on both tree root and shoot growth (Figure 10). This illustrates the paramount importance of weed control in such cases.

Figure 9. Effect of increasing nitrogen on the volumetric (ml) root to shoot ratio of newly planted transplants on china clay wastes.

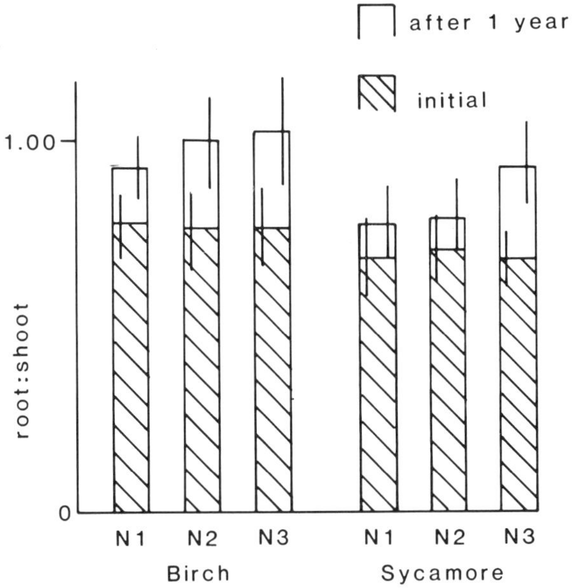

Vertical bars = 95% confidence limits.

Figure 10. Volumetric root and shoot increments (ml) of transplants showing the interaction of weed control and fertiliser.

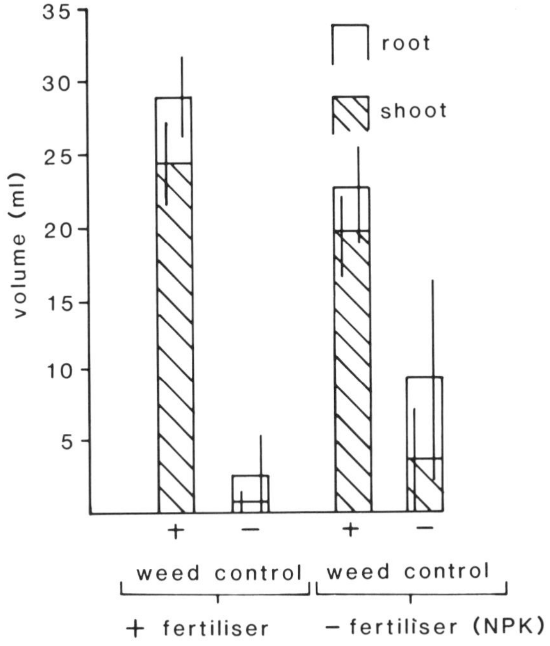

Vertical bars = 95% confidence limits.

Following weed control, response to added fertiliser may be small. Not only are the weeds no longer competing for soil reserves of water and nutrients, there is commonly a flush of released nutrient ions from the killed sward biomass, either or both of which may make additional fertiliser less necessary in the summer of the planting on reasonable soils. On industrial or urban derelict land, in the absence of some overriding limiting factor, it is almost inevitable that added nutrients will produce a positive response following weed control.

Conclusions

Most of these concepts are far from new. Indeed they have been the key to good cultural practices for centuries. But recently the attitude has been taken that they can be ignored. In the current high levels of amenity tree planting in the UK, insufficient attention is clearly being given to basic plant needs, resulting in poor average performance. Yet simple management practices are available to improve water and nutrient supply to newly planted trees. At the same time it must be remembered that the availability of these resources is strongly influenced by the functional rooting volume of the tree in relation to the requirements of the canopy. Hence particular attention must be given to the merits of early planting, pruning, watering, fertilising and weed control for their effects on root, rather than shoot growth. It is our intention to quantify these effects further.

Acknowledgements

This work has been jointly funded by NERC, SERC, ESRC, and ECC.

References

ATKINSON, D. and WILSON, S.A. (1980). The growth and distribution of fruit tree roots: some consequences for nutrient uptake. In, *Mineral nutrition of fruit trees*, eds. D. Atkinson, Jackson, J.E., Sharples R.O. and Waller, W.M. Studies in the Agricultural and Food Sciences. Butterworths, London.

BINNS, W.O. and FOURT, D.F. (1981). Surface workings and trees. In, *Research for practical arboriculture*. Forestry Commission Occasional Paper 10, 60–75. Forestry Commission, Edinburgh.

BRADSHAW, A.D. (1981). Growing trees in difficult environments. In, *Research for practical arboriculture*. Forestry Commission Occasional Paper 10, 93–106. Forestry Commission, Edinburgh.

BRADSHAW, A.D. (1985). Ecological principles in landscape. In, *Ecology and design in landscape*, eds. Bradshaw, A.D., Thorp, E. and Goode, D.A. Blackwell, Oxford.

DAVIES, R.J. (1987). Weed competition and broadleaved tree establishment. In, *Advances in practical arboriculture*. Forestry Commission Bulletin 65, 91–99. HMSO, London.

DUTTON, R.A. and BRADSHAW, A.D. (1981). *Land reclamation in cities*. HMSO, London.

FAYLE, D.F.C. (1968). Radial growth in tree roots. *Faculty of Forestry, Technical Report 9*. University of Toronto, Canada.

GILBERTSON, P. and BRADSHAW, A.D. (1985). Tree survival in cities: the extent and nature of the problem *Arboricultural Journal* **9**, 131–142.

LEVITT, J. (1972). *Responses of plants to environmental stresses*. Physiological Ecology series, Academic Press, New York.

RUSSELL, R.S. (1977). *Plant root systems, their function and interaction with the soil*. McGraw Hill, Maidenhead.

Discussion

T.H.R. HALL (Oxford University Parks).

Does the timing of fertiliser application affect root growth?

A.D. BRADSHAW Limited resources have not allowed this to be tested so we have no evidence as yet. However, it is clear that a first year application is definitely useful.

J.B.H. GARDINER (Forestry Commission).

From my experience, if you add large quantities of N and K at planting, conifers may be killed. The experts tell us that on lowland clay soils, P and K are seldom deficient.

A.D. BRADSHAW This is generally so, but on some artificial soils, such as demolition sites, Ca^{++} induced deficiencies may arise.

S.J. MORRIS (Devon Tree Services).

How much nitrate should be applied at planting?

A.D. BRADSHAW Typically, 50 kg/ha will be sufficient. 50, 100 and 150 kg/ha have been tried but effectiveness varies according to site.

R.I. FAIRLEY (Countryside Commission for Scotland).

N increases the root/shoot ratio but is this effect disrupted by planting?

A.D. BRADSHAW If no nutrients are present then roots will not grow. Roots do not grow to seek nutrients.

D. GILBERT (Ministry of Agriculture, Fisheries and Food).

In my experience I have seen no response to N or P on ordinary nursery soils, though in some circumstances slow release fertilisers have had an effect.

A.D. BRADSHAW Leaching is a dominant factor, particularly on the china clay spoil sites and here a slow release formulation is most applicable. Our experience shows that 'good' soils, such as bought top soils, are frequently depleted and poorly structured.

Tree Shelters

J. Evans
Forestry Commission

Summary

Tree shelters are transparent or translucent tubes placed around newly planted trees. They aid establishment and early growth, provide good protection and permit easier and safer weeding of broadleaved species with herbicide. Almost all broadleaved and most coniferous species benefit from tree shelters and some normally slow growing ones, e.g. oak, beech and lime, show greatly increased early growth.

Since their introduction in 1979 the use of tree shelters has grown very rapidly; about 1.5 million trees were planted with them in 1985. Present research aims to monitor their use and identify ways of further improving this promising technique. The paper also comments on practical aspects of using tree shelters.

Introduction

It is remarkable that only 5 years ago at the Preston Conference 'Research for Practical Arboriculture' tree shelters were virtually unheard of and there is not a single reference to them in the entire proceedings. Today their use is an important establishment technique, with an estimated 1.5 million shelters used in 1985.

A tree shelter is simply a transparent or translucent tube inside which the newly planted tree is grown. The idea was developed by Graham Tuley, hence the colloquial name of 'Tuley tube' for the tree shelter, beginning with a single experiment on oak trees in 1979. Accounts of this early work and subsequent research will be found in papers by Tuley (1982, 1983, 1984a and 1985).

The object of this paper is to assist potential users of tree shelters by (a) describing tree shelters and the effects they have, (b) reporting on the very wide ranging research already carried out, (c) making recommendations regarding their use, and (d) summarising present research developments.

Types of Shelter and the Environment Created

Types of tree shelter

The original tree shelter was a sleeve of polythene placed over plastic netting to form a tube 1.2 m tall and about 8 cm in diameter. Since 1979 a great many different kinds of materials, sizes and shapes have been used for shelters and it is clear that any reasonably transparent or translucent material formed into a tube will function as a tree shelter.

Today, three main kinds of materials are used, corrugated polypropylene, folded into square or hexagonal shaped tubes, extruded polypropylene tubing either ribbed or smooth, and PVC sheeting, usually reinforced with polyester, made into open cones or tubes. Most shelters used are 1.2 m tall but both shorter and taller ones are available to meet differing protection requirements to prevent browsing damage (Pepper, 1983). A list of suppliers of tree shelters is appended.

Microclimate inside tree shelters

Tree shelters protect trees from herbicide, mammal and mechanical damage but one of their main advantages is that the 'mini-greenhouse' environment appears to stimulate tree growth. The reasons for this are complex including shelter from wind, lack of stem movement, even possible suppression of side light – see the work of Donssal (1975) and the Manchon effect reported by Roussel (1972), but clearly the internal microclimate in which the tree grows is likely to be important.

The microclimate of tree shelters is under investigation at Wye College and the data portrayed in Figures 1 and 2 derive from this work (Rendle, personal communication). Both figures compare microclimatic data collected inside a corrugated polypropylene shelter with conditions nearby

Figure 1. Solar radiation during 25th August 1984 measured in the open ———— and inside a corrugated polypropylene tree shelter - - - - - -

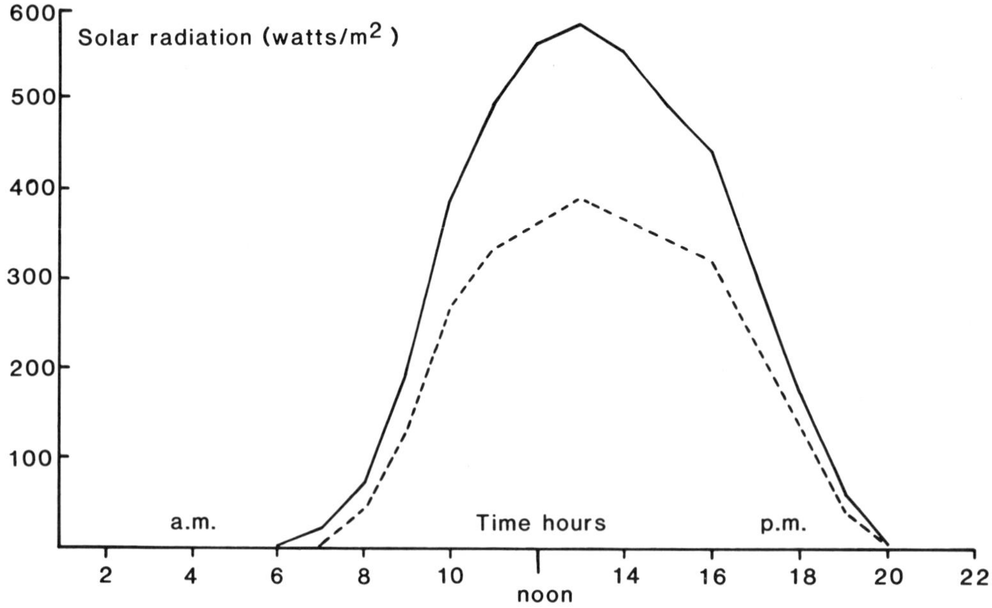

Figure 2. Air temperature and relative humidity measured during 25th August 1984

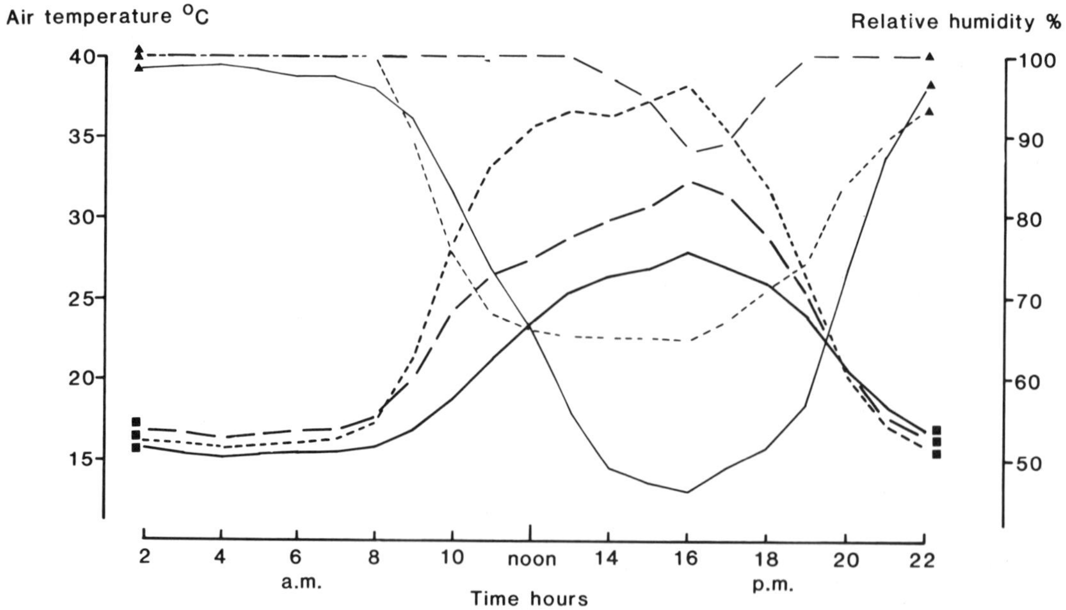

Key: ■ = temperature °C ———— field — — — tube with tall tree out of top plus leaves inside

▲ = relative humidity - - - - - - - - tube with small tree

in the open. They show that solar radiation within this type of shelter is somewhat depressed in the *middle* of the day but that total day length is little affected (Figure 1) and that air temperatures are elevated (Figure 2). However, the greatest difference is the sustained high relative humidity inside a shelter, especially when fully occupied by a tree and its foliage (Figure 2). This combination of elevated temperature and relative humidity may enhance photosynthesis and explain the generally better growth and good survival of newly planted trees. Such conditions are certainly not damaging, despite signs of occasional leaf scorch, and have not been found to increase disease or pest problems.

Overview of Research with Tree Shelters

Since 1979, and including work currently in hand, some 96 experiments have been carried out with tree shelters by the Forestry Commission's Research Division. This considerable body of research, almost wholly under Graham Tuley's leadership, falls into four main categories.

1. Response of different tree species to shelters.
2. An examination of materials and methods of erecting shelters.
3. The interaction of the tree shelter technique with other establishment practices such as plant type, weed control and fertilising.
4. The practical application of tree shelters, i.e. the sites and situations where they are of most value.

Response to tree shelters by species

The first species to be tried in shelters was Sessile oak (*Quercus petraea*), which is normally expensive to establish and slow growing. Performance of Sessile oak grown in tree shelters compared with that in tree guards and in the open, but protected from mammal damage, is shown in Figure 3. The data are from the oldest experiment and are representative not only of oak but of the kind of response of most broadleaved species; the shelters were removed after 3 years, a practice not now recommended. Figure 3 shows that height growth is rapid during the first few years *while the tree is in the shelter* but thereafter settles down to a more normal rate; the shelter has not changed the growing potential (fertility) of the site but accelerated the tree through the initial, slow and expensive early establishment stage. In terms of stem volume and basal area development (Figure 4) trees in shelters are advanced by about 2 years compared with their 'unsheltered' counterparts.

Figure 3. Mean height of oak transplants

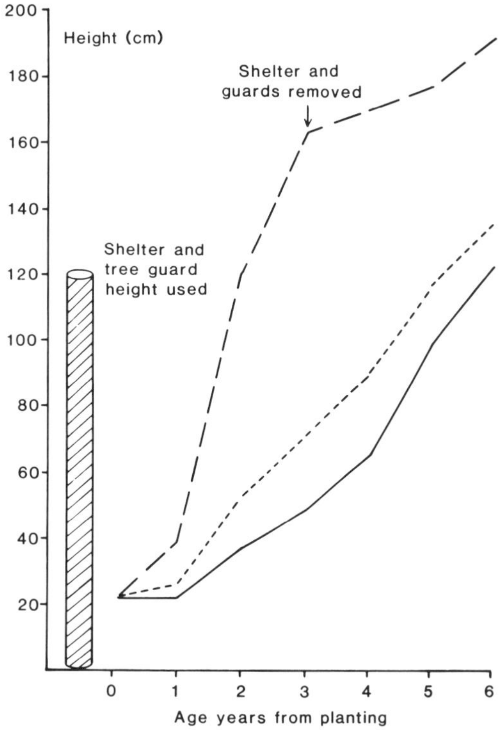

Key: ——— tree shelters for 3 years
- - - - - - - tree guards for 3 years ——— unprotected

Figure 4. Mean stem diameter at 5cm ■ and stem volume ▲ growth of oak transplants

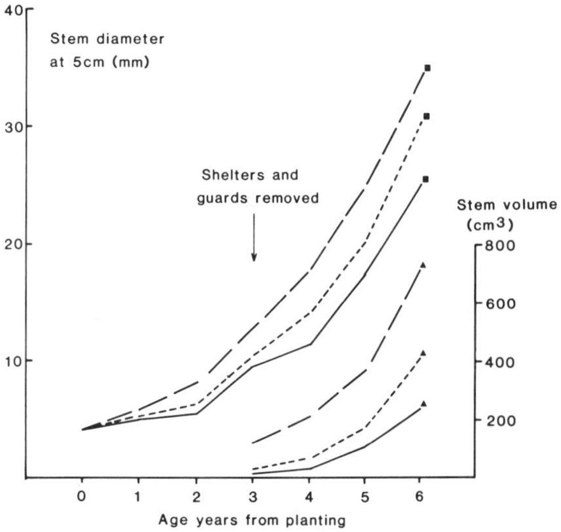

Key: ——— in tree shelters for 3 years
- - - - - - in tree guards for 3 years ——— unprotected

Table 1. Effects of tree shelters on growth by species

Species[a] Common name	Scientific name	No. of experiments where present[b]	Overall growth response[c] (1) very good	(2) good	(3) initial	(4) some	(5) none	Comments
BROADLEAVES								
Alder, Common	*Alnus glutinosa*	7			x			
Alder, Italian	*A. cordata*	2				x		
Ash, Common	*Fraxinus excelsior*	3		x				few early experiments
,, , Narrow leaved	*F. angustifolia*	3						
Beech	*Fagus sylvatica*	9	x					occasionally slow or poor response
Birch	*Betula pendula*	10			x			
Cherry	*Prunus avium*	4			x			rapidly grows out of shelter
Crab apple	*Malus sylvestris*	3		x				
Eucalypt	*Eucalyptus gunnii*	1					x	develop oedema on leaves
Hawthorn	*Crataegus monogyna*	5	x					
Holly	*Ilex aquifolium*	2		x				
Hornbeam	*Carpinus betulus*	3		x				variable, site sensitive
Horse chestnut	*Aesculus hippocastanum*	1				x		
Lime, Large-leaved	*Tilia platyphyllos*	7		x				often very good response
,, , Small-leaved	*T. cordata*	1	x					
Maple, Field	*Acer campestre*	5		x				variable
,, , Norway	*A. platanoides*	2		x				variable
Oak, Pedunculate	*Quercus robur*	2	x					
,, , Sessile	*Q. petraea*	many	x					one or two trees often fail to respond
,, , Holm	*Q. ilex*	1		x				
Rowan	*Sorbus aucuparia*	6			x			
Southern beech								
,, , Dombeys	*Nothofagus dombeii*	3		x				variable
,, , Roble	*N. obliqua*	3		x				} very variable, often die back then
,, , Rauli	*N. procera*	8				x		} good recovery. Site sensitive.

Sweet chestnut	*Castanea sativa*	4	x					tending to rapid initial response only
Sycamore	*Acer pseudoplatanus*	8		x				
Walnut, Black	*Juglans nigra*	3		x				} both species very site sensitive
,, , Common	*J. regia*	3		x				}
Whitebeam	*Sorbus aria*	3				x		
Wingnut	*Pterocarya x rehderana*	3				x		
CONIFERS								
Douglas fir	*Pseudotsuga menziesii*	3		x				
Grand fir	*Abies grandis*	5				x		
Japanese larch	*Larix kaempferi*	5			x			
Pine	*Pinus*							
,, , Corsican	*P. nigra* var. *maritima*	8		x				branches constricted
,, , Bishop	*P. muricata*	5		x				site sensitive
Western red cedar	*Thuja plicata*	8		x				site sensitive
Spruce, Norway	*Picea abies*	3		x				} both very variable in their response
,, , Sitka	*P. sitchensis*	5		x				}
Western hemlock	*Tsuga heterophylla*	8					x	significant response on only one site
Yew	*Taxus baccata*	3		x				still very slow growing!

Footnotes to Table 1.

a. Omission of a species from the list should not be interpreted as being unsuitable for growing in tree shelters; it simply has not been formally evaluated.

b. Mostly experiments specifically comparing species' performance in tree shelters. There are many other experiments with shelters and now a considerable amount of field experience but mostly with the main forest species.

c. Overall growth response.

1. Very good. Species showing consistently good response to shelters, usually more than doubling rate of height growth in first 2–3 years after planting.
2. Good. Generally a significant improvement in growth on most sites but not as marked as in 1.
3. Initial. Species which initially respond well to shelters but, because of early emergence from the top (end of first or during second year) and naturally fast growth anyway, do not sustain significant improvement beyond the third year.
4. Some. On average growth appears somewhat improved by shelters but either there is great variability or, in the experiments in question, the improvement was not statistically significant.
5. None. Shelters confer little advantage, or may even be detrimental.

Many species, both broadleaves and conifers have been included in trials with shelters. Table 1 indicates how different species can be expected to respond to tree shelters based on our experiments.

No species has grown significantly slower or suffered higher mortality as a result of being enclosed in a tree shelter. Indeed, there have been several reports of shelters improving post-planting survival even with difficult species such as Corsican pine (*Pinus nigra* var. *maritima*). As a recommendation, tree shelters merit consideration when planting any species in categories 1 and 2 in Table 1 and category 3 species if individual tree protection is justified on economic grounds (Pepper *et al.*, 1985).

Materials and methods of erection

There are several different types of tree shelter commercially available (Appendix 1). This range represents only a small proportion of the total number of different materials and plastics which have been evaluated as possible shelter materials. All transparent or transluscent materials when formed into a tube appear to give much the same kind of benefit. However, not all materials last for the same period of time and this has led to considerable differences in their potential usefulness. Several points have emerged:

1. To ensure enhanced tree growth, shelters need to last for at least 3 years.

2. Ideally, however, a shelter should last at least 5 years to protect the stem from rabbit and deer damage and to give support to the tree while its crown builds and stem thickens and so minimise risk of snap or bending once the shelter disintegrates.

3. To achieve a 5-year shelter life the plastic must incorporate an ultra violet inhibitor to prevent premature breakdown. In early experiments the use of ordinary polythene sleeves and PVC sheeting, such as Jetran, proved unsatisfactory since they disintegrated or became brittle within about 2 years.

4. As tree shelters have become more widely used other factors shortening their useful life have become evident. The main one, apart from actual stake breakage, is splitting of the corners of corrugated polypropylene shelters. Splitting is sometimes observed to begin after 2 or 3 years and is probably initiated by trees hitting the corners once they grow out of the shelter. This weakness develops earlier on exposed sites which suggests that wind fluttering the shelter is a contributory factor. The folding process and transport of tubes flat may additionally weaken the corners (Jeffs, personal communication). However, it is clear from many experiments laid down in 1981 and 1982 that corrugated polypropylene shelters with large fluting are more prone to corner splitting than those with small fluting. Manufacturers of corrugated polypropylene now use small fluting and, indeed, are also seeking other ways of eliminating the problem.

5. The useful life of round, smooth or ribbed, tubular tree shelters made of extruded plastic may be greatly shortened if they are supported by a metal rod or only a short stake. Under such conditions they are frequently observed to deform and bend and become set in a curved shape. This is obviously useless; such shelters require a stake for at least two-thirds of their length.

Apart from the factor of useful life significant differences have been observed between shelter types in their effect on tree growth. However, overall, no firm conclusions can yet be drawn to say that for example Gro-cones (reinforced PVC sheet) are superior or inferior to Correx tubes (corrugated polypropylene). This type of comparison is currently in hand.

In recent years coloured shelters, notably tints of brown or green, have been widely used. There is a slight suggestion that they may be marginally inferior for tree growth to white or clear shelters but possibly superior for life span.

Although individual experiments have shown significant differences between types of stake, e.g. wooden versus metal rods, overall there is no general trend to suggest one method of erection and fastening is superior to another *from the point of view* of tree survival and growth. There are, of course, big differences between methods; if weak or knotty stakes are used or where sites are exposed or suffer considerable livestock pressure then robust external stakes are necessary.

Research into shelter size has examined both diameter and, to some extent, shelter height. Early work showed that tube diameters greater than 12 cm lead to aerodynamic problems owing to the large surface exposed and exaggerated fluttering. The optimum size would appear to be between 8 and 10 cm across. Height of shelter is mainly a function of the need to protect against mammal damage but it is emphasised that growth enhancement is confined to the period a tree is within a shelter. Short shelters will give a smaller boost to growth than large and tall shelters.

Tree Shelters in Relation to Other Establishment Practices

Tree shelters on their own appear to benefit many species. Nevertheless, it is important to evaluate the benefit in relation to other establishment practices such as plant quality and planting, weed control, fertilising, and the whole subject of where shelters can be used – open ground, woodland glades, enrichment of scrub. Many experiments have been laid down looking at these aspects of tree establishment and the conclusions below draw on these.

Planting and plant quality

Tree shelters aid establishment and early growth of newly planted trees but *they will not resuscitate poor plants* (spindly, damaged, diseased) or *compensate for the effects of bad handling* (physical damage, desiccation, bad or late planting). Tree shelters should be viewed as an aid to further improving good establishment practices not a compensation for poor workmanship.

Shelters are best used with healthy, sturdy, transplants (less than 0.5 m tall); container stock appear to be of no advantage at all and shelters are of little benefit to whips or larger planting stock. Using shelters with direct sowing or with trees which are stumped back is feasible but in both instances shoot growth may develop outside the shelter; thus enclosing a tree which you can see appears the safest course.

Weed control

This subject is amply covered by Davies (1987) but the point is stressed that though shelters will still enhance growth when trees are in a weedy environment good weed control is additionally beneficial (Tuley, 1984b). Moreover, since the tree is enclosed it is relatively easy to apply herbicide at medium to high volume to control competing weeds, say in a 1 m diameter spot around the tree, without risking damage.

Weeds will develop and grow inside a shelter if allowed to do so. Several experiments have investigated the value of very small plastic mats or other mulches to prevent this internal weed growth. Although weeds are prevented, rather surprisingly no significant improvement in tree growth has been recorded.

Fertilisers

Since a tree shelter accelerates early growth greater demands must be made on soil for nutrients, thus fertilising could be worthwhile. On some sites small but significant responses to compound fertilisers (N:P:K) have been demonstrated but in most cases nutrients do not seem limiting to growth. Where weeding treatments have been included in such experiments, achieving a good standard of weed control has always been far more important.

Sites and situations

The individual protection afforded by tree shelters allows the possibility of planting trees in virtually any situation without fear of mechanical or browsing damage. This is a relatively new opportunity for the forester and thus a number of experiments have investigated the effects of amount of overhead shade, density of scrub, the use of shelters on exposed man-made waste sites, and in places frequented by the public such as golf courses and recreation grounds. No unexpected results have emerged but the following are of note.

1. Planting in dense thickets where there is no access to overhead light will fail. Shelters genuinely provide some advantage but do not obviate the need for some cleaning to provide sufficient light for tree growth. The more the opening of the scrub around the trees in shelters the better is the growth.
2. Planting trees with shelters under the canopy of other trees is feasible but results from the three experiments are somewhat conflicting. Two show a steady fall-off in growth with increasing overhead shading; the third shows no significant difference between trees planted in the open or under a tree canopy.
3. In public places tree shelters attract considerable attention but very little vandalism. There appears little preference for colour. Shelters arouse interest which is almost wholly favourable because they show that tree planting is being carried out.

Present Research

Tree shelters have been one of the most exciting developments in silviculture in recent years and research is actively continuing.

Firstly, there is continued monitoring of the performance of tree shelters both in a 'work study' sense by comparing different manufacturers' products and by following up possible problems, especially reports in 1984 of birds becoming trapped and dying inside shelters. Happily, a survey showed such deaths to have been extremely uncommon in all our research trials but clearly monitoring must continue. This year five experiments have been established throughout southern England, comparing the principal types of tree shelter commercially available, on the growth of ash (*Fraxinus excelsior*), beech (*Fagus sylvatica*) and oak.

Secondly, research is continuing into making the most of the shelter environment. Foresters have comparatively little influence over the environment which their trees experience – in marked contrast to crops grown under glass – but the shelter offers virtually complete enclosure and hence control of a tree's environment for the first 2 or 3 years. How can this best be used? Is the environment described earlier the most favourable that can be achieved? One avenue currently being investigated is to modify the light regime. One series of experiments is designed to evaluate the effects of different colours; Correx shelters have been specially printed for this research by Corruplast Limited to cover the entire range of the visible spectrum and into the ultra violet. Another series of experiments compares the effects of increasing degrees of shading by the shelter itself. The manipulation of shelter colour offers intriguing possibilities because it is well known that different wavelengths influence different aspects of plant growth, in particular the near-red wavelengths may be inhibitory, green wavelengths are the least well absorbed by leaves (that is why leaves are green), and ultra violet is known to interfere with auxin synthesis. Whether simply colouring a shelter can have sufficient influence, especially if it results in further reduction in total light supply, is unknown. Nevertheless, in Israel and in Kenya growers have found that yellow film provides the best light environment for growth of many vegetables and tender house plants destined for the European market.

A third area of research is the question of irrigation. The tree shelter is part of a high input:high output establishment package for individual trees and irrigation may afford further substantial benefit. Treatments of trees in shelters will range from a single watering after a long spell of dry weather in early summer to weekly watering throughout the growing season. The latter, very high intensity of irrigation, is not expected ever to have practical application but, as with all research, the range of response needs determining and extreme ideas are always worth trying – after all that is how tree shelters first began!

There is much opportunity for further research with tree shelters notably on the question of size, design, and incorporation of other features which may further enhance initial tree performance such as ability to sway, carbon dioxide enrichment, and so on. There also remain many questions unanswered, in particular are stem wood properties significantly altered when trees are grown in shelters and will side branch and epicormic branch development be affected?

Shelters in Practical Use

This paper has surveyed research with tree shelters but it will be helpful to conclude by summarising the main recommendations regarding use of shelters in practice – further elaboration of these points will be found in Evans (1984) and Evans and Shanks (1985) and in an instruction describing their use, which is available from the Forestry Commission's Research Station, Alice Holt Lodge.

1. When growing broadleaved trees individually or at wide spacing in areas of less than one hectare consider using tree shelters to achieve full tree protection and rapid establishment.
2. Ensure a high standard in all aspects of planting practice – plant quality, plant handling, and timing and methods of planting.
3. Rigorously control competing weeds around each tree, both before planting and subsequently by localised application of herbicide.
4. Use a shelter which has at least a 5–year life, and leave the shelter around the tree until it disintegrates or breaks down.
5. Ensure stake strength and fastening will withstand the stresses the site will be subject to.
6. At the present time there is no evidence that one make of shelter is superior to another.

Acknowledgements

The data used in Figures 1 and 2 were kindly provided by Miss E. Rendle who is undertaking post-graduate research at Wye College on a Forestry Commission research grant. Mr K.D. Jeffs (K.M. Packaging Services Limited) provided some of the information about corner splitting of corrugated polypropylene shelters; he is a plastics consultant to Munro-Alexander Limited.

References

DAVIES, R.J. (1987). Weed competition and broadleaved tree establishment. In, *Advances in practical arboriculture*, ed. D. Patch. Forestry Commission Bulletin 65, 91–99. HMSO, London.

DONSSAL, C. Le (1975). *Eléments d'écologie forestière – la chêne et la chênaie*. Centre Regional de Recherche et de Documentation Pedagogiques d'Orleans.

EVANS, J. (1984). Tree shelters – notes on their use. *Timber Grower* 93, 29–33.

EVANS, J. and SHANKS, C.W. (1985). *Treeshelters*. Arboriculture Research Note 63/85/SILS. DOE Arboricultural Advisory and Information Service, Forestry Commission.

PEPPER, H.W. (1983). *Plastic net tree guards*. Arboriculture Research Note 5/83/WILD. DOE Arboricultural Advisory and Information Service, Forestry Commission.

PEPPER, H.W., ROWE J.J. and TEE, L.A. (1985). *Individual tree protection*. Arboricultural Leaflet 10. HMSO, London.

ROUSSEL, L. (1972). *Phytologie forestière*. Mason et Cie.

TULEY, G. (1982). Tree shelters increase the early growth of broadleaved trees. In, *Broadleaves in Britain: future management and research*, eds. Malcom, D.C., Evans, J. and Edwards, P.N., 176–182. Institute of Chartered Foresters, Edinburgh.

TULEY, G. (1983). Shelters improve the growth of young trees in the forest. *Quarterly Journal of Forestry* **77**, 77–87.

TULEY, G. (1984a). Trees in plastic tubes – taking the greenhouse to the tree. *Paper presented to Forestry Section, British Association for the Advancement of Science*, Norwich.

TULEY, G. (1984b). Trees in shelters do need to be weeded. *Aspects of Applied Biology* **5**, Weed control on vegetation management in forests and amenity areas, 315–318.

TULEY, G. (1985). The growth of young oak trees in shelters. *Forestry* **58**, 181–195.

Discussion

F.R. CLAXON (Epping Forest District Council)

(i) Were shelters successful when used on natural regeneration of beech?
(ii) Have you experienced vandalism of shelters?
(iii) How does Netlon improve growth over the control trees?

J. EVANS

(i) The trial is only 2 years old, though some seedlings have emerged from their 60 cm shelters. The 'control' seedlings have not been very successful, possibly due to browsing by deer.
(ii) On three experiments sited so as to almost invite vandalism, interference has been surprisingly minimal though much curiosity has been aroused. A good stake may prevent casual damage.
(iii) This may be due to the shelter effect of the stake.

G. ANDREWS (Somerset County Council)

Has any work been undertaken on the optimum diameter of shelters?

J. EVANS

Shelters up to 20 cm across have been tried though 12 cm is the dimension widely used. Trees respond well if enclosed in a shelter which is over 8 cm diameter but there is no advantage if this is increased above 12 cm. The narrower sizes have produced some branch distortion in conifers.

G. ANDREWS

Have there been any deaths due to high temperature inside the shelters?

J. EVANS

The air temperature in the shelters may be high, but recorded leaf temperatures have not been critical.

S.A. COOMBES (Raven Tree Services)

Does vigorous weed growth in shelters harm trees?

J. EVANS

Weed growth, like tree growth, is enhanced by the shelter effect and weed control around the tree remains generally important, but where weeds have been seen within the shelter this has not been found to be significantly detrimental.

R.G. TAYLOR (Wealden Woodlands Ltd)

The results so far seem very good, but the technique needs to be tried on oak regeneration.

J. EVANS

All the evidence suggests that the response will be good, but it should be remembered that not every individual oak will respond similarly due to natural variation within the species.

Appendix

Corrugated polypropylene

Correx Tree Shelters (welded)

Corruplast Ltd., Correx House, Moreland Trading Estate, Bristol Road, Gloucester GL1 5RZ (Tel. Gloucester (0452) 31893).

Scottish Distributor:
Plasboard Plastics Ltd., Unit 8, Broomfield Industrial Estate, Montrose DD10 8SY (Tel. Montrose (0674) 76006).

Somerford Sheltatree (stapled)

Monro, Alexander & Co. Ltd., Newleaze, Great Somerford, Nr. Chippenham, Wiltshire (Tel. Seagry (0249) 720442).

Smooth extruded tube

Greenleaf Tree Tubes

Stanton Hope Ltd., 11 Seax Court, Southfields, Laindon, Basildon, Essex SS15 6LY (Tel. Basildon (0268) 419141).

Ribbed extruded tube

Tree Cells

Argival Plastics Ltd., 47–49 Cowley Street, Methil, Fife KY8 3QQ, Scotland (Tel. Kirkcaldy (0592) 713801).

Wire in small quantities and staple guns are easiest to purchase at local hardware shops or garden centres.

Twin wall extruded shelter

Tubex Treeshelter

Tubex Ltd., Littlers Close, Colliers Wood, London SW19 2TF (Tel. 01-542-9767).

PVC sheet and shelter pieces

Andrew Mitchell & Co. Ltd., Hainault Road, Little Heath, Romford, Essex RM6 5ST (Tel. 01-590-6070).

PVC manufactured cones and mulching mats

Gro-cones

Acorn Planting Products, Mornington, Walnut Hill, Surlingham, Norwich NR14 7DQ (Tel. Surlingham (05088) 279).

Trouble at the Stake

D. Patch
Forestry Commission

Summary

Early recommendations for tree planting recognised that, after transplanting, large trees needed support to hold them upright until a new root system developed. Staking has become the method of providing this support and is almost universally adopted in landscape situations irrespective of the size of tree being planted. Absence of after-care means supports, especially stakes and ties, are retained for many years. The result is physical and morphological damage to the tree. This paper reviews research into the effects of support on stem development, reasons for supporting trees and recommends modified practices of support.

Introduction

John Evelyn (1678) recognised that natural regeneration was not occurring sufficiently rapidly to replace trees which died or were felled. In order to meet this shortfall he recommended a policy of growing trees in a nursery and transplanting them to a final position. Today the need for continued tree planting in the landscape is acknowledged by both national and local governments (Anon., 1978). The provision of legal and fiscal measures emphasises the importance of tree planting; but statutes and finances alone do not guarantee success and as Peters (1987) suggests early losses in contemporary tree planting schemes may be in excess of 50 per cent.

In his treatise Evelyn (*op. cit.*) emphasised the need for thorough site preparation before planting takes place – a requirement often reiterated today (Binns, 1983; Thoday, 1983) but rarely adequately practised. In contrast, Evelyn's recognition that when planted into cultivated soil nursery grown trees have inadequate natural anchorage to hold them upright and his suggestion that artificial support should be provided has become widely accepted. Evelyn suggested brambles around the base of trees in rural areas could provide adequate natural support. In addition he described a technique of utilizing a piece of rope (natural fibre) tied round a pad of straw high up on the stem and secured to a peg in the ground. This might be appropriate for sites where wind from only one direction is a problem.

Today Evelyn's recommendation for newly planted trees to be supported has evolved to the use of a stout wooden stake, often pressure treated with preservatives to ensure a long life and one, two or more ties to "secure the tree to the stake to prevent excessive movement" (Anon., 1969). This specification appears to be increasingly used irrespective of the size of tree being planted. However, Evelyn and his contemporaries would have been able to give regular after-care and attention to newly planted trees. During the routine inspections and after-care, gardeners would have recognised when support was no longer needed and removed it before the tree suffered damage. This is in contrast to the present day when money and manpower are unavailable for ongoing cultural practices to ensure that newly planted trees establish and subsequently flourish without being damaged.

People responsible for contemporary amenity tree management in both rural and urban situations are concerned about the possibility of stems of newly planted trees being broken either by wind or vandals. Indeed, the sight of standard trees with stems broken at the position of the attachment to the support is not uncommon. Unfortunately surveys of the occurrence of such physical damage in Britain have not been made.

Casual observations of amenity trees indicate that the support systems themselves cause mechanical damage to trees. The only report of young trees which identifies and describes the damaging agent was done by Foster and Blaine (1978). This survey, carried out on the street trees of three districts of Boston, Massachusetts, showed that stakes and ties were the most common cause of damage to newly planted street trees, while vehicles and vandalism were subordinate as causes of damage. Furthermore, physical damage was visible as little as 3 months after planting.

The work reported in this paper was undertaken to assess the effects of currently recommended support methods on the development of newly planted broadleaved trees.

Materials and Methods

In order to minimise the possibility of variation in the plant material used, trees of the poplar clone T x T 32 (now *Populus* 'Balsam Spire') were used. These were approximately 1.8 m tall rooted plants and unrooted sets which had been graded for uniformity of height and stem diameter within each plant type. The trees were planted 3 m x 3 m in stratified blocks on an exposed site north of the Hogs Back, Surrey, and supported with four intensities of stout wooden stakes and proprietary all-plastic ties. The treatments were:

(a) control – no stake (O);
(b) stake and one tie at the root collar (GL);
(c) one-third stem length supported using one tie ($\frac{1}{3}$);
(d) half-stem length supported using two ties ($\frac{1}{2}$);
(e) stem fully supported using three ties (1/1).

In each treatment a tie was positioned at the top of the stake to prevent contact between the tree and the stake at that point. As the planting material was straight stemmed there was no need to prevent the lower stem rubbing on the stake. The ties were nailed to the stakes to prevent them slipping down the stake.

One year after planting the support was removed from half the trees in each treatment. Where possible this included the removal of the stakes, otherwise they were cut off at ground level. The stakes on the fully supported trees, where the support was to be retained, were raised to the tip of the leading shoot to minimise any effect of stem movement during the second year's growth.

Results

Assessments were made of height increment and stem diameter after one and two growing seasons. After one growing season there were no significant differences between the intensity of staking treatments (Figure 1).

After two growing seasons assessments showed:

i) *Height*
Trees staked for 2 years had greater height growth in the second growing season than trees staked for only the first year after planting. The differences between duration of staking and staking intensity were significant in both rooted and unrooted trees (Table 1).

Figure 1. Stem form based on radius after 2 years' growth. Stems supported for one year. (Stem form after first year shown hatched.)

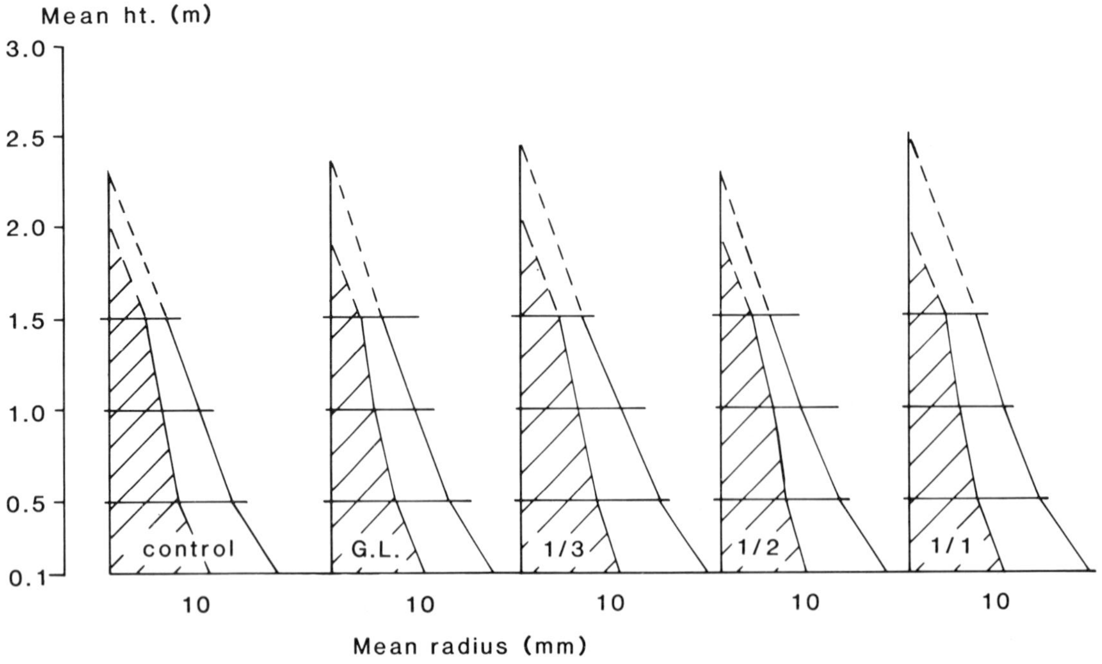

Table 1. Mean height increments (metres)

	All plants													
	Increment during first year				Increment during second year					Total increment				
Staking treatment	Rooted plants		Unrooted setts		Rooted plants staked 2 yr	Rooted plants staked 1 yr	Unrooted setts staked 2 yr	Unrooted setts staked 1 yr		Rooted plants staked 2 yr	Rooted plants staked 1 yr	Unrooted setts staked 2 yr	Unrooted setts staked 1 yr	
Control	0.096	0.105	0.138	0.073	0.407	0.230	0.386	0.352		0.503	0.335	0.524	0.425	
GL	0.072	0.114	0.028	0.035	0.682	0.343	0.226	0.463		0.754	0.457	0.294	0.498	
30	0.104	0.083	0.172	0.042	0.648	0.400	0.513	0.414	*	0.752	0.483	0.685	0.456	*
50	0.060	0.040	0.033	0.118	0.696	0.328	0.450	0.348		0.756	0.368	0.483	0.466	
100	0.120	0.185	0.062	0.080	1.028	0.343	0.606	0.315		1.148	0.528	0.688	0.395	
Σ	0.452	0.527	0.433	0.348	3.461	1.644	2.181	1.892		3.913	2.171	2.674	2.24	
1 year stake mean		0.105		0.069		0.329		0.378			0.434		0.448	
2 year stake mean	0.090		0.087		0.692		0.436			0.783		0.535		
					***		***			**		**		

Figure 2. Stem form based on radius after 2 years' growth. Stems supported for 2 years. (Stem form after first year shown hatched.)

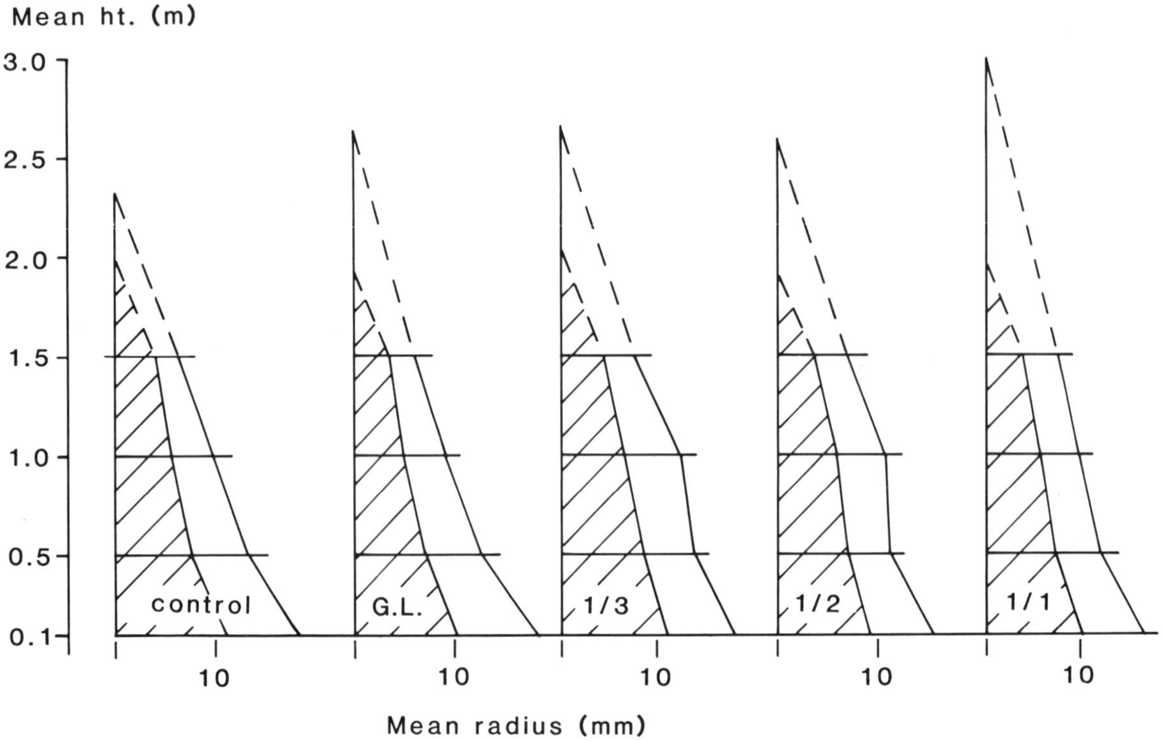

ii) *Stem diameter*
Stem diameter was measured at 0.1 m, 0.5 m, 1.0 m and 1.5 m above ground level. At each point two measurements were taken at right angles using a vernier calliper and means calculated for each treatment. The treatment means differed with height of assessment above ground level. As shown in Figure 2 the second year diameter increment at ground level (0.1 m) and 0.5 m were greater in trees staked for one third, one half and totally for 1 year only. In contrast, diameters measured at 1.0 m did not show any significant difference between deviation or intensity of staking.

iii) *Root development*
Assessment of the roots was attempted but there was no detectable effect on the size or architecture of the root systems.

Discussion

Each year trees grow above ground in height and girth by forming a cone of new wood outside the core of existing wood. The size of the cone is determined by the physiological condition of the tree, including the area of leaf it is able to support (Büsgen *et al.*, 1929). By altering the aerial conditions in which trees grow foresters manipulate the proportions of height and diameter growth (Baker, 1950). That is, in woodlands where trees are close together and there is competition for light, height growth is encouraged at the expense of stem diameter growth. In contrast, hedgerow and parkland trees tend to be shorter in stature than woodland trees of the same age but with larger breast height girths, larger branches, wider crowns and more root buttressing (Hummel, 1951).

Competition for light is not the only criterion influencing the growth of a tree. Jacobs (1954) working with an established plantation of Monterey pine (*Pinus radiata*) in Australia investigated the effects of thinning on stem development. He held the lower 6 m (20 feet) of stem firm with guys secured to cup hooks secured into the stems of trees exposed to different thinning regimes. Measurement of the trees after 2 years showed that the supported trees were taller than the unsupported trees. However, the supported trees had less diameter growth below the support but were fatter than the unsupported trees above the support. That is, the unsupported trees were more strongly tapered than the supported trees.

After 2 years Jacobs removed the artificial support with the result that trees, which had stood upright when originally isolated by thinning, bent away from the vertical while others snapped in the vicinity of the support. From this work Jacobs deduced that in addition to competition for light, movement of the stem contributed to the stimulation of diameter growth but suppressed height growth.

Büsgen *et al.* (*op. cit.*) suggest that movement of a freely swaying stem had maximum effect on diameter growth at ground level. However, the number of branches on the stem also influences stem diameter growth and this effect is greater at the base of the live crown of a tree (Büsgen *et al.*, *op. cit.*).

Harris *et al.* (1971 and 1972) examined staked landscape trees growing in containers standing packed tightly together in nursery conditions and detected similar effects on stem growth after as little as 3 months. However, a microscope was needed to detect the differences. The recent work (Patch, 1980) with clonal poplars used widely spaced trees with branches retained to ground level. Maximum diameter growth both from the influence of branches and, in unsupported trees, sway of the stem should have been at ground level.

Results from Patch (1980) are consistent with those of both Jacobs and Harris *et al.* working with older woodland grown trees and landscape trees in a nursery respectively. However, the four diameter measurements made by Patch on the poplar were too widely spaced to indicate the precise effect of the staking on stem form. In addition the apparent lack of effect of the treatments on the diameter measurement taken at 1.0 m appears anomalous: the position of the ties, which was determined by the tree height, could have coincided with this measurement point.

Why stake trees?

Three reasons are usually quoted for supporting the stems of newly planted landscape trees. These are:

a) to hold the stem upright until it is thick enough to support the crown;

b) to reduce physical damage to the stem, including vandalism;

c) to hold the tree steady until a new anchorage has developed.

(a) *Supporting the crown*

Examination of British grown nursery stock (Anon., 1980) used on amenity sites suggests that generally, in spite of being clean pruned, stems of standard trees are adequately developed to support the crown without artifical support. Provision of support after planting, as currently practised and required by British Standard 4428 (Anon., 1969), will suppress stem diameter growth between ground level and the upper support point, promote height growth and result in stem thickening above the tie. Although these effects are rarely noticed they are detectable with callipers after only two growing seasons. The duration of the imbalance is not known.

Trees released from artificial support after two or more growing seasons may lean. This can be unsightly and prove a serious hazard where amenity trees are growing in public places. Harris and Hamilton (1969) suggest this lean is the result of the shadow cast by the stake, influencing movement of auxins through the stem creating differential growth on the exposed and shaded sides of the stem. This explanation is not confirmed by the results of Jacobs (1954) who supported his trees with guy wires connected to hooks secured to the trunk. As a result there was no shading effect on the stems and physical damage resulting from insertion of the hooks would have been too localised to affect the movement of hormones. Observations in Britain suggest that when trees that have been supported for many years are released they lean away from the prevailing wind.

(b) *Physical damage*

Physical damage, whether caused by wind or vandals, is difficult to research because it is unpredictable. However, trees supported in the traditional manner (Anon., 1969) are not immune from distortion and damage. Large trees planted in windy locations and staked to hold them upright may lean away from the prevailing wind above the upper support point. This distortion persists for many years and it may not 'grow out', resulting in trees which are a hazard in public places. If these trees have to be removed in the interest of public safety it is unlikely there will be an opportunity to establish replacement trees which could be managed better.

Stems snapped just above or just below the upper support point are often seen. Leiser and Kemper (1968) in considering this damage were able to show mathematically that a tree with a branch-free stem supported to the base of the live crown was more easily broken than a similar unsupported tree. The upper support acts as a fulcum, minimising flexing of the lower stem so that the effectiveness of any force applied to the top of the tree is, suggested Leiser and Kemper, six times greater than the same load applied to an unsupported stem.

If the stem of the tree was cylindrical it would have uniform strength throughout its length and equal resistance to bending (Metzger's beam of uniform strength (Büsgen *et al.*, *op. cit.*)). A cone, in contrast, increases in strength from its apex to its base and the effect of any force applied to the cone will therefore depend on the point around which the force acts, i.e. the length of the cone. A tree stem is more or less conical, the degree of taper depending on species and cultural practices in the nursery.

A force applied to the leading shoot of a planted tree with an unsupported stem will be dissipated throughout its stem length. Supporting the stem as recommended in the British Standard (Anon., 1969) reduces the length of stem over which the load can be dispersed. At the same time the effective cone to which the load is applied is much smaller in diameter than at the base of the stem, and therefore less able to resist the force. Leiser and Kemper's modelling has not been tested with plant material – confirmatory data would be invaluable in assisting understanding of physical damage and minimising its occurrence.

Furthermore, a survey of young landscape trees may indicate whether stem snapping is commonest in the first year after planting. If this is the case it is possible that nursery practices which grow trees close together and remove their lower branches, resulting in stems with a tendency to be cylindrical, will require rethinking. Trees grown at wider spacing with branches retained to ground level, at least until the trees are lifted for sale, may prove more appropriate for landscape use even if the practice is less economically acceptable.

Evidence that stem fracture occurs mainly in the second or subsequent year after planting could suggest stem form and strength are being adversely affected only by the support method.

Support techniques using a stake and ties (Anon., 1969) involve a band of inert material around the stem. Such bands are usually pulled tight to minimise stem movement and the possibility of abrasion of the bark. Glenn (1964) assessed tie materials for effectiveness and cost only. Our understanding of the effects of physical pressure exerted on the tissue of the stem by both proprietary and bespoke ties is poor. Brown (1987) reports early work to assess the significance of pressure on stem growth.

(c) *Anchorage*

The roots of standard trees are usually severely shortened when trees are lifted from a nursery bed. Once planted into cultivated soil, especially if bulky organic matter has been added (Anon., 1969), such root systems do not provide anchorage to resist toppling in a wind. Furthermore, if such a tree does remain upright without artificial support, the movement of the crown and stem would tear developing roots and slow establishment. Fayle (1968) reports that an unsupported tree has more root diameter growth than a supported tree although the effect is restricted to about 0.6 m radius from the root collar. Thickening of roots in this region corresponds with maximum stem diameter growth at ground level and results in the familiar buttressing seen on parkland and open grown trees. Buttressing is much less marked or it may be absent in woodland trees which experience less crown movement. Temporary artificial support of newly planted standard trees is needed.

The aim of any support should be to hold the root collar steady until a new anchorage has developed. Theoretically a support attached only to the root collar should be adequate, but in practice a force applied to the leading shoot could pivot the tree around the support point and the inadequately held roots would come out of the soil. A higher support appears justified therefore but it need not extend to the base of the live crown let alone into the head of the tree.

Conclusions

Visible mechanical damage and invisible morphological damage caused by stakes and ties may be attributed to poor practices and lack of aftercare. Both should be avoidable, but there is need for more research to determine the effects of support and pressure on stem form and structure.

Support for newly planted trees usually follows BS 4428 (Anon., 1969) and is designed to "hold stems firmly". Further research is needed to test Leiser and Kemper's mathematical model as applied to plant material and to assess the possible role of flexible materials in place of wooden stakes. In addition, the question of materials for tying trees to the support must be appraised so that appropriate properties are used and morphological damage to the stem is minimised.

Recommendations

The philosophy of John Evelyn (1678) appears to be justified insofar as the need to support newly planted trees is concerned but the recommendation in British Standard 4428 and currently practised needs rethinking.

Evidence from Australia, America and British research suggests that currently practised techniques for supporting newly planted trees can be physically damaging to the tree and, if allowed to remain more than one growing season, will adversely affect development of the tree. It is recommended, therefore, that:

1) Newly planted trees should be as small as possible so that support is not needed.

2) Where large trees (over 2 m tall) cannot be avoided support should be provided but:

   i) it should not extend more than 1/3rd total tree height and should be attached to the stem only at the top of the stake;

ii) if adverse effects of support on stem development are to be avoided the support must be removed after one growing season. In practice the most appropriate time to remove the support would be at the onset of the second growing season after planting. This means that the unsupported tree is not immediately exposed to winter gales and has a growing season to become 'balanced'. If the tree has not developed an adequate anchorage by this time the soil physical conditions are probably inappropriate as a result of inadequate cultivation prior to planting. It is then unlikely that an adequate anchorage will ever develop.

3) Where trees have been supported for many years, removal of the support could allow the tree to lean, fall over, or the stem to snap. It is necessary, therefore, to 'wean' the tree from its support. This can be achieved by releasing the tie and, if the tree leans, finding the position on the stem which just holds the crown upright. The tree should be refixed at that point and the excess stake should be cut off just above the tie. After another growing season the procedure should be repeated until the support can be removed completely.

Footnote

The above principles are based on research, mathematical modelling and field observation, but it is far from easy to test them in the field because damaging agents are transient. It is suggested therefore that the efficacy of these recommendations should be tested on a small percentage of trees in large planting schemes. To date, practitioners who have tried the technique report no greater physical damage to the trees than occurs to the traditionally supported trees. Furthermore, practitioners comment that "trees with short stakes look better" and suggest that "they are growing better". If this is the case then the trees are likely to achieve the objectives of the original planting with a minimum of damage.

Acknowledgements

The work on which this paper was based was funded initially by Surrey County Council and subsequently by the Forestry Commission. My particular thanks go to Professor J.D. Matthews of Aberdeen University who supervised the work.

Discussion

For the discussion on this paper see page 00.

References

ANON. (1969). Recommendations for general landscape operations. *British Standard 4428 : 1969*. British Standards Institution, London.

ANON. (1978). Trees and forestry. *DOE Circular 36/78*. HMSO, London.

ANON. (1980). Specification for nursery stock – trees and shrubs. *BS 3936 Part 1 : 1980*. British Standards Institution, London.

BAKER, F.S. (1950). *Principles of silviculture*. McGraw-Hill, New York.

BINNS, W.O. (1983). Establishing trees on damaged soils. In, *Proceedings of Tree Establishment*. A Symposium at University of Bath, July 1983. University of Bath.

BROWN, I.R. (1987). Suffering at the stake. In, *Advances in practical arboriculture*, ed. D. Patch. Forestry Commission Bulletin 65, 85–90. HMSO, London.

BÜSGEN, M., MÜNCH, E. and THOMPSON, T. (1929). *The structure and life of forest trees*. 3rd edition (translated by Thomson). Chapman and Hall Ltd., London.

EVELYN, J. (1678). *Sylva, or a discourse of forest trees and the propagation of timber*. 4th edition. Arthur Doubleday & Co. Ltd., London.

FAYLE, D.C.F. (1968). Radial growth in tree roots. *Technical Report No. 9. Faculty of Forestry, University of Toronto*, Canada.

FOSTER, R.S. and BLAINE, J. (1978). Urban tree survival: trees in the side-walk. *Journal of Arboriculture* **4**, 14–17.

GLENN, E.M. (1965). A trial of tree ties. *East Malling Research Station Annual Report 1964*.

HARRIS, R.W. and HAMILTON, W.D. (1969). Staking and pruning young *Myoporum laetum*. *Journal of American Society of Horticultural Science* **94**, 359–361.

HARRIS, R.W., LEISER, A.T., NEEL, P.L., LONG, D., STICE, N.W. and MAIRE, R.G. (1971). Tree trunk development: influence of spacing and movement. *Combined Proceedings of International Plant Propagators Society 1971* **21**, 149–161.

HARRIS, R.W., LEISER, A.T., NEEL, P.L., LONG, D., STICE, N.W. and MAIRE, R.G. (1972). Spacing of container-grown trees in the nursery. *Journal of American Society of Horticultural Science* **97**, 33–51.

HUMMEL, F.C. (1951). Increment of 'free-grown' oak. *Report on Forest Research 1950*, 65–66. HMSO, London.

JACOBS, R.M. (1954). The effects of wind sway on the form and development of *Pinus radiata*. *Australian Journal of Botany* **2**, 33–51.

LEISER, A.T. and KEMPER, J.D. (1968). A theoretical analysis of a critical height of staking landscape trees. *Journal of American Society for Horticultural Science* **92**, 713–720.

PATCH, D. (1980). *The influence of staking on the growth of the Poplar clone T × T 32.* MSc thesis, University of Aberdeen.

PETERS, J.C. (1987). Introduction. In, *Advances in practical arboriculture*, ed. D. Patch. Forestry Commission Bulletin 65, 7–10. HMSO, London.

THODAY, P.R. (1983). Tree establishment on amenity sites. In, *Proceedings of Tree Establishment.* A Symposium at University of Bath, July 1983. University of Bath.

Suffering at the Stake

I.R. Brown
Department of Forestry, Aberdeen University

Summary

Improperly used tree stakes may do as much harm as good. This paper describes attempts to determine the effects that pressure and wounding have on the cambium and stem wood. In birch, pressures of only 17 to 50 g/mm^2 cause the formation of structurally weak, abnormal wood and similar tissues were found in elm. When stems are wounded they may respond by producing callus and then one or more abnormally wide growth rings. In many species this 'healed' wound is intrinsically weaker than normal tissue. These mechanical and developmental defects are of course additional to those due to fungal invasion.

Introduction

Everywhere amenity trees are staked – in streets, parks, road verges and private gardens. Common sights include weedy-stemmed standards, firmly strapped between three posts by large elastic bands, feathered trees flanked on two sides by stout 4″ × 4″ posts and heavy, vandal-proof standards succumbing to a variety of restraints ranging from rubber and iron contraptions to lengths of canvas webbing. An anonymous article in ILAM Journal (1983) stated that "Official research may have come to the conclusion there is much in favour of half-staking but would it have done so . . . if (research) had been carried out . . . where the vagaries of the human race are much more a force to be contended with than our capricious weather?" The problem with this philosophy is that it assumes a greater degree of wilful damage than that deriving from cultural neglect. Certainly on the basis of casual observation I would dispute this point. Until both people at large and particularly the arborists' attitudes to staking and tying change, these practices, through lack of aftercare until the trees are self-sufficient, will continue to result in restricted growth, increased susceptibility to disease and mis-shapen trees and stem fracture. Could it be that a bio-degradable tree-tie that expands as the tree grows is the answer?

It was speculation about this idea that led to thoughts concerning the degree of pressure the cambium of a young tree can resist. It is known that explants of cambial tissue, cultivated in agar, require applied pressure before the new cells they produce approximate in shape to normal wood elements (Brown, 1974; Makino *et al.*, 1983). It is also a matter of common observation that cambial growth in roots can withstand and overcome the pressure applied by paving stones, for example, but a piece of garden twine tied around a stem soon becomes engulfed in swollen tissue. So far as I am aware, there is little if any precise data concerning the exact degree of pressure that will restrict cambial activity and provoke such increased radial growth above and below the point of pressure.

Seedling Pressure Experiments

The plants used in these experiments were greenhouse-grown birch, one year old, between 400 and 500 mm tall, and with a diameter of between 5 and 10 mm at about one third of their height. Various weights from 10 g up to 1280 g were applied to the stems at this height via a hard rubber pad, 50.5 mm^2 in area and thus pressure varied from 0.2 to 25.3 g/mm^2. The experimental set-up is shown in Figure 1. Each treatment was applied to four trees and, each week, stem diameter was measured using a micrometer along two axes, parallel and at right angles to the applied force, at the pressure points and at points 1 cm above and below the upper and lower pads.

Results

All treatments showed growth curves similar to those in Figure 2 and there were no consistent differences in rates

Figure 1. Birch seedling pressure experiment.

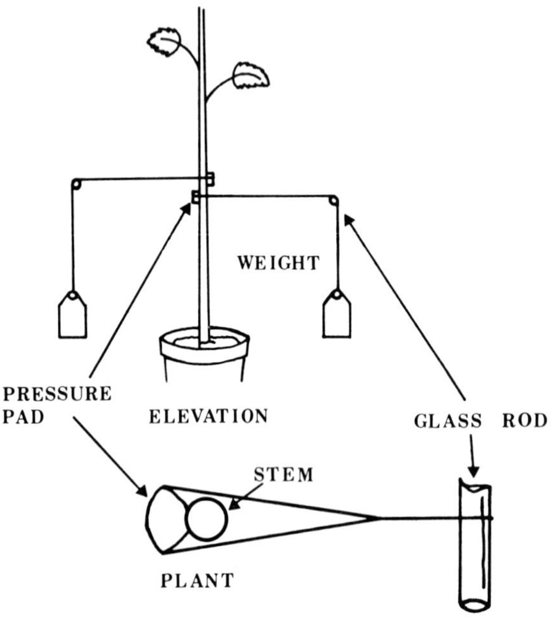

of diameter increment along the two axes measured at the four points on the stems. Thus it appeared that none of the treatments affected cambial activity or else that the expected localised effect was replaced by one that influenced the whole length of stem under observation.

To investigate the matter further the stems were sectioned and, as is shown in Plates 1a and 1b, lens-shaped areas of abnormal tissues were seen in transverse sections of the xylem. Unlike normal vessel elements, the cells were not elongated parallel to the stem (Plate 1c), appeared to be filled with starch granules and were similar to callus or parenchyma cells. It also became clear, however, that this tissue was not limited to the pressure points but was to be found in varying amounts on four radii just where measurements had been taken! At this stage it was somewhat belatedly realised that the culprit could be the micrometer and when its clutch mechanism was investigated it was found that clutch slippage, while tightening up the jaws, occurred when the jaws applied forces to the stems of between 100 and 300 g. A more accurate figure than this could not be achieved since the point of slippage seemed to be affected by the speed at which the knurled screw was tightened and the type of stem surface over which the face of one jaw rotated.

Figure 2. Birch pressure experiment (25 g mm^{-2}).

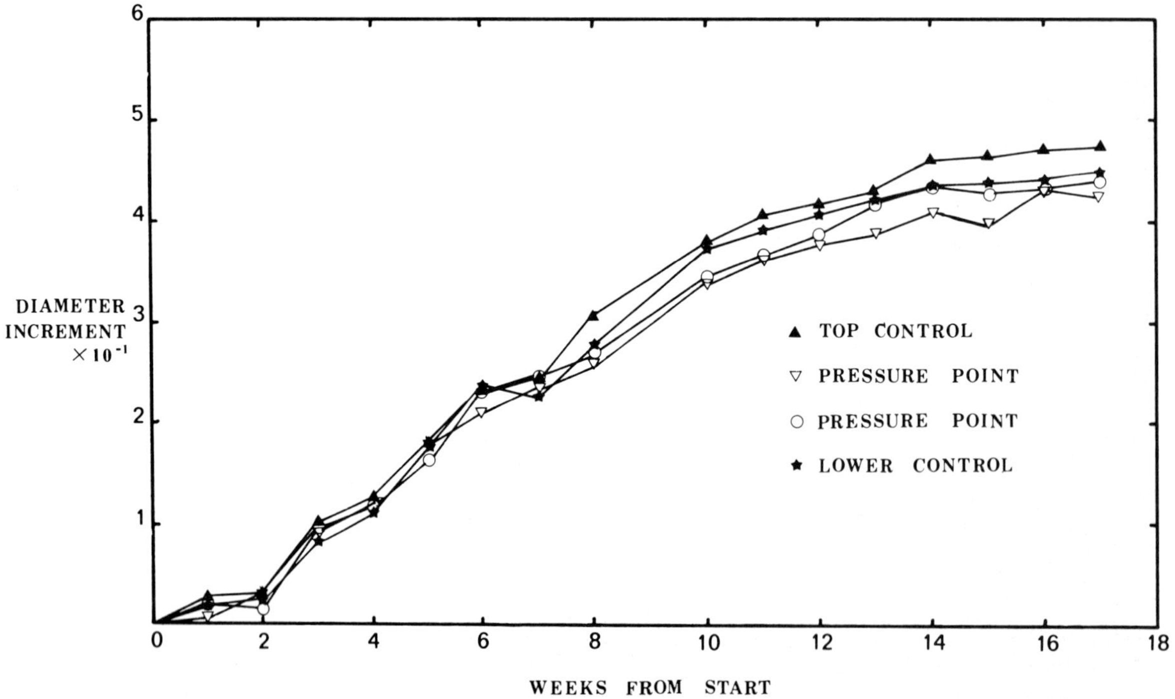

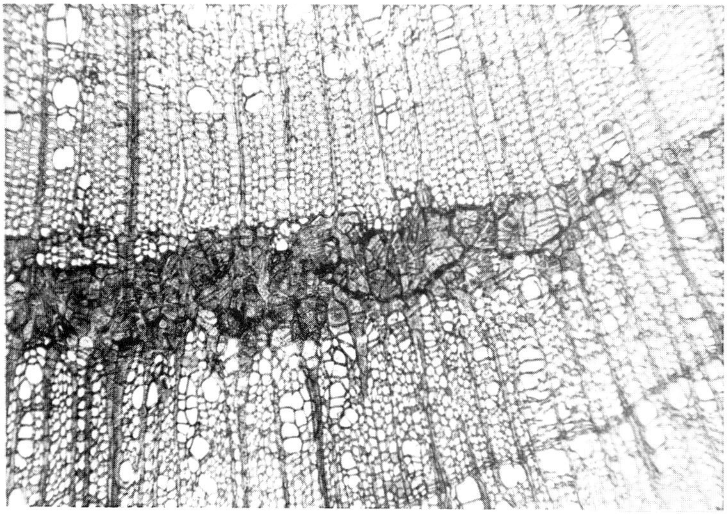

(a) Low power transverse section × 122

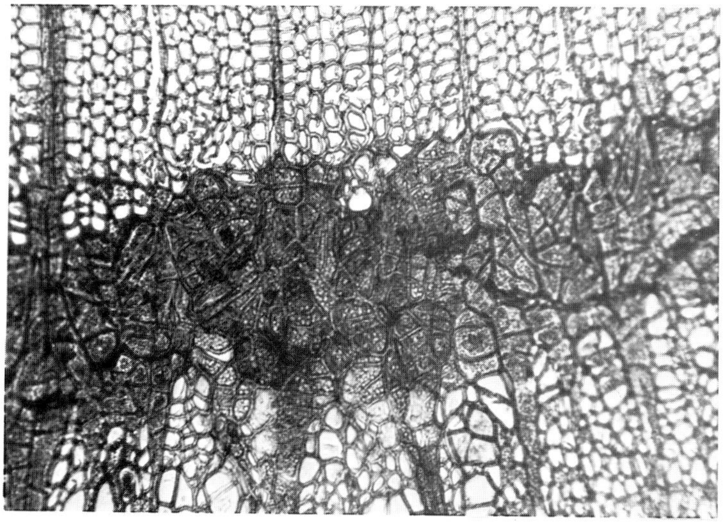

(b) Medium power transverse section × 337

(c) Low power radial section × 204

Plate 1. Abnormal tissue developed in birch seedling pressure experiments.

The area on the stem in contact with the micrometer was estimated to be 6 mm vertically by 1 mm wide thus the pressure applied was between 17 and 50 g/mm^2. This is up to twice the treatment pressure but, of course, applied only once each week rather than continuously. Despite the pressure being applied mainly vertically the effect (if that is what it was) spread horizontally to produce a lens-shaped area in cross section. In addition it appears that the amount of abnormal tissue, at all the measuring points, increases with the experimental pressure applied just as if one treatment was reinforcing the effect of the other.

Discussion

While the above experiment was in some respects a failure there are still some lessons to be learned. Just because a tree shows no external signs of damage or reduced growth that does not mean that it is growing normally. In some of the seedlings, up to 25 per cent of the increment was replaced by abnormal tissue. An amenity tree suffering from similar pressures, either intermittent or continuous, could also develop such tissue resulting in a reduction in both stem strength and water conducting tissue. The former effect making breakage by vandalism easier, for example, and the other increasing the risk of water stress and its attendant ills in the first few years after planting.

Effects of Pressure on Older Trees

It was intended to examine the wood formed under the ties of conventionally staked and neglected trees but these, for a variety of reasons, proved remarkably difficult to obtain locally. The most suitable material that became available were branches from the crown of a Wych elm (*Ulmus glabra*) that had been braced with wire ropes many years ago. Initially the rope had been prevented from chafing the bark by pieces of wood inserted between the rope loops and stem. Eventually, however, the wood packing and the rope became embedded in the living wood of the branches.

Having split the limbs at the point of attachment of the rope and sectioned the wood, three features were observed. Firstly there had been an increase in radial growth when the ropes were first braced, even though pressure was applied to the tree only under the pieces of wood. Secondly this accelerated growth soon resulted in the branch expanding to meet the rope, at which point, all growth ceased below the rope while diameter increment continued above. Finally, the wood produced in the last few years was abnormal in that there was a reduction in vessel elements, tracheids and fibres and an increase in disorganised parenchyma tissue (Plate 2).

Thus under pressure, similar abnormal tissue is produced in elm as in birch and the wood is weakened by its presence. Since elm is ring porous the accelerated growth mentioned above would tend to produce stronger wood than normal. The cessation or reduction of growth below a constriction, even if it resumes following removal of the cause will alter stem form and so will enhance the abnormal distribution of wood brought about by staking *per se* that has been described by Patch (1987).

While it is always unwise to extrapolate from one circumstance or species to another, the experiments and observations described above point, if nothing else, to the need for further research to quantify the effects of pressure on stems in terms of both water and assimilate conduction and strength.

Wounding

So far nothing has been said about the more obvious wounding of stems by abrasion against ties or tops of posts. Usually in young stems, once the cause has been removed, 'healing' is rapid. Firstly callus forms in which a new cambium develops which then usually quite vigorously goes on to produce new wood and bark and the scars soon disappear.

This process was seen in larch (*Larix* spp.) that became available by courtesy of the Forestry Commission. These trees, growing in the larch seed orchard at Research Branch Nursery, Newton near Elgin, had been treated to induce flowering with two half girdles in which bands of bark and cambium about one inch wide had been removed from the stems. As can be seen in Figure 3, callus was formed in the first growing season following treatment and, the next year, an abnormally wide increment was laid down before normal growth was resumed. Externally, five or so years after wounding, few signs of damage can be seen but the callus and fast grown wood remain areas of structural weakness. In large stems where such areas take up proportionately little of the total area this may not matter but in young trees where scars often occupy half the stem circumference and cut well into the wood the relative proportion of weakened wood will be high, especially in the first few years following removal of the causative agent, and fracture may be facilitated. In addition exposed wood would serve as an entry point for decay or fresh wound parasites.

Conclusions

It appears that at present it is generally assumed that if stakes and ties damage stems then this is a regrettable consequence of undermanning in parks or leisure and

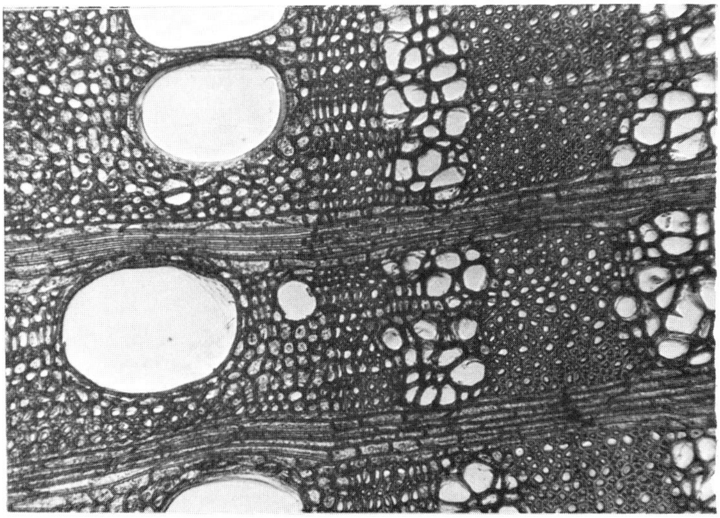

(a)
× 122

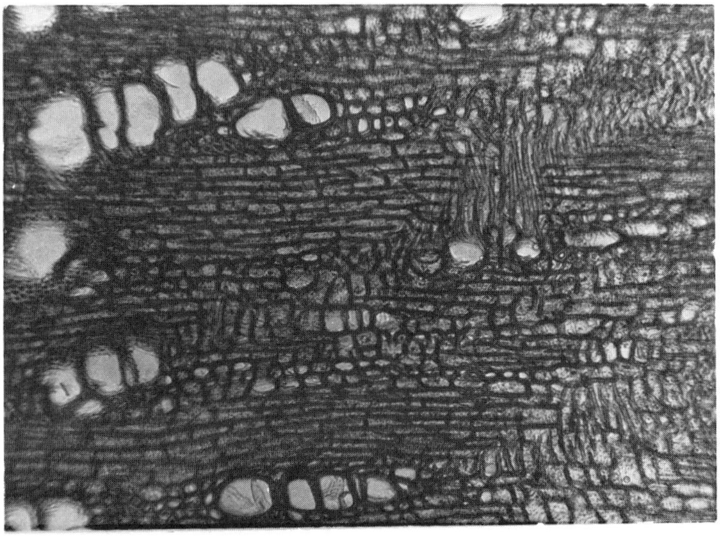

(b)
× 122

Plate 2. Transverse section of Wych elm (a) normal and (b) abnormal wood developed above a pressure point.

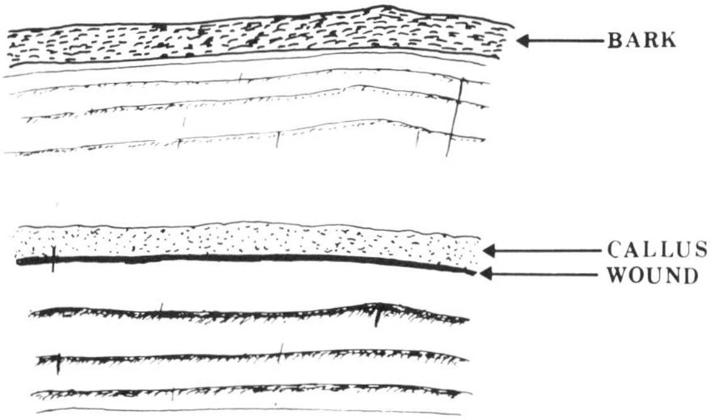

Figure 3. Transverse section of *Larix* showing girdling wound and callus.

recreation departments. Besides, once the tie is removed and the scars and dents painted with bitumen all will eventually be well! The work described in this paper suggests that such a sanguine attitude may not be justified and that damage to young trees stores up troubles for the future.

References

ANON. (1983). In support of trees. *ILAM Journal* (Sept.), 51–53.

BROWN, C.L. (1964). The influence of external pressure on the differentiation of cells and tissues cultured *in vitro*. In Zimmerman M.H., *The formation of wood in forest trees*. Academic Press, NY.

MAKINO, R., JURODA, H. and SHIMAJI, K. (1983). Callus formation and effects of applied pressure in the cultured cambial explant of sugi (*Cryptomeria japonica* D. Don). *Wood Research, Kyoto University*, No. 69, 1–11.

PATCH, D. (1987). Trouble at the stake. In, *Advances in practical arboriculture*, ed. D. Patch. Forestry Commission Bulletin 65, 77–84. HMSO, London.

Discussion

T. RENDELL (Gwynedd County Council).

Can 1½″ hessian strap ties be recommended and is there any likelihood of damage from them?

I.R. BROWN Good for one year but then fell to pieces in the second year.

D. PATCH A limited life tie is of little value if the stake remains and the tree sways freely to damage itself on the stake. A short stake is less likely to rub on a tree than a long stake.

M.J. HAXELTINE (Domestic Recreation Services).

Does bark wounding result in strengthening of the stem?

I.R. BROWN The idea has been tested by half girdling trees which later snapped at the wound. As such the idea appears unreasonable.

A.D. BRADSHAW (Department of Botany, University of Liverpool).

Believed that short staking and removal of support after only one year would only work if trees were growing vigorously. He suggested one year was too soon and that possibly five years should elapse before support was removed.

Weed Competition and Broadleaved Tree Establishment

R.J. Davies
Forestry Commission

Summary

The results of eight experiments are used to illustrate the mechanisms and effects of weed competition on broadleaved tree establishment. Studies of soil moisture tension and foliar nutrient concentrations indicated that competition was primarily for moisture and nutrients. Cutting weeds above ground level did not relieve this competition, and in one experiment increased the sward's vigour and reduced tree growth. Hoeing, herbicides and polythene mulches relieved trees from competition, increasing their survival, root and shoot growth.

Introduction

Many arborists do not recognise the damage caused by weeds to their trees and consequently do not control them effectively. Often the 'weeds' are grasses and clovers, sown to reduce soil erosion and improve site appearance. 'Weeds' may be a misleading term because they were established deliberately, 'weeds' generally being defined as plants growing where they are not wanted. Nevertheless, these plants compete strongly with newly planted trees and will be called 'weeds' in this paper.

This paper reports results from eight experiments which demonstrate that weeds reduce the survival and growth of newly planted trees. How weeds damage trees is discussed. It is hoped that this will stimulate more arborists to employ appropriate weeding methods.

Eight Weeding Experiments

Three hoeing experiments

In the 1920s and 1930s many Forestry Commission experiments in the south of Britain compared the effects of hoeing and hook weeding on the establishment of broadleaved transplants. Hook weeding reduces the height of the weeds without disturbing their roots. Controls were not included. Table 1 summarises the results of three of these experiments.

The Friston site was a deep loam over chalk dominated by grasses; the experimental treatments were first applied to slow-growing beech (*Fagus sylvatica*) 4 years after planting, and probably ceased 3 years later. The other two experiments were on felled broadleaved woodland sites

Table 1. The effect of hook weeding and hoeing on tree height in three experiments

Forest name and experiment number		Friston 14	Micheldever 1	Dean 36
Species		Beech	Oak	Ash
Season planted		1926/27	1927/28	1930/31
Years hoed, inclusive		1931–(33?)	1928–30	1931–33
Assessment date		November 1937	Late 1934	Late 1934
Height (cm)	Hook	99	106	44
	Hoe	188	119	80

Table 2. The effect of different sized paraquat weed control spots on the survival and growth of hawthorn (*Crataegus monogyna*), Italian alder (*Alnus cordata*) and sycamore (*Acer psuedoplatanus*) transplants over 3 years. Ripley.

Diameter of paraquat spot (m)	0	0.54	0.76	1.06	Significance [2]
	Hawthorn				
Survival (%) [1]	97	100	100	100	**
Height growth (cm)	78	96	100	104	***
Basal area growth (mm^2)	146	238	265	303	***
	Italian alder				
Survival (%) [1]	62	86	94	94	***
Height growth (cm)	96	123	121	145	***
Basal area growth (mm^2)	386	601	571	792	***
	Sycamore				
Survival (%) [1]	78	97	96	91	**
Height growth (cm)	58	75	86	100	***
Basal area growth (mm^2)	101	160	210	287	***

[1] Survival transformed to angles for statistical analysis.

[2] NS = Effect not significant at 5% level.
* = Effect significant at 5% level.
** = Effect significant at 1% level.
*** = Effect significant at 0.1% level.

with a wide variety of weed species including some coppice regrowth.

In this series of experiments, of which these three form a part, survival was generally good with no significant differences between treatments. It was noted that grasses were the strongest competitors, response to hoeing being greatest where these were dominant. Hoed ash (*Fraxinus excelsior*) had larger buds and larger darker leaves than unhoed ash.

Hoeing seldom increased oak (*Quercus petraea* or *Q. robur*) height growth. Only height was measured in these experiments. However, more recent experiments (e.g. the experiment at Thorpe reported below) indicate that weeding may increase diameter growth while height growth is unaffected. In many experiments it was noted that the trees were cut back by rabbits, deer or frost. The Dean experiment, for example, suffered all three of these agents. This would reduce height differences between treatments. (All the experiments reported below were fenced to exclude rabbits and in some cases deer.)

Three roadside herbicide spot experiments

These three experiments were planted on sites with disturbed soils beside recently constructed roads. Each site was seeded with the standard Department of Transport (1976) grass/clover mixture before planting. Each experiment compared the effect of different sized herbicide spot treatments on the establishment of broadleaved transplants.

Half of the experiment planted near Ripley, Derbyshire in April 1979 was on cuttings and half on embankments. The herbicide spots were sprayed with paraquat once each summer for 3 years. Survival, height and basal area growth were related to the herbicide spot size (Table 2). Better weed control than that achieved by the annual paraquat applications would probably have further increased growth.

The other two experiments were planted in early 1982 at the M3/M25 motorway interchange near Thorpe, Surrey, and alongside the M11 near Cambridge, with oak (*Quercus petraea*) and sycamore (*Acer pseudoplatanus*) transplants. Each experiment compared four weeding treatments each with or without tree shelters – eight treatments in all. The four weeding treatments were herbicide spots of 0.25, 0.5 and 1 m diameter and a control. The spots were kept largely weed-free using paraquat, glyphosate and propyzamide when necessary.

Table 3 gives results from Thorpe after two growing seasons. Tree shelters produced large increases in height

Table 3. The effect of different sized spot herbicide treatments and tree shelters on the establishment of oak and sycamore transplants over 2 years. Thorpe, M3/M25.

		Treatment means				Significance (2)		
Diameter of herbicide spot (m)		0	0.25	0.5	1	Shelter	Herbicide	S × H interaction
Oak								
Two years height growth (cm)	No shelter	22	25	28	25	***	*	NS
	With shelter	51	64	69	78			
Two years basal area growth (mm^2)	No shelter	23	35	39	54	*	***	NS
	With shelter	13	31	34	42			
Volume index (1) two years growth	No shelter	31	42	45	60	***	***	NS
	With shelter	34	67	75	100			
Survival	No shelter	97	100	98	95	*	NS	NS
	With shelter	91	92	97	95			
Vole damage (% of trees)	No shelter	9	3	0	0	***	***	NS
	With shelter	2	0	0	0			
Sycamore								
Two years height growth (cm)	No shelter	40	45	50	63	***	***	NS
	With shelter	70	86	105	114			
Two years basal area growth (mm^2)	No shelter	41	60	97	168	***	***	NS
	With shelter	25	44	62	86			
Volume index (1) two years growth	No shelter	23	34	53	100	NS	***	NS
	With shelter	25	39	60	80			
Survival (%)	No shelter	81	66	92	97	***	**	*
	With shelter	89	98	100	95			
Vole damage (% of trees)	No shelter	86	86	75	47	***	NS	NS
	With shelter	14	11	11	8			

(1) Index of stem volume derived by multiplying tree height and basal area.

(2) See note (2) below Table 2.

growth, but depressed basal area growth. Weed control, by contrast, had a greater beneficial effect on basal area than height growth. Indeed the height growth of oak without shelters was not improved by weeding (a similiar result to the early hoeing experiments). There is a progressive response to weed control with growth improving as spot size is increased. Both weeding and shelters aided sycamore survival, largely by their influence on vole (*Microtus agrestis*) damage; shelters protected the sycamore and weeding discouraged the voles, since they dislike crossing open ground.

In October 1984, after three growing seasons, four trees from each treatment, a total of 32 out of the original 512 trees of each species, were excavated from the Cambridge experiment. Tree shelters had little effect on rooting depth or spread, but weeding increased root spread considerably and depth slightly. Despite the small sample size it appears that weeding increased root and

Table 4. The effect of different sized spot herbicide treatments and tree shelters on root and shoot growth of oak and sycamore transplants after 3 years. Cambridge, M11.

		Treatment means				Significance [2]		
Diameter of herbicide spot (m)		0	0.25	0.5	1	Shelter	Herbicide	S × H interaction
	Oak							
Root dry weight (g)	No shelter	15	22	181	261	NS	***	**
	With shelter	39	68	130	127			
Shoot dry weight (g)	No shelter	8	9	146	207	NS	***	NS
	With shelter	31	83	172	197			
	Sycamore							
Root dry weight (g)	No shelter	46	58	219	440	NS	***	NS
	With shelter	40	131	199	266			
Shoot dry weight (g)	No shelter	23	38	125	288	*	***	NS
	With shelter	30	134	199	269			

[1] See note (2) below Table 2.

shoot growth to a similar extent, this increase being related to the herbicide spot size (Table 4). Tree shelters resulted in more shoot growth per unit weight of root growth with both species (Figure 1).

The Thorpe and Cambridge experiments are two of a series of identical experiments in various parts of England. Despite differences in soil and climate weeding improved growth similarly on each site. On some sites herbicide damage almost cancelled out relief from competition of the non-sheltered trees with 0.25 m diameter herbicide spots.

First weeding experiment at Alice Holt

It was believed that competition for soil moisture might be responsible for some experimental results, such as those reported above. An experiment was therefore established at Alice Holt to study the effect of commonly used sward control methods on soil moisture. For a detailed account of this experiment, which also investigated soil temperatures, see Insley (1982).

The site, which is almost flat with a heavy, silty loam was sown with the standard Department of Transport seed mixture in June 1979. Gypsum soil moisture blocks were placed at various depths in the soil. The electrical resistance of the blocks is closely related to the moisture tension of the surrounding soil. In April 1980 five treatments were applied:

(a) sward left uncut;
(b) sward mown regularly leaving the cuttings on the plot;
(c) paraquat applied when necessary to keep plots weed-free;
(d) sheet mulch of bituminised felt laid down after vegetation killed with paraquat; and
(e) 150 mm depth of pulverised bark mulch laid down after vegetation treated with paraquat.

In April 1982 Wild cherry (*Prunus avium*) trees were planted at 1 × 1 m spacing. Foliage was collected in July 1982 for analysis.

Moisture block resistance at 100 mm depth under the five treatments through the summers of 1981 and 1982 are shown in Figure 2. In 1981 the two mulch treatments remained at field capacity, the bare soil dried slightly, and the two sward treatments dried the soil considerably. Cutting the grass increased its soil drying ability, presumably through increased transpiration.

From July 1982 on, the bare soil dried out rapidly; this probably resulted from transpiration by the vigorous cherry in these plots.

In the mulch treatments many trees died in 1982, and growth of the survivors was erratic. Investigation of soil oxygen availability using steel rods (Carnell and Anderson, 1986) demonstrated that conditions were too anaerobic to support roots below a few centimetres. This was

Figure 1. The relationship between root and shoot dry weights of trees grown without (triangles and unbroken line) or with (circles and broken line) shelters after three seasons; four weeding treatments combined. Cambridge, M11.

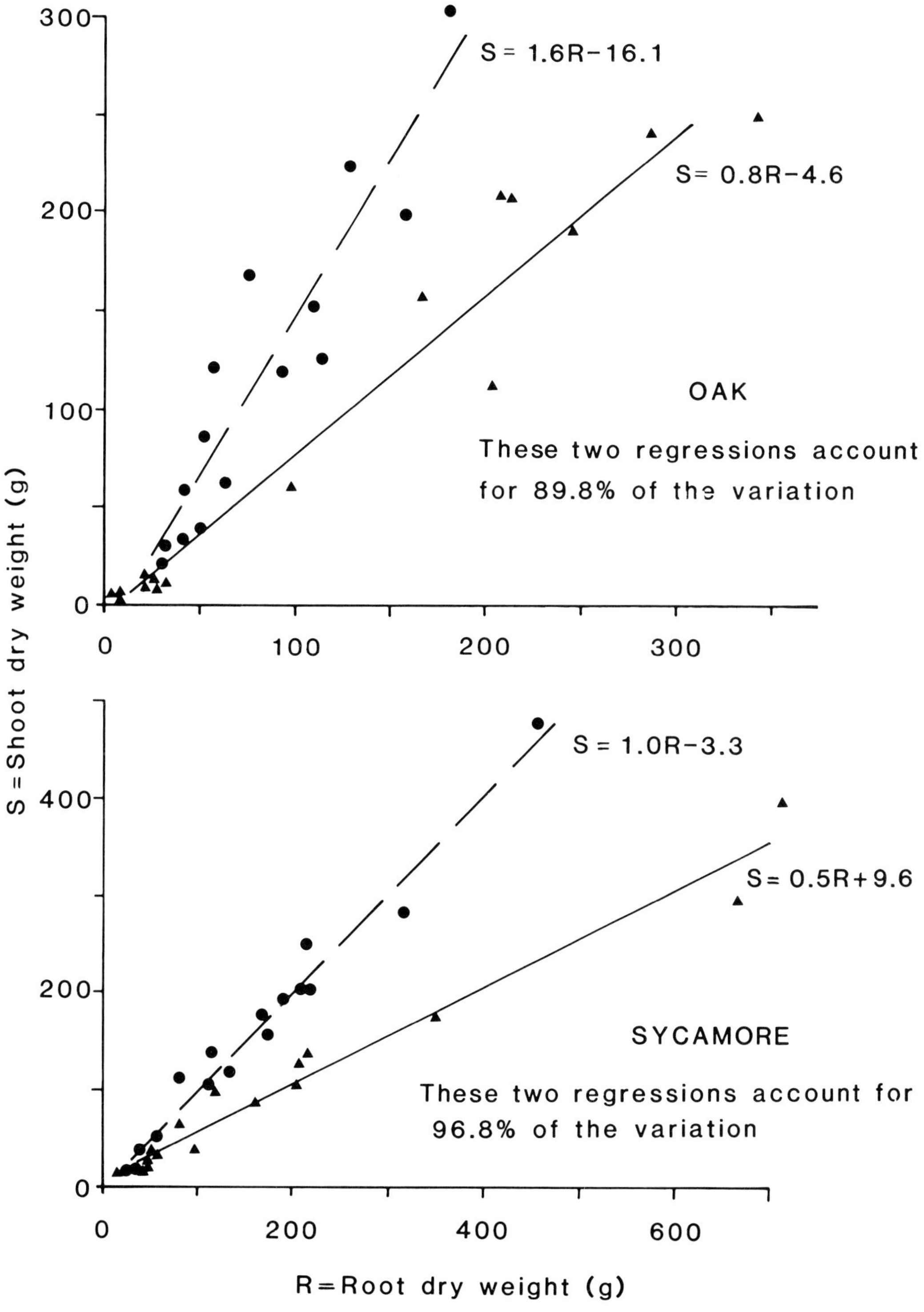

Figure 2. Moisture block resistances at 100 mm depth under five ground covers in summer 1981 and 1982. Alice Holt.

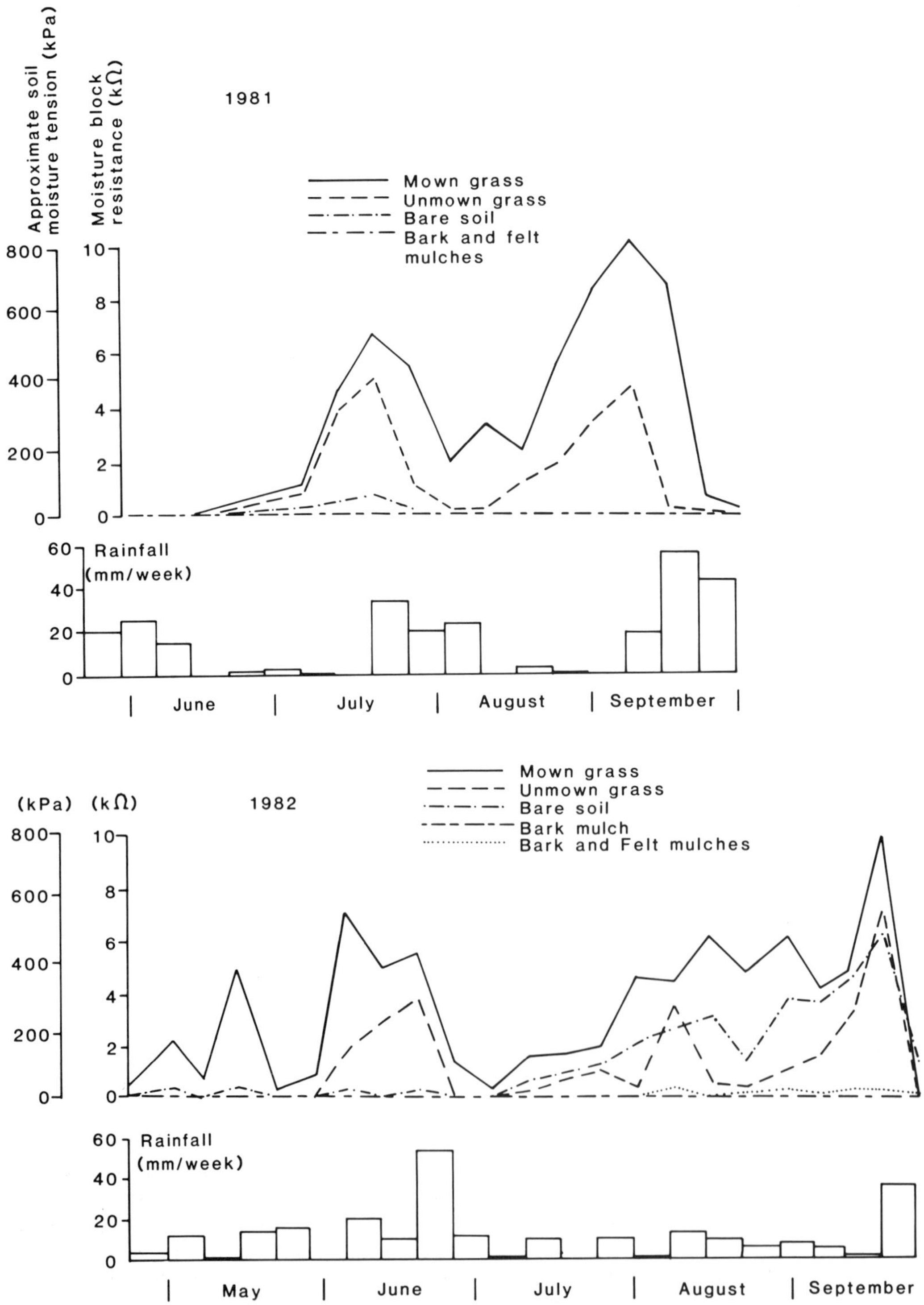

caused by waterlogging. The mulches, applied 24 months before planting, had caused the soil to become stagnant. Detrimental effects with mulches applied at time of planting are rare, although Davies (1983) reported them in two experiments on poorly drained sites.

All the cherry in the remaining three treatments survived. Table 5 gives their first season's growth, relative leaf sizes and foliar nutrient concentrations. Tree growth was best in the bare soil plots where soil moisture tension was lowest; growth was poorer in the unmown plots where moisture tension was higher; and growth was poorest in the mown plots where moisture tension was highest. Although growth was related to soil moisture availability, the significant differences in foliar nutrient concentrations indicate that nutrient uptake was also affected by the treatments.

Second weeding experiment at Alice Holt

The site and soil of this experiment were similar to the first experiment at Alice Holt. Most of the area was sown with the standard Department of Transport seed mixture in June 1983. Some plots were sown with White clover (*Trifolium repens* 'Huia') only. The experiment compared the effects of six treatments on the establishment of Silver maple (*Acer saccharinum*) transplants planted in February 1984:

(a) grass/clover sward left uncut;
(b) grass/clover sward mown regularly, leaving the cuttings where they fell;
(c) pure clover sward;
(d) mown grass/clover sward with 1 m diameter herbicide spots around the trees;
(e) mown grass/clover sward with 1 m diameter black polythene mulching mats around the trees; and
(f) grass/clover sward over the whole plot killed with herbicide.

Paraquat and glyphosate were applied as needed to maintain weed-free spots and plots in treatments c and f. The polythene mats were secured by inserting their edges into the soil and pegging them with bent wire 'staples'.

Table 6 gives the first year height and diameter growth. Trees with weed-free areas around their bases grew better than those growing intimately with vegetation. The pure clover sward and the grass/clover swards were equally detrimental. Early in the growing season trees with polythene mats grew faster and had larger darker leaves than trees in the herbicide treatments, but they ceased growth sooner, so that by the end of the year they were not significantly larger than trees with herbicide spots, and trees in the whole plot herbicide treatment had overtaken them.

Table 5. First year's growth and foliar analysis results for cherry at Alice Holt (1982).

		Mown grass	Unmown grass	Bare soil	Significance [1]
Height growth (cm)		9	31	80	***
Diameter growth (mm)		3	7	14	***
Leaf size index [2]		29	53	100	***
Foliar nutrient concentrations (% of dry weight)	N	2.6	2.5	3.4	***
	P	0.19	0.17	0.20	**
	K	1.1	1.1	1.6	***

(1) See note (2) below Table 2.

(2) Leaf size index obtained by multiplying leaf lengths and widths.

Table 6. First season height and diameter growth in the second Alice Holt weeding experiment.

	Height growth (cm)	Diameter growth (mm)
Grass/clover, unmown	2 a	−0.1 a
Grass/clover, mown	−1 a	0.3 a
Pure clover sward	0 a	0.4 a
Grass/clover + herbicide spots	38 b	5.0 b
Grass/clover + polythene mats	41 b	5.8 b
Whole plot herbicide	58 c	7.8 c

In each column treatment means with the same letter following do not differ significantly ($p < 0.05$).

Discussion

In the first Alice Holt experiment competition for light was relatively unimportant: otherwise mowing would have been beneficial. In that experiment tree growth was related to soil moisture availability but nutrient uptake was also affected by weeding treatments. It therefore appears that competition for moisture and nutrients was largely responsible for the growth differences in that experiment, and probably in others also. These two factors are interrelated. Plants are unable to extract nutrients from dry soil. Plant growth can be checked when the surface soil (where most of the available nutrients are) dries, despite adequate moisture availability at depth (Garwood and Williams, 1967).

It follows that effective weeding must eliminate competition from weed roots. This can be done by hoeing, herbicides or mulching. Cutting grass swards above ground level is ineffective. In the first Alice Holt experiment mowing increased the sward's vigour and reduced tree growth, but in the second experiment mown and unmown swards were equally detrimental. This difference may be related to the age of the swards, the older sward in the first experiment producing a self-mulching effect when left uncut.

Most weeds other than grasses are weakened by cutting. However, the principle that effective weed control must eliminate root competition still applies. Lund-Høie (1984), for example, working with Norway spruce (*Picea abies*) on restocking sites in Norway showed that cutting woody weeds increased tree growth slightly and reduced survival, probably through accidental damage to the spruce; by contrast weeding with glyphosate increased growth considerably, especially diameter growth.

Legumes are often recommended as ground cover species because of their ability to fix atmospheric nitrogen. But it is clear from the second Alice Holt experiment that clover, a legume, can reduce tree growth. Even on pulverised fuel ash, a medium devoid of nitrogen, Buckley (1977) found that the growth and nitrogen uptake of poplar (*Populus tacamahaca* × *trichorcarpa*) cuttings was inversely proportional to the productivity of undersown clover (results given by Insley and Buckley, 1980).

In the Ripley experiment survival and growth were related to the weed-free area around the tree. Weed-induced stress levels were presumably lower in other experiments where survival was less clearly related to weed competition. Others have related performance to the degree of control achieved within the herbicide spots or strips. For example, Hanschke (1968) found a correlation between the survival and growth of Scots pine (*Pinus sylvestris*) seedlings and the percentage cover and dry weight of competing weeds. The optimal weeding regime balances predicted costs and benefits. The major element in the cost of herbicide treatment is usually labour; doubling the area treated per tree does not double the cost per tree. In general it is more cost-effective to treat a larger area per tree, accepting a lower standard of weed control within that area, rather than more thorough control (achieved with more applications per year) over a smaller area. A one metre diameter herbicide spot is often appropriate for transplants, although larger areas usually give more growth, as in the second Alice Holt experiment.

The soils of many landscape tree planting sites are compacted and their structure damaged by machinery or storage of building materials; such sites are usually sown with vigorous agricultural grasses and clovers. So the sward used in most of the experiments reported here and the soil conditions of the three roadside experiments are representative of many sites. Soil compaction restricts rooting depth and impedes water movement through soil. Thus soil moisture tension around newly planted tree roots rises rapidly in dry weather, especially if vigorous weeds are present. Weed competition is likely to be more harmful under these conditions than with undisturbed soils and less vigorous weeds.

The first Alice Holt experiment showed that most soil moisture is lost by transpiration and relatively little by evaporation; also that mulches reduced evaporational moisture losses. Davies (1985) found that the moisture-conserving effect of a sheet mulch greatly improved the survival and growth of Italian alder (*Alnus cordata*) transplants and willow (*Salix daphnoides*) cuttings on a site with very poor soil moisture retention. Parfitt and Stott (1984) also examined soil moisture at various depths under different sward control treatments and found that straw and polythene mulches concentrated moisture in the upper 75 mm of soil, the horizon that dried most rapidly in bare soil conditions or under grass. As plant-available nutrients are concentrated near the soil surface, mulching maintains nutrient availability.

Although weeds and trees compete for nutrients, fertilising alone may stimulate weeds to compete more strongly for soil moisture. Davies (1987) found that fertilising without weeding reduced the growth of oak transplants; fertilising with weed control increased their growth.

The hoeing experiments conducted over 50 years ago showed that cutting weeds above ground level was relatively ineffective; but because of its expense hoeing was never adopted by foresters. Many arborists, particularly those with forestry training, continue to 'mow' weeds. This is of little or negative benefit to the tree. Herbicides are now available which provide an inexpensive and effective alternative to hoeing.

Conclusions

Weed competition for soil moisture and nutrients reduces survival and growth of young trees. These effects are greatest on disturbed sites with vigorous weeds. Effective weed control must eliminate root competition; cutting weeds above ground level is insufficient. Mulching helps the tree by reducing evaporation from the soil surface and concentrating moisture in the upper soil horizons.

Acknowledgements

Many past and present colleagues have been involved in these experiments. I particularly wish to thank Mr J.B.H. Gardiner, Mr T.J. Davis, Mr R.E. Warn, Mr I.H. Blackmore, Mr P.R. Barwick, Mr S.E. Malone, Mr T.D. Cooper for help with the experiments and Mr R.C. Boswell for the statistical analyses. Most of the work was funded by the Department of the Environment.

References

BUCKLEY, G.P. (1977). *The establishment of tree species on pulverised fuel ash.* PhD thesis, University of Leeds.

CARNELL, R. and ANDERSON, M.A. (1986). A technique for extensive field measurement of soil anaerobism by rusting of steel rods. *Forestry* **59** (2), 129–140.

DAVIES, R.J. (1983). Transplant stress. In, *Tree establishment,* Proceedings of a symposium at Bath University, 40–50.

DAVIES, R.J. (1985). The importance of weed control and the use of tree shelters for establishing broadleaved trees on grass dominated sites in England. *Forestry* **58** (2), 167–180.

DAVIES, R.J. (1987). Fertilising broadleaved landscape trees. In, *Advances in practical arboriculture.* Forestry Commission Bulletin 65, 107–114. HMSO, London.

DEPARTMENT OF TRANSPORT (1976). *Specifications for road and bridge works.* 5th edtn. HMSO, London.

GARWOOD, E.A. and WILLIAMS, T.E. (1967). Growth, water use and nutrient uptake from the subsoil by grass swards. *Journal of Agricultural Science,* Cambridge **69**, 125–130.

HANSCHKE, D. (1968). Weed competition and its quantitative evaluation. *Forst und Holzwirt* **21**, 434–439. (Translation by W. Linnard.)

INSLEY, H. (1982). *The effects of stock type, handling and sward control on amenity tree establishment.* Unpublished PhD thesis, Wye College, University of London.

INSLEY, H. and BUCKLEY, G.P. (1980). Some aspects of weed control for amenity trees on man-made sites. In Proceedings of the conference on *Weed control in forestry,* Nottingham University, 189–200.

LUND-HØIE, K. (1984). Growth responses of Norway spruce to different vegetation management programmes – preliminary results. In, Weed control and vegetation management in forests and amenity areas. *Aspects of Applied Biology* **5**, 127–133.

PARFITT, R.I. and STOTT, K.G. (1984). Effect of mulch covers and herbicides on the establishment, growth and nutrition of poplar and willow cuttings. In, Weed control and vegetation management in forests and amenity areas. *Aspects of Applied Biology* **5**, 305–313.

Discussion

K. ROWTON (Tunbridge Wells Borough Council).

The techniques described don't seem new. What account have you taken of existing practice?

R.J. DAVIES Many of the results had been previously found, but it is the variability of current practice that necessitates the work.

MR COLEMAN Is there an effect of herbicide directly on the growth of the tree?

R.J. DAVIES Atrazine can affect the nutrient regime; others can be directly damaging.

T. OTENIYA (Sterling District Council).

The damaging effects of weeds are known by managers but limited resources restrict practices.

R.J. DAVIES In my experience there remains much ignorance on the effects of weeds.

J. THOMAS (Janie Thomas Associates)

Have there been any experiments comparing basal area growth with the progression of tube disintegration?

J. EVANS (Forestry Commission Research Division).

Diameters have been measured all along the stem before and after the shelter effect. At 10 cm height the diameter is narrow, but higher up the diameter is greater than normal. The stem has become more cylindrical and less conical.

Water Relations and Irrigation Methods for Trees

J. Grace
Department of Forestry and Natural Resources,
University of Edinburgh

Summary

Routine irrigation of intensively cultivated crops is becoming common place. In contrast newly planted trees, other than those in nurseries, rarely receive any watering and as a result they die or grow only slowly. Plant response to water stress, reduction of water stress at planting and irrigation needs of trees are discussed.

Introduction

The tree is a hydraulic system of considerable complexity. Most of the time it functions under tension, water being lost at the leaf surface and the supply being drawn through the soil and into the roots, stem and leaves. This flow of water is initiated by the climate at the leaf surface, which drives evaporation and so reduces the water content of the leaf tissues, inducing the gradient of water potential between the leaf and soil (Jarvis, 1975).

The pathway for water flowing through the plant includes cellulose cell walls in the leaf and the roots, and the fine tube system of the vessels and/or tracheids in the stem. The mechanism of movement is purely physical, with myriads of rather tortuous and interconnecting water columns being drawn through the plant in response to a water deficit in the leaves. There is much resistance to flow and the system functions only because water molecules have the property of clinging together. If the tension becomes too great, cavitation occurs: that is to say some of the columns break. Breakage can be detected as sound or ultrasound (Milburn and Johnson, 1966; Sandford and Grace, 1985). Once broken, it is generally held that water columns do not easily repair.

The hydraulic resistance of the pathway is considerable, and as the evaporation at the leaf surface increases the water content and water potential of the leaf tissues decline. In large trees there are significant stores of water in the trunk. These are drawn upon, on a seasonal and diurnal basis, causing shrinking and swelling of the stem (Zimmerman, 1983), but the bulk of the water supply must be from the soil via the roots. All biochemical processes in the plant occur in a medium of water, and any reduction in the availability of water will inevitably affect their rate. Cell division is thus very sensitive to a shortage of water. Once plant cells are formed by cell division, they expand greatly as a result of the positive pressure inside them. This too is sensitive to water deficit. Thus, even the smallest reductions in tissue water content, such as occurs even in a well-watered plant, cause a measureable reduction in growth (Kozlowski, 1982).

The hydraulic system must function adequately. If too many breakages in the water columns occur then the overall resistance may be so great as to lead to large enough tissue water deficits to be lethal. In most normal conditions this does not happen because there is an appropriate balance of leaf area, conducting wood and root surface. Superimposed upon a basic 'ground plan', any plant during its development displays adaptability and adjusts to the prevailing conditions. Parameters which have been widely studied experimentally include the ratio of root:shoot and the ratio of leaf area:leaf weight. The plant adjusts these ratios by subtly sensing the environment and displays a degree of internal homeostasis in relation to acquisition of basic resources like light, water and nutrients (Hunt, 1975).

The trees of our parks and gardens should display this homeostasis and grow satisfactorily if planted as seeds in the place where they are intended to grow, or even if planted as seedlings. More often, however, they are grown for several years elsewhere and planted, with considerable disturbance and loss of roots, in a completely different environment.

Commonly, they suffer water stress and die soon after planting or remain at risk for 2 or 3 years afterwards simply because their root systems are inadequate as a water conducting pathway. The sight of young trees dying through lack of water has evoked letters to *The Times* (Pollard, 1976), and represents a financial loss which varies according to the prevailing summer rainfall.

Although the lack of a properly developed root system is an obvious reason for concern, it should be pointed out that there are other features of the aerial environment of amenity trees which may lead to high rates of water loss. In natural vegetation, a tree is to some degree sheltered by its neighbours. In particular, the flux of solar radiation to the leaves is reduced by shading, and the humidity of the air surrounding the leaves is generally much higher than that of the atmosphere above the vegetation. Amenity trees on the other hand are not sheltered from radiation except perhaps by buildings, and if they are planted near south-facing walls among paving stones then the flux of radiant energy to them may cause a desert environment. The humidity of the air around an isolated tree is likely to approximate more closely to that of the atmosphere as a whole rather than that of a forest microclimate. There are possibilities in an urban environment of many kinds of pollution, especially near the roads. As this subject is considered in a Symposium volume elsewhere (Colwill *et al.*, 1979), it is perhaps necessary here only to point out that the oxides of sulphur and nitrogen may have significant effects on plant growth, and, in relation to water stress, have been found to influence directly the stomatal control of transpiration.

Water is Important Even to Established Trees

In trees, the formation of the wood (increment) is a useful record of cell division, extending over many decades. The width of the annual radial increment has often been shown to be highly correlated with rainfall or a water deficit calculated retrospectively from climatological data (Zahner and Donnelly, 1967; Zahner, 1968). In Britain, this sort of work is less common than elsewhere, though Brett (1983) has demonstrated good correlations between summer rainfall and radial increment of elms (*Ulmus* spp.) in London parks. Even in extremely dry years, it is rare to see established trees in this country suffering from dieback or wilting, though browning of the leaf margin is common and splitting of sapwood has been reported (Day, 1954). Elsewhere, severe droughts do sometimes result in phenological disturbances and the shedding of plant parts (Hinckley *et al.*, 1979).

There are several studies of the effect of irrigation and fertilisers applied to plantations in north-temperate parts of the world (Brix, 1972, 1979; Holstener-Jørgensen and Holmsgaard, 1975; Aussenac, 1980; Polge, 1982; Linder and Axelsson, 1982). In nearly all cases timber production responds somewhat to both irrigation and nutrient supply, but the effect of both factors together exceeds the sum of their separate effects. In many cases involving plantations on poor soils, the yield could probably be increased several-fold by irrigation combined with fertiliser application. It should be borne in mind that isolated trees are likely to respond more strongly to irrigation than are trees in plantations.

Reducing Water Stress at Establishment

In relation to newly-transplanted material, it has been of paramount importance to nurserymen, foresters and landscape tree planters to adopt effective planting procedures. Water stress is often severe following planting-out, and recovery depends strongly on the development of the root system, particularly the growth of fine roots (Baldwin and Barney, 1976). The exposure of roots to the air during handling and planting nearly always reduces survival (Kozlowski and Davies, 1975; Kozlowski, 1975; Coutts, 1980, 1981) and its effect may be apparent for years.

There are several practices which have been proposed from time to time in order to prevent these difficulties, or to reduce them to a more acceptable level. These include cold storage (Stone *et al.*, 1962, 1963; Rook, 1973), root-pruning to stimulate the development of a fibrous root ball (Rook, 1971; Bacon and Hawkins, 1979; Beckjord and Franklin, 1980; Toliver *et al.*, 1980), the application of an inert slurry to the roots to form a waterproof skin (Bacon *et al.*, 1979), the use of plant growth substances (Hartwig and Larson, 1980), antitranspirants (Rabensteiner and Tranquillini, 1970) and chemical defoliants (Insley, 1981). Conclusions from such studies should not be taken as prescriptions for planting schemes elsewhere as much depends on species, timing and location. The use of an inert slurry or paste to protect roots from desiccation during handling does seem widely effective and cheap. In this country a commercial product made from seaweed 'Alginure' is widely used by Local Authorities.

Climatological Controls of Water Loss

The transpiration rate depends on climatological variables – notably solar radiation, humidity and wind speed. It also depends on plant variables: the stomatal and aerodynamic resistance. The biophysical principles of evaporation from

Figure 1. The effect of climatological variables on transpiration rate of broadleaved temperate trees. The four columns refer to different radiation loads (roughly equivalent to dawn, early morning, late morning and noon). The three rows refer to different conditions of humidity expressed as saturation vapour pressure deficit. The lines drawn in each window represent 5 values of stomatal opening, the upper line being 0.3 s mm^{-1} and the lower line being 1.1 s mm^{-1} (these cover the range from wide open to partially shut).

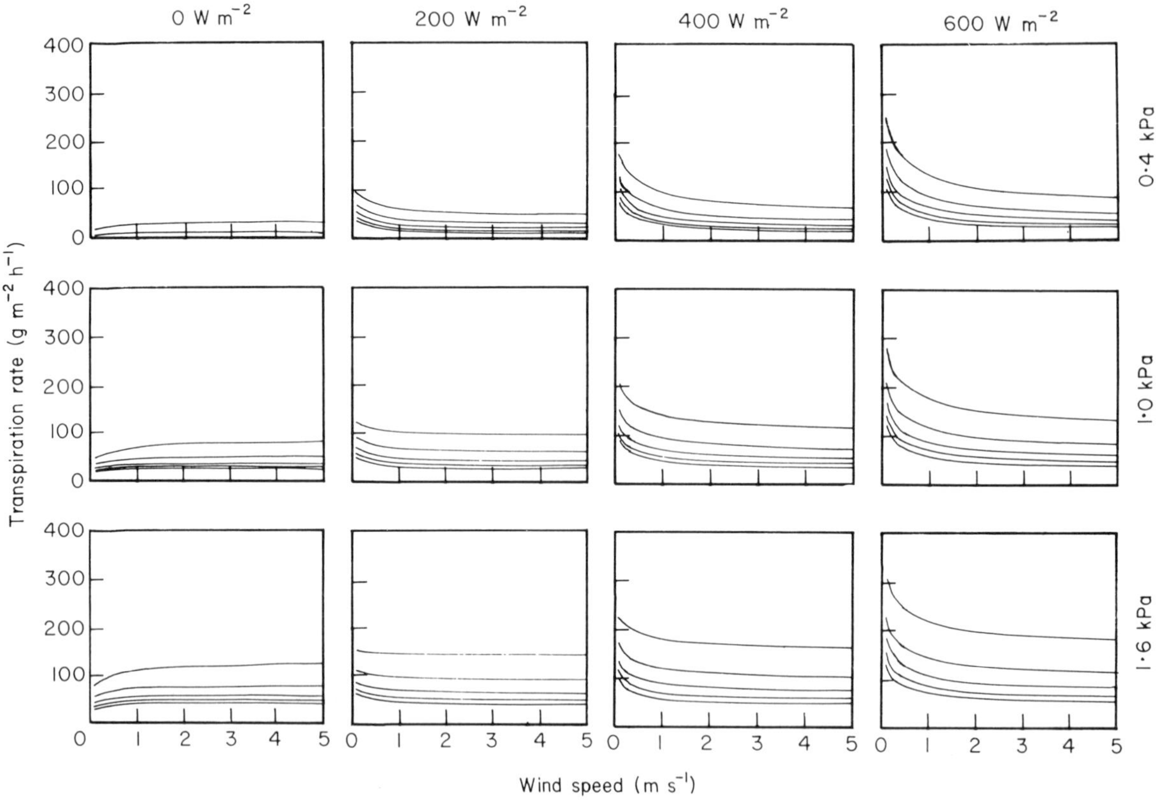

vegetation are well understood and transpiration rate can be calculated in cases where the values of these variables are known. This leads naturally to applications in hydrology where the aim is to predict the effect of vegetation on the water yield of catchments. It also leads naturally to a calculation of irrigation needs for crops, based on the idea of adding just enough irrigation water to make up for that lost by transpiration.

The uptake of water by individual trees can however be measured. The most direct methods involve using tracers in the trunk such as radioisotopes or a heat pulse (Stewart, 1984). In one study, large trees were found to use 100 litres of water a day when their stomata were open (Doley, 1981).

The effects of environmental variables on the transpiration rate of individual leaves can be studied experimentally; and also theoretically by calculating the separate components of the heat balance (Grace, 1984). The result of such a study, applied to temperate broadleaved species, showed that the most important environmental variable was solar radiation, and that humidity was also important (Figure 1). Wind speed is relatively unimportant, and contrary to what is said in many text books, an increase in wind speed causes a decline in the rate of transpiration in undamaged leaves (Dixon and Grace, 1984). In gales, however, the cuticle on the leaf surfaces becomes abraded and then the plant looses its ability to retain water.

Since the total transpirational surface of a tree is conceptually the collection of leaves, it ought to be possible to extend this type of calculation to estimate the water use by individual trees in any environment. This has been achieved by Thorpe *et al.* (1978) and elaborated by Landsberg and McMurtrie (1984). The principle difficulty here is that the calculation of solar energy absorbed by the leaves is a fairly complex problem in geometry and the result bears only a weak relationship to the solar radiation that is recorded by a solarimeter at a meteorological station. For amenity trees, the calculation would be even more complex, because radiative transfers with walls and paving material add to the energy flux to the leaves. Nevertheless, the work of Thorpe *et al.* (1978) does provide an interesting example of a calculated and measured quantity of water which was transpired by a small apple tree (*Malus* spp.) on a sunny summer's day. The size of the tree was typical of container-grown amenity trees (canopy diameter 1.0 m, canopy height 1.6 m, leaf area 1.4 m^2), and so the magnitude may be of interest to arboriculturists. The tree in question used 4.6 kg of water (4.6 litres, or rather more than a gallon) over a 16 hour day.

In principle it would be possible to compute the irrigation needs of isolated trees by this method. In practice it would be of questionable value as the water added to the soil would not all be taken up by the plant, particularly if the plant had a restricted root mass, and would flow away from the point of application to the unirrigated, drier, surrounding soil.

There is an alternative approach to irrigation which does not depend on adding water to the soil to replenish the deficit, and which does seem more applicable to isolated trees. This is mist irrigation. So far, among tree crops, it has been used only in orchards.

Mist Versus Soil Irrigation

The principle of mist irrigation is to reduce water loss at the leaf surface instead of increasing water supply at the roots. If a fine spray is released in the centre of the canopy, a proportion of the leaves at any time can be maintained in a wet state, whilst the remainder is exposed to air which is considerably humid. This system was recently tested in an apple orchard in Kent (Brough *et al.*, 1986). Trees were irrigated by misting the canopy or spraying the soil.

Trees irrigated by both methods suffered some degree of water stress, as any transpiring plant does, but the mist-irrigated trees suffered least (Figure 2). The water potentials in these trees were consistently above those of the ground irrigated plants and substantially above those of the controls. The water content of the stems was also much higher. Over the season, the soil water content of the controls fell markedly, but that of the misted trees remained close to field capacity (Brough *et al.*, 1986). Thus, although the water was applied to the leaves, the treatment was successful in conserving soil moisture, presumably because some of the applied water fell on the soil, and the misted tree took up far less water from the soil anyway.

At the end of the season, it was the misted tree that produced the most apples and the most new growth.

Figure 2. Evaporation rates E and water status of apple trees. Trees were irrigated by mist (●), irrigated by watering the soil (△) and unirrigated (○). Climatological variables: solar radiation (broken line) and air temperature (solid line). The water potential defines the availability of water within the plant. Values less than zero are detrimental to growth, and growth is likely to cease in the range −2 to −3 megapascals.

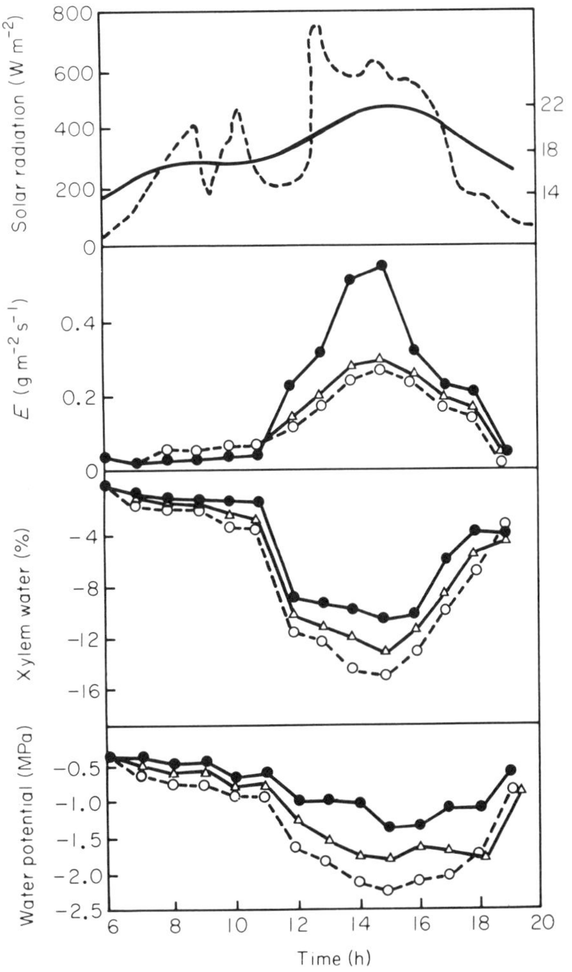

Recommendations

1. Planting

(a) Any irrigation is likely to assist the establishment of trees, but it should be done before visible injury is widespread. By then, the hydraulic system has probably failed and no amount of watering is likely to be successful. Concentrate your effort during the phase of leaf expansion.

(b) Where trees are scattered over parkland, water could be taken to them in a petrol driven tanker of the type sometimes seen on golf-courses. Where trees are in large populations in a single area, a water-pipe could be run to them. If rapid growth is required, liquid fertiliser should be added to the water.

(c) In large schemes, where there is a water-pipe, mist irrigation is best. The mist need not be run continuously, but turned on by a sensor-actuated valve when solar radiation reaches a preset value, probably in the range 100–200 watts m^{-2}.

(d) Where irrigation is impossible, consideration should be given to the careful preparation of the plants in the nursery and the timing of the plantings and the use of slurries to form a water-permeable skin on the exposed roots. Judicious root-pruning in the nursery bed might be a good method to stimulate the growth of new roots. Pruning of branches reduces the transpiring area and thus is likely to be beneficial.

(e) Mulching and weeding is recommended as a means of reducing water consumption by competing vegetation.

2. After-care

Local authorities may have insufficient manpower for aftercare. It may be possible to pay the supplier for aftercare. As the cost-effectiveness of such irrigation has never been investigated, both local authorities and contractors should play safe and use small trees to obtain a high probability of survival.

3. Isolated established trees

Although such trees suffer water stress, they do not usually die unless diseased. In a park or garden, fast growth is not required (indeed, slow growth is normally preferred) and so irrigation would be pointless.

4. Woodlands

Irrigation of woodlands, especially if coupled with fertiliser application, is likely to increase timber production. Some indication of the response, in an unknown situation, can be obtained by examination of cores taken with an increment borer, and inspection of the correlation between ring width and summer rainfall or calculated water deficit.

5. Research

There has been little research on the water relations of amenity trees, despite their economic importance. Studies of the hydraulic architecture, transpiration and water potential would be a very useful aid in formulating a code of practice.

References

AUSSENAC, G. (1980). Premiers resultats d'une étude de l'influence de l'alimentation en eau sur la croissance des arbres dans un peuplement de Douglas (*Pseudotsuga menziesii* (Mirb.) Franco). *Revue Forestière Française* **32,** 167–172.

BACON, C.J. and HAWKINS, P.J. (1979). Intensive root wrenching: a pre-requisite for the successful establishment of 1–0 *Pinus caribaea* Mor. seedlings. *Tree Planters Notes* **30** (1), 5–7.

BACON, G.J., HAWKINS, P.J. and JERMYN, D. (1979). Comparison of clay slurry and agricol root dips applied to 1–0 slash pine seedlings. *Tree Planters Notes* **30** (1), 34–35.

BALDWIN, V.C. and BARNEY, C.W. (1976). Leaf water potential in planted Ponderosa and Lodgepole pines. *Forest Science* **22,** 344–350.

BECKJORD, P.R. and FRANKLIN, C.C. (1980). Effect of various methods of root pruning on the establishment of transplanted red oak seedlings. *Tree Planters Notes* **31** (4), 10–11.

BRETT, D.W. (1983). Records of temperature and drought in London's park trees. *Arboricultural Journal* **7,** 63–71.

BRIX, H. (1972). Nitrogen fertilisation and water effects on photosynthesis and earlywood-latewood production in Douglas-fir. *Canadian Journal of Forest Research* **2,** 467–478.

BRIX, H. (1979). Moisure-nutrient inter-relationship. In, *Proceedings of the Forest Fertilisation Conference*, Sept. 25–27, 1979, pp. 48–52. eds. Gessel, S.P., Kenady, R.M. and Atkinson, W.A. University of Washington, Institute of Forest Resources, Contribution 40, Seattle, WA, USA.

BROUGH, D., JONES H.G. and GRACE, J. (1986). Diurnal changes in the water content of the stems of apple trees, as influenced by irrigation. *Plant, Cell and Environment* **9** (1), 1–7.

COLWILL, D.M., THOMSON, J.R. and RUTTER, A.J. (1979). *The impact of road traffic on plants.* TRRL Supplementary Report 513, Department of Transport, Crowthorne, Berks.
COUTTS, M.P. (1980). Control of water loss by actively growing Sitka spruce seedlings after transplanting. *Journal of Experimental Botany* **31,** 1587–1597.
COUTTS, M.P. (1981). Effects of root or shoot exposure before planting on the water relations, growth and survival of Sitka spruce. *Canadian Journal of Forest Research* **11,** 703–709.
DAY, W.R. (1954). *Drought crack in conifers.* Forestry Commission Forest Record 26. HMSO, London.
DIXON, M. and GRACE, J. (1984). Effect of wind on the transpiration of young trees. *Annals of Botany* **53,** 811–819.
DOLEY, D. (1981). Tropical and subtropical forests and woodlands. In, *Water deficits and plant growth* Vol. VI, 209–320. Academic Press, London.
GRACE, J. (1984). *Plant-atmosphere relationships.* Chapman & Hall, London.
HARTWIG, R.C. and LARSON, M.M. (1980). Hormone root-soak can increase initial growth of planted hardwood stock. *Tree Planters Notes* **32** (1), 29–33.
HINCKLEY, T.M., DOUGHERTY, P.M., LASSOIE, J.P., ROBERTS, J.E. and TESKEY, R.O. (1979). A severe drought: impact on tree growth, phenology, net photosynthetic rate and water relations. *American Midland Naturalist* **102,** 307–316.
HOLSTENER-JØRGENSEN, H. and HOLMSGAARD, E. (1975). Fertilization and irrigation of young Norway spruce on sandy soil. *Det Forstlige Forsøgsvaesen i Danmark* **34,** 263–270.
HUNT, R. (1975). Further observations on the root-shoot equilibria in perennial ryegrass (*Lolium perenne* L.). *Annals of Botany* **39,** 745–755.
INSLEY, H. (1981). Moving plants safely. In, *Research for practical arboriculture*. Forestry Commission Occasional Paper 10. Forestry Commission, Edinburgh.
JARVIS, P.G. (1975). Water transfer in plants. In, *Heat and mass transfer in the plant environment,* eds. D.A. de Vries and N.G. Afgan, 369–394. Scripta Book Co., Washington, D.C., USA.
KOZLOWSKI, T.T. (1975). Effects of transplanting and site on water relations of trees. *American Nurseryman* **141** (9), 84–94.
KOZLOWSKI, T.T. (1982). Water supply and tree growth. I. Water deficits. *Forestry Abstracts* **43,** 57–95.
KOZLOWSKI, T.T. and DAVIES, W.J. (1975). Control of water balance in transplanted trees. *Journal of Arboriculture* **1,** 1–10.
LANDSBERG, J.J. and McMURTRIE, R. (1984). Water used by isolated trees. *Agricultural Water Management* **8,** 223–242.
LINDER, S. and AXELSSON, B. (1982). Changes in carbon uptake and allocation patterns as a result of irrigation and fertilization in a young *Pinus sylvestris* stand. In, *Carbon uptake and allocation in sub-alpine ecosystems as a key to management,* ed. Waring, R.H., 38–44. Oregon State University, Corvallis, USA.
MILBURN, J.A. and JOHNSON, R.P.C. (1966). The conduction of sap. II. Detection of vibration by sap cavitation in *Ricinus* xylem. *Planta* **69,** 43–52.
POLGE, H. (1982). Influence de la competition et de la disponibilité en eau sur l'importance de l'aubier du douglas. *Annales des Sciences Forestières.* **39,** 279–298.
POLLARD, R.S.W. (1976). Letter to the editor. *Arboricultural Journal* **2,** 488.
RABENSTEINER, G. and TRANQUILLINI, W. (1970). The importance of antitranspirants and root dips for the survival of forest trees after transplanting. *Allgemeine Forstzeitung* **81,** 319–320.
ROOK, D.A. (1971). Effect of undercutting and wrenching on growth of *Pinus radiata* D. Don. seedlings. *Journal of Applied Ecology* **8,** 477–490.
ROOK, D.A. (1973). Conditioning radiata pine seedlings to transplanting by restricted watering. *New Zealand Journal of Forestry Science* **3,** 54–59.
SANDFORD, A.P. and GRACE, J. (1985). The measurement and interpretation of ultrasound from woody stems. *Journal of Experimental Botany* **36,** 298–311.
STEWART, J.B. (1984). Measurement and prediction of evaporation from forested and agricultural catchments. *Agricultural Water Management* **8,** 1–28.
STONE, E.C., JENKINSON, J.L. and KRUGMAN, S.L. (1962). Root regenerating potential of Douglas-fir seedlings lifted at different times of the year. *Forest Science* **8,** 288–297.
STONE, E.C., SCHUBERT, G.H., BENSELER, R.W., BARON, F.J. and KRUGMAN, S.L. (1963). Variation in the root-regenerating potentials of ponderosa pine from four California nurseries. *Forest Science* **9,** 253–256.
THORPE, M.R., SAUGIER, B., AUGER, S., BERGER, A. and METHY, M. (1978). Photosynthesis and transpiration of an isolated tree: model and validation. *Plant, Cell and Environment* **1,** 269–277.
TOLIVER, J.R., SPARKS, R.C. and HANSBROUGH, T. (1980). Effects of top and lateral root pruning on survival and early growth – three bottomland hardwood tree species. *Tree Planters Notes* **31** (3).
ZAHNER, R. (1968). Water deficits and growth of trees. In, *Water deficits and plant growth* Vol. 2, ed. T.T. Kozlowski, 191–254. Academic Press, New York.
ZAHNER, R. and DONNELLY, J.R. (1967). Refining correlations of water deficits and radial growth in young red pine. *Ecology* **48,** 525–530.
ZIMMERMAN, M.H. (1983). *Xylem structure and the ascent of sap.* Springer-Verlag, Berlin.

Discussion

P.R. THODAY (Department of Biological Sciences, University of Bath).

Are data available on decreasing water loss from leaves depending on previous (stressed/unstressed) treatment?

J. GRACE Leaves from stressed plants differ from those on unstressed plants in that both mass and stomatal conductance are likely to be lower.

P.A. HEMSLEY (Askham Bryan College of Agriculture and Horticulture).

What level of soil water deficit indicates a need for water?

J. GRACE Even in bogs trees may suffer from water stress. Ideally, irrigation would respond to solar radiation once this reaches about 200 watts m^{-2}.

Fertilising Broadleaved Landscape Trees

R.J. Davies
Forestry Commission

Summary

Ash, Norway maple and Norway spruce planted into subsoil in a factorial NPK trial did not benefit from fertiliser whether surface applied or mixed in; some treatments reduced survival or growth.

NPK increased first and later years' growth and foliar nutrient concentrations of oak and ash planted in herbicide-killed grass on London clay; fertilising without herbicide reduced oak growth. Bituminised felt mulch was less effective than herbicide.

NPK fertiliser increased foliar nitrogen concentrations but had little effect on growth of 10-year-old ash growing slowly in dense grass beside a motorway; whereas herbicide increased both foliar nitrogen concentrations and growth.

Fertilising had no effect on height or diameter growth of 25-year-old lime (N or K), 50-year-old lime (N, P or K) and 40-year-old Horse chestnut and London plane (NPK). The 50-year-old lime showed a significant uptake of phosphorus.

Soil structure and weeds appeared more limiting than nutrients in these trials. Fertilisers only improved the growth of trees in the establishment stage.

Introduction

Why fertilise landscape trees? Maximising timber production is not the objective.

1. Large trees have higher landscape values than small trees. Fertilising may speed the attainment of these values, thus increasing net discounted landscape value.
2. Fertilising may enhance the colour or quality of foliage.
3. Fertilising may aid recovery from shock (e.g. site disturbance or defoliation) or increase the tree's resistance to future shock.
4. Fertilising may prolong the life of over-mature trees.

These are reasonable objectives; but will fertilising achieve them? It would be difficult to test the achievement of objectives 3 and 4, and to a lesser extent objective 2, in empirical fertiliser trials. Better understanding of the nutrition and growth of landscape trees is needed to assess the likelihood of fertilisers achieving these objectives.

Fertiliser trials with large landscape trees are difficult. To obtain statistically sound results many trees of one species and age on a uniform site are required. Avenues provide suitable sites but, after every other tree has been rejected (because of overlapping root systems) and some more rejected because they appear abnormal, few avenues have sufficient trees.

This paper reports six fertiliser trials using trees of various species and ages, some newly planted and others up to 50 years old. Response was assessed by measuring height and stem diameter, foliar nutrient concentrations and leaf size. Attempts were made to assess leaf colour by comparison with colour charts; this proved subjective and unreliable. It was impossible to measure the heights of the older trees, which had lost much of their apical dominance, with sufficient accuracy to detect small responses. Total shoot extension growth or leaf area might be good indicators of fertiliser response, but are difficult to assess accurately with large trees.

Six Fertiliser Experiments

i. Fertiliser experiment at Milton Keynes

The site of this experiment was a 10 m tall heap of subsoil. The soil contained some building rubble, but was predominantly fine textured, poorly drained and had little structure. Most of the site had a gentle slope.

The experiment compared the effects of factorial combinations of nitrogen, phosphorus and potassium, confounded with two application methods, on the establishment of ash (*Fraxinus excelsior*), Norway maple (*Acer platanoides*) and Norway spruce (*Picea abies*) transplants. Nitram (supplying 2.5 g of nitrogen), triple superphosphate (1.3 g of phosphorus) and muriate of potash (2.5 g of potassium) were spread close to the tree or mixed into the soil at time of planting in March 1983. Periodic glyphosate applications controlled weeds.

After two growing seasons average survival and growth of all trees, fertilised and controls, were:

	Ash	Maple	Spruce
Survival (%)	99	56	16
Height growth (cm)	11.6	−6.6	0.9
Diameter growth (mm)	0.6	0.1	0.5

These disappointing figures were depressed rather than improved by fertilising. Nitrogen reduced spruce survival ($p<0.01$) and ash ($p<0.01$) and maple ($p<0.05$) diameters. Potassium reduced ash diameters ($p<0.05$). Phosphorus reduced ash heights and maple heights and diameters (these effects were not quite significant at $p = 0.05$). There were no significant interactions. Fertilisers reduced spruce survival more when mixed into the soil than when applied as a top dressing ($p<0.01$) – probably because difficulties in mixing fertilisers with heavy soil left pockets of fertiliser which damaged adjacent tree roots.

ii. Fertiliser × weeding experiment at Wormwood Scrubs, west London

The site had been regularly mown and used as a recreation area for many years prior to the planting of this experiment in March 1982 with oak (*Quercus petraea*) and ash transplants. The soil was a moderately compacted London clay.

Three weeding treatments, each with or without fertiliser – six treatments in all – were used. Each fertilised tree received 45 g of a 9:25:25 fertiliser in April 1982 and 40 g of a 27:5:5 fertiliser in April 1983, spread over a 300 mm diameter area around the tree. The first application supplied 4.1, 4.9 and 9.3 g, and the second 10.8, 0.9 and 1.7 g per tree of nitrogen, phosphorus and potassium respectively. The weeding treatments were a control, 0.5 × 0.5 m bituminised felt mulching mats ('Tree Spats') and 0.5 m diameter herbicide spots. Glyphosate in March 1982, paraquat in June 1982, June 1983 and July 1984, and propyzamide in November 1982 and February 1984 were applied to the herbicide spots.

Figure 1 depicts height and diameter growth over the first three growing seasons. Table 1 summarises the analyses of foliage collected in August 1982 and July 1983.

Fertilising increased foliar nutrient concentrations in both the first and second seasons. There was a significant fertiliser × weeding interaction with oak: the increase in foliar nitrogen concentrations produced by fertilising was dependent on the weeding treatment. In general weeding had as much effect on foliar nutrient concentrations, and a larger effect on leaf size, height and diameter growth, than fertilising. Despite their greater area 'Tree Spats' (0.25 m^2) were less effective than herbicide spots (0.20 m^2); this was probably owing to grasses, particularly *Agropyron repens*, rooting under the spats.

Fertilising only improved oak growth when accompanied by herbicide; without herbicide it was detrimental. Fertilising barely increased ash growth unless 'Tree Spats' or herbicide were also applied.

iii. Motorway experiment near Leeds

This experiment was superimposed on slow-growing, 10-year-old ash. The site had been disturbed and compacted during motorway construction. It is almost level, has a fine textured soil and supports strong grass growth. It is waterlogged after prolonged rain and was single-furrow ploughed prior to planting.

Although soil compaction and poor drainage were thought to be the main problems this experiment compared the effects of fertiliser, with or without chemical weed control, on the ash. (The correct time to ameliorate the soil was before planting.) Each fertilised tree received 80 g of a 21:14:14 fertiliser in June 1983, repeated in April 1984: in total 33.6, 9.8 and 18.6 g per tree of nitrogen, phosphorus and potassium respectively. Paraquat in June 1983 and propyzamide in November 1983 were applied to 1.2 m diameter spots around half the trees.

Figure 2 shows height and basal area (measured 0.5 m above ground) over the first two seasons of the experiment. Table 2 summarises the analyses of foliage collected in late August 1983.

Both fertilising and herbicide increased foliar nitrogen concentrations, and the increase when both were applied together was greater than the sum of the increases due to each separately. Surprisingly fertilising depressed foliar phosphorus concentrations, but all of the phosphorus concentrations appear adequate for healthy growth. Herbicide increased growth more than fertilising; only the herbicide effect was significant.

Figure 1. The effect of fertiliser (F) and three weeding treatments (W) on height and diameter of oak and ash transplants over their first three seasons. Wormwood Scrubs.

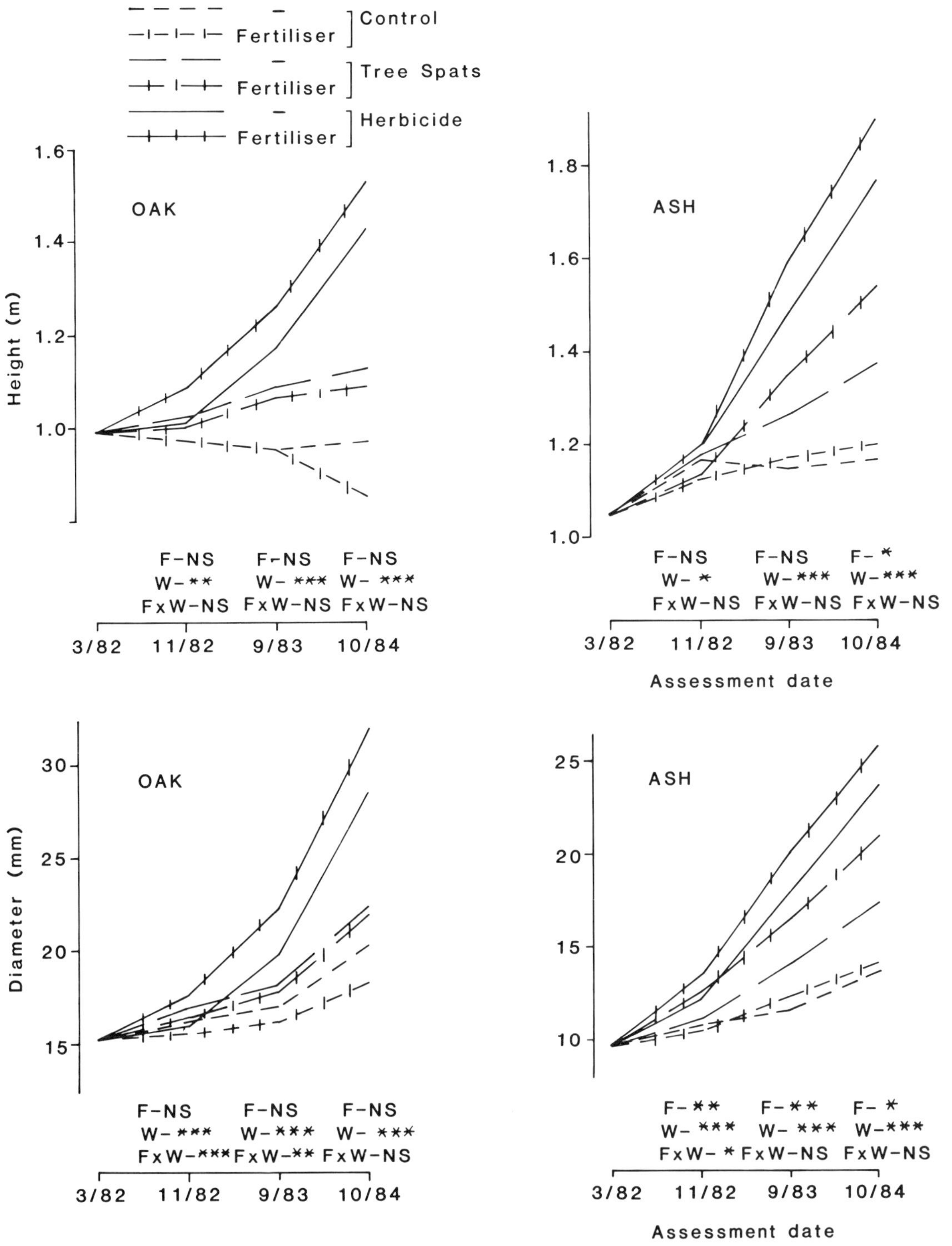

All heights and diameters have been adjusted for covariates using initial heights and diameters. See Note (1) below Table 1 for explanation of the significance levels, which refer to the six treatment means on each assessment date.

Table 1. The effect of three weeding treatments (WC = control; WT = 'Tree Spats'; WH = herbicide spots) with (F) or without (No F) fertiliser on foliar nutrient concentrations (as % of dry weight) and leaf size index (L)

				Significance [1]					Significance [1]		
		NoF	F	F	W	F&W	NoF	F	F	W	F&W
		1982 Ash foliage					1983 Ash foliage				
	WC	2.13	2.37				2.14	2.15			
N	WT	2.06	2.47	***	***	NS	2.00	2.71	***	**	*
	WH	2.75	3.18				2.37	3.26			
	WC	0.145	0.146				0.114	0.104			
P	WT	0.167	0.162	NS	***	NS	0.115	0.112	NS	**	NS
	WH	0.189	0.181				0.125	0.128			
	WC	1.20	1.50				1.36	1.52			
K	WT	1.21	1.28	**	NS	NS	1.46	1.41	NS	NS	NS
	WH	1.15	1.27				1.36	1.51			
	WC	122	125				122	131			
L[2]	WT	140	133	NS	*	NS	163	202	**	***	NS
	WH	164	163				212	269			
		1982 Oak foliage					1983 Oak foliage				
	WC	2.87	2.77				2.26	2.77			
N	WT	2.74	2.70	NS	***	**	2.27	2.87	***	NS	NS
	WH	2.88	3.33				2.56	2.95			
	WC	0.187	0.192				0.143	0.166			
P	WT	0.174	0.185	NS	NS	NS	0.124	0.152	*	NS	NS
	WH	0.184	0.184				0.149	0.156			
	WC	0.764	0.810				0.767	0.800			
K	WT	0.772	0.809	NS	NS	NS	0.748	0.867	NS	NS	NS
	WH	0.857	0.801				0.870	0.868			
	WC	238	235				411	433			
L[2]	WT	315	302	NS	***	NS	437	432	NS	*	NS
	WH	377	407				546	540			

Notes: [1] NS = effect not significant at 5% level.
* = effect significant at 5% level.
** = effect significant at 1% level.
*** = effect significant at 0.1% level.

[2] Leaf size index (L) obtained by multiplying leaf (oak) or leaflet (ash) lengths and widths measured in millimetres.

Figure 2. The effect of fertiliser (F) and herbicide (H) on ash height and basal area. Leeds, M62.

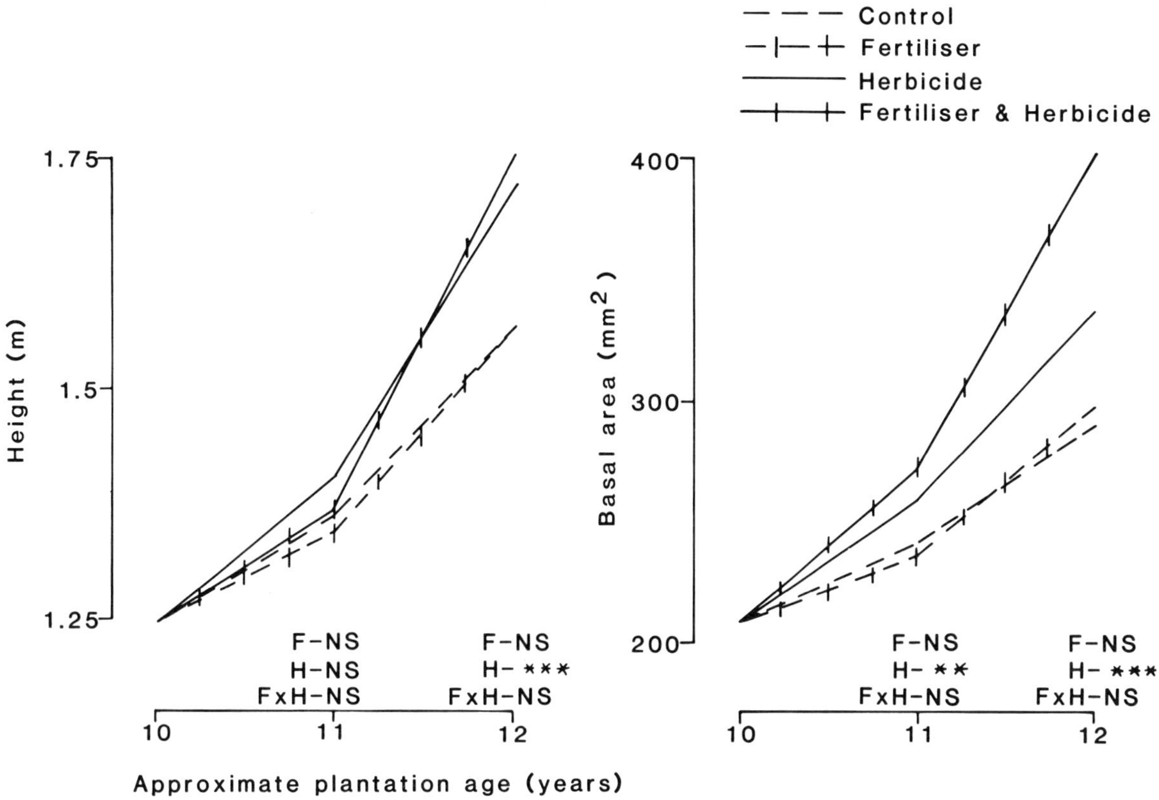

All heights and diameters have been adjusted for covariates using initial heights and diameters. See Note (1) below Table 1 for explanation of the significance levels, which refer to the four treatment means on each assessment date.

Table 2. The effects of fertiliser and weed control on foliar nitrogen, phosphorus and potassium concentrations (as % of dry weight) and leaf weight. Leeds.

	Control	Fertiliser	Herbicide	Both		Significance[1]	
	—	F	H	F+H	F	H	F×H
N	1.91	1.99	2.05	2.57	***	***	***
P	0.25	0.22	0.29	0.23	***	NS	NS
K	1.02	0.96	0.95	0.84	NS	*	NS
Leaf weight (g)	0.171	0.167	0.185	0.188	NS	NS	NS

(1) See note (1) below Table 1.

iv. Avenue experiment at Blickling, Norfolk

This experiment used 44 Common lime (*Tilia x europaea*) approximately 25 years old, growing in a double avenue on the National Trust's Blickling Estate in Norfolk. The four experimental treatments were a control, 100 g and 200 g of nitrogen (supplied as urea) and 100 g of potassium (supplied as muriate of potash) per tree spread on the grass sward beneath the trees' crowns. These rates are roughly equivalent to 63 and 126 kg of nitrogen and 63 kg of potassium per hectare. Before fertilising the trees averaged 5.6 m tall and 154 mm diameter at 1.4 m. The treatments had no detectable effect on height or diameter growth, foliar nutrient concentrations or leaf size in the 4 years following application.

v. Avenue experiment at Hardwick Hall, Derbyshire

A similar experiment on another National Trust double avenue of Common lime at Hardwick Hall, Derbyshire used 160 trees approximately 50 years old. The 10 experimental treatments included a control and various rates and combinations of nitrogen, phosphorus and potassium, applied as urea, Gafsa rock phosphate and muriate of potash spread on the grass within a 3 m wide ring under the trees' drip lines. The highest rates supplied 3 kg of nitrogen, 2 kg of phosphorus and 1.7 kg of potassium per tree. These maximum rates are roughly equivalent to 420, 280 and 240 kg per hectare of each element assuming the trees were growing in a plantation with crowns touching. Before fertilising the trees averaged 12.4 m tall and 364 mm diameter. In the year after fertilising trees which had received phosphorus had significantly higher foliar phosphorus concentrations (0.21 per cent) than trees receiving no phosphorus (0.19 per cent). No other effects on height or diameter growth, foliar nutrient concentrations or leaf size were detected in the 4 years following application.

vi. Avenue experiment in Windsor Great Park, Berkshire

This experiment used 60 Horse chestnut (*Aesculus hippocastanum*) from the inner, and 60 London plane (*Platanus x hispanica*) from the outer rows of 'The Long Walk', a double avenue approximately 40 years old. Both species, especially the Horse chestnut, lacked vigour and were a poor colour. The three experimental treatments were a control, 2.1 kg of a 27:5:5, and 2.4 kg of a 9:24:24 fertiliser per tree spread on the grass within a 3 m radius of the tree, the total quantity being split equally between two consecutive years. The fertiliser treatments were roughly equivalent to 200 or 76 kg of nitrogen, 16 or 89 kg of phosphorus and 31 or 160 kg of potassium per hectare. Before fertilising the mean heights and diameters at 1.3 m of the plane were 12.8 m and 300 mm and of the chestnut 10.5 m and 343 mm respectively. The soil along the avenue had been compacted by horse riding, and parts of the avenue were poorly drained, suffering periodic waterlogging. The treatments had no significant effects on height or diameter growth, foliar nutrient concentrations or leaf size in the 4 years following the first fertiliser application. Irrespective of treatment, all the plant foliar nitrogen, phosphorus and potassium concentrations are close to those reported by Binns *et al.* (1983) as indicating possible deficiencies. The Horse chestnut foliar nitrogen (1.5 per cent) and phosphorus (0.11 per cent) concentrations were very low.

Discussion

Adverse soil physical conditions probably caused the poor tree growth at Milton Keynes. Although the soil was not badly compacted and there was sufficient slope to prevent large puddles forming, the fine textured soil's lack of structure resulted in slow drainage and long periods of waterlogging. This will have reduced gaseous diffusion, producing anaerobic soil conditions and restricting root growth. Since the surviving trees grew little or died back their nutrient requirements will have been small. Fertilisers do not improve soil structure and their osmotic effect probably damaged the trees.

Fertilising improved tree growth at Leeds, despite this site also being poorly drained. The 10-year-old trees at Leeds had more time to develop a root system than the newly planted trees at Milton Keynes, and the longer period following engineering works at Leeds allowed some soil structure to develop.

The time of fertilising in relation to planting has been debated. Patch *et al.* (1984), for example, suggest that because trees absorb nutrients mainly through the fine roots which die during transplanting, fertilising should be delayed until the start of the second growing season. However, fertiliser applied a few weeks after planting at Wormwood Scrubs increased ash foliar nutrient concentrations and diameter growth. The first season fertiliser response of the oak was confused by the fertiliser-weeding interaction. Fertiliser at planting may benefit trees, provided soil air and water regimes and careful plant handling facilitate early root growth. However, at both Wormwood Scrubs and Leeds weeding increased tree growth more than fertilising. Fertilising was most beneficial when accompanied by weed control. In one case fertilising without weeding depressed growth, probably because weeds were stimulated to compete more strongly for moisture.

Fertilisers improved tree growth at Wormwood Scrubs and Leeds, where small trees were used; but in the three experiments using larger trees (Blickling, Hardwick and Windsor) little response was detected despite high fertiliser rates. This is not surprising assuming that the guiding concepts proposed by Miller (1981) for forest fertilisation also apply, albeit in modified form, to landscape trees. The newly planted tree has a rapidly increasing crown size and total nutrient content which is fed by a restricted root system; consequently the concentration of available soil nutrients may be critical. At a later stage of growth the tree has a larger root system and because its crown is not enlarging so rapidly, a large proportion of its nutrient requirements can be met by recycling within the tree.

After the establishment stage landscape trees are unlikely to benefit from fertilisers. Severely pruned or defoliated trees are possibly exceptions to this generalisation; their nutrient requirements may temporarily increase while they rebuild their crowns. This hypothesis is difficult to test in empirical fertiliser trials, estate owners being understandably reluctant to permit a large number of trees to be uniformly mutilated.

However, detailed site studies could yield nutrient cycling models with which to assess the likelihood of fertilisers benefiting severely pruned or defoliated trees. They could also test the likelihood of mature tree fertilising achieving some of the other objectives which are impossible to test in empirical fertiliser trials. Annual leaf-sweeping, lawn-mowing or grazing could be built into such models. These studies should be relatively straightforward in parks, but the presence of sewers and drains with their varied contents would complicate matters in streets.

If the fertiliser treatments used at Blickling, Hardwick or Windsor accelerated tree growth slightly, the benefits may only become apparent many years later. We could remeasure these trees at a later date; but the site studies suggested in the previous paragraph would probably reveal that the fertiliser rates used were small in relation to the nutrient content of soil, ground vegetation and trees, despite being high relative to forest and arboricultural practice. It therefore seems unlikely that responses will be detected in the long term when none were detected in the few years after fertilising.

In plantation forestry it appears that fertiliser responses can be adequately described in terms of reduced rotation length (Miller, 1981). Typically the response lasts of the order of 5 to 10 years after fertilising, the stand then resuming the growth pattern that it would have exhibited at a later date without fertilising. Thus fertilising may reduce landscape trees' life expectancy; but it may prolong life expectancy if it helps the tree recover from an otherwise fatal disturbance. Fertiliser responses during the establishment stage though will shorten the vandal vulnerable stage, and speed the attainment of the high landscape value of a mature tree. Fertiliser responses are both more likely and more desirable during the establishment stage than later.

Fertilising did not benefit the Windsor trees despite some of the foliar nutrient concentrations being below those reported by Binns *et al.* (1983) as indicating possible nutrient deficiencies. The low concentrations and the trees' lack of vigour were probably caused by soil compaction and poor drainage restricting root growth. This emphasises the need for care, advocated by Binns *et al.*, when interpreting foliar analyses to diagnose deficiencies.

Growth and foliar nutrient concentrations of the trees at Blickling and Hardwick did not appear to be deficient. This may explain the lack of response. These avenues were used because I have not found nutrient deficient avenue trees with which to experiment. Many trees superficially appear nutrient deficient, as at Windsor, but investigations have invariably revealed other adverse factors; soil compaction and poor drainage being common. In searching for experiment sites in England I have gained the impression that simple nutrient deficiency of landscape trees is rare.

The fertiliser application method used in the avenue experiments might be criticised; we spread fertilisers on top of the grass rather than close to the trees' roots. Arborists frequently place fertilisers in holes augered in the soil around well established trees. However, grasses can root deeply, and Garwood and Williams (1967) demonstrated that grasses readily absorb nutrients injected directly into the subsoil. To prevent competition for nutrients grasses must be killed. If, as suggested previously, soil compaction is a more frequent cause of poor tree growth than nutrient deficiency then augering hundreds of holes in the soil may in itself improve growth. Experiments examining the effects of sward removal with herbicides, and soil augering on well established landscape trees have recently been established.

Conclusions

Fertilising is unlikely to benefit established landscape trees. During the establishment stage fertilising may reduce tree survival and growth unless adverse soil air and water regimes are corrected by cultivation and drainage and weeds are adequately controlled. Poor health of established trees is frequently associated with soil compaction or poor drainage; simple nutrient deficiency is rare. Fertilising may possibly help the well established landscape tree rebuild its crown or foliage following severe pruning or defoliation.

Acknowledgements

Milton Keynes Development Corporation, The Greater London Council, West Yorkshire Metropolitan County Council, The National Trust and the Crown Estate Windsor provided sites for experiments. J.B.H. Gardiner, K.A.S. Gabriel, A.L. Sharpe, I.H. Blackmore, P.R. Barwick, T.J. Davis and R.E. Warn did much of the experimental work and R.C. Boswell conducted statistical analyses. This work was initiated by Dr. H. Insley and funded by the Department of the Environment.

References

BINNS, W.O., INSLEY, H. and GARDINER, J.B.H. (1983). *Nutrition of broadleaved amenity trees. I. Foliar sampling and analysis for determining nutrient status*. Arboriculture Research Note 50/83/SSS. DOE Arboricultural Advisory and Information Service, Forestry Commission.

GARWOOD, E.A. and WILLIAMS, T.E. (1967). Growth, water use and nutrient uptake from the subsoil by grass sward. *Journal of Agricultural Science*, Cambridge **69**, 125–130.

MILLER, H.G. (1981). Forest fertilization: some guiding concepts. *Forestry* **54**, 157–167.

PATCH, D., BINNS, W.O. and FOURT, D.F. (1984). *Nutrition of broadleaved amenity trees. II. Fertilisers*. Arboriculture Research Note 52/84/SSS. DOE Arboricultural Advisory and Information Service, Forestry Commission.

Discussion

M. HAXELTINE (Domestic Recreation Services).

Would the use of superphosphate accelerate root growth and improve callus formation?

R.J. DAVIES Neither triple nor superphosphate gave a response in the experiments and it appeared that nutrients were not the limiting factor. The influence of fertilisers on callus formation was not tested.

A. KING (Tree and Landscape Consultant)

Your experiments have been conducted on avenue trees but would fertilisers benefit trees affected by building operations?

R.J. DAVIES The need for adequate replications for experiments made it very difficult to find suitable sites which include building works.

T.H.R. HALL (Parks Department, Oxford University).

There are avenues of trees on roads leading out of Oxford and Manchester which might be suitable for experiments.

R.J. DAVIES Roadside trees presented difficulties for experiments because of extraneous effects of tarmac, drains, sewers and other services.

J. EVANS (Forestry Commission, Research Division).

(1) Nutrients were only one of several factors limiting tree growth – others such as compaction or drainage are more important on many sites.
(2) In experiments in fertilising pole stage broadleaved woods nutrient uptake has been detected in the foliage but the only growth rate affected had been diameter increase of ash following application of nitrogen.
(3) Though the data for the Windsor experiment showed low nutrient levels, they were in balance (N:P:K – 10:1:5).

The Control of Epicormic Branches

J. Evans
Forestry Commission

Summary

Epicormic branches which develop on the trunk of well established trees are a nuisance to arboriculturists and foresters. Oak, elm, Common lime, poplar and willow are mainly affected. Epicormic shoots are stimulated to develop by stress to a tree or sudden change in its environment; crown pruning, waterlogging of the soil and root severance during site development are all common causes of the problem in amenity trees.

Control by pruning is expensive and tends to increase subsequent production of epicormics. Several chemicals will control epicormic shoots in the year of application but maleic hydrazide, applied to newly emerging shoots in May or June, will also suppress development in the following year. Maleic hydrazide appears to depress trunk diameter growth.

Introduction

Side branches which develop low down on the trunks of well established trees can be a problem on amenity and forest trees. On amenity trees such epicormic branches, although never becoming very large thick limbs, are unsightly and can be a nuisance to traffic and pedestrians. On forest trees they cause knots in the wood, a defect which lowers timber quality and strength.

This paper briefly reviews the subject of epicormic branching and reports research carried out by the Forestry Commission on the control of epicormics on oak. Much of the work is of relevance to arboricultural practice.

Species Affected

In temperate regions the problem of epicormic branching is largely confined to five genera of broadleaves – oak (*Quercus*), elm (*Ulmus*), lime (*Tilia*), poplar (*Populus*), and Willow (*Salix*). Some species outside these genera can be seriously affected (e.g. eucalypts) but in general profuse epicormic development is the exception rather than the rule.

Within the genera mentioned and between different seed origins of one species the propensity to produce epicormic shoots varies greatly. For example, among the oaks Red (*Q. rubra*) and Turkey oak (*Q. cerris*) generally produce far fewer epicormics than either of our native species. But even among the native oaks Sessile oak is less prone to epicormics than Pedunculate (Table 1). Table 1 also shows highly significant differences in numbers of epicormics between different origins of the *same* species (Sessile oak).

Among other species Common lime (*Tilia* x *europaea*) will throw epicormics profusely as well as basal suckers while the Small-leaved lime (*T. cordata*) is often free of them. Among poplar, aspen (*P. tremula*) and Grey poplar (*P. canescens*) very rarely produce epicormics while varieties of Black cottonwood (*P. trichocarpa*) show great variability from profuse production of them in a Canadian provenance of *P. trichocarpa* to weak production in *P. trichocarpa* 'Scott Pauley'. The Black poplars are also variable.

Clearly one way of reducing epicormic branching is to select species, provenances or cultivars which are relatively free of the problem.

Initiation and Causes of Epicormic Development

Anatomically, nearly all epicormic shoots arise from dormant buds in the bark which were originally laid down when the young leading shoot first grew. Following

Table 1. Variation in number of epicormics in relation to oak seed origin

Species and origin	No. epicormics between 1.3 and 2.3 m counted on all origins in 1980 – Dean (Penyard) 1 P53	As previous column, but assessment in 1984 of three Sessile origins used to make four blocks[1] for experiment Dean 135/81
Pedunculate oak		
A Lodgehill, New Forest	27.5	
H South Side Wood, Powys	57.8	
Sessile oak		
B Pritchard's Hill, Dean (gen)	16.9	
C Pritchard's Hill, (tree No. 1)	13.2	8.2
D Hadnocks, Dean (gen)	15.3	
E Hadnocks, Dean (tree No. 1)	19.0	12.4
F Bonds Wood, Dean	15.3	
G Skiltons Paddock, Alice Holt	20.9	
J Vestmorland, Norway	12.3	
K Lot 1, Sweden	11.4	5.1 and 5.0[2]
L Lot 2, Sweden	12.3	
Statistical significance	Very high (***)	Very high (***)

Notes: (1) Note similar relative ranking for these three origins as previous column.
(2) Two blocks laid down in large area of this origin.

formation these barely differentiated buds grow sufficiently each year not only to maintain their position in the inner bark but sometimes to divide into two or more buds. In some species, such as Pedunculate oak (*Q. robur*), this process can lead to a 'pad' of epicormic bud tissue which in older trees often takes the shape of a 'cat's paw'.

Epicormic buds can survive but remain dormant for many decades or even hundreds of years in long lived species. They are stimulated to grow and produce shoots in many ways all of which reflect some stress or sudden change to a tree's environment. In forestry practice thinning operations in a stand are much the most frequent cause of renewed production of epicormics on oak. For amenity trees factors such as crown dieback or heavy pruning, root death from disease or severance caused during building development, change in water-table, cold damage to tender species can all lead to epicormic growth.

The physiological mechanisms governing epicormic development are not wholly understood but change in the hormonal environment of the epicormic buds must be the immediate cause of breaking dormancy. There is dispute over whether it is change in a tree's light, temperature, water or nutrient environment which is the trigger to the process, though change in water relations, at present, seems the most likely. Discussion of this will be found in Evans (1982) and is the subject of research being carried out on contract for the Forestry Commission at the East Malling Research Station.

Although the mechanisms of initiating epicormic development are unclear at present it is clear that once epicormic shoots have emerged there must be adequate light for them to grow and develop into branches. Thus, keeping a trunk in heavy shade, as is traditionally achieved by growing an understorey of beech (*Fagus sylvatica*) in oak woodland, will prevent epicormic shoots becoming sizeable woody branches. In amenity planting the manipulation of shade in this way is rarely feasible apart from using shrubs as an understorey in group plantings.

Research into Epicormic Branching

Two aspects of the epicormic problem have been under investigation by the Forestry Commission:

a. factors which cause or stimulate epicormics;
b. methods of direct control.

Stimulation of epicormics

As mentioned, thinning is the main cause of epicormic development in normal forestry work, but from the point of view of epicormic control, both the intensity of thinning (degree of opening up of a stand) and the timing or season of thinning may be significant. An experiment in the Forest of Dean examined the effect of thinning intensity in an oak stand which had been planted at 1.2 x 0.9 m and not thinned for 27 years. Selected trees were released by one of three thinning treatments:

a. no thinning;
b. removal of all adjacent trees to a radius of 2 m, and
c. removal of all trees to a radius of 3 m, i.e., just over double the growing space affected by thinning treatment b.

The thinning was carried out in March 1981 and after 3 years the numbers of live epicormics present in the assessment zone were significantly different between treatments. Table 2 shows the assessment data for both July 1983 and July 1984.

Table 2. Average number of epicormics between 1.3 m and 2.3 m above ground (by thinning treatment)

Date assessed	Thinning treatment		
	No thinning	Moderate thinning (trees within 2 m removed)	Heavy thinning (trees within 3 m removed)
July 1983	5.4	2.9	17.4
July 1984	7.1	5.3	21.3

Table 2 shows that the heavy thinning greatly stimulated epicormic development but, perhaps surprisingly, the moderate thinning produced the lowest numbers of epicormics. This is probably explained by the phenomenon of 'agony shoots' — the development of weak epicormics on trees becoming suppressed or under severe competition in very dense stands. The very crowded nature of the unthinned stand gave conditions leading to 'agony shoot' development where thinning was still not done (the 'no thinning' treatments).

Two recent experiments, at Alice Holt and Dean Forests, are investigating the effects of season of thinning on epicormic production. They were laid down following observation of rapid response (new epicormic shoots emerging within days) to thinning carried out in July (Wignal, personal communication) and even after thinning in September in an oak/Norway spruce mixture at Rockingham Forest. No data are available from these new experiments but if production of epicormic branches is related to the season of thinning then this offers one approach to reducing the problem for foresters.

It may have application in arboricultural practice regarding the timing of any crown pruning of amenity trees.

Direct control of epicormic branches

Six experiments have investigated methods of direct control of epicormics including pruning treatments, wrapping the stems and applying chemicals. In most cases prior to application of treatments trees were high pruned to 5 m and new epicormics stimulated to emerge by heavy thinning, usually removal of all trees within 3 m radius of selected trees.

Results of the most comprehensive experiment, laid down in early 1982, are portrayed in Figure 1 where epicormic development over three growing seasons is compared. Information on costs of application of each treatment are in Patch *et al.* (1984).

Based on the data in Figure 1 and the other experiments laid down the following interim conclusions can be made about control methods.

Pruning

Pruning removes epicormic branches but in all experiments this operation alone did not help to reduce the number of new epicormics which *emerged in later years.* Indeed, in most cases, pruning led to a significant increase in new epicormics in the year following treatment. Thus, where epicormic branches need to be controlled from time to time no hope can be held out that pruning itself will gradually reduce the problem.

Stem wrapping

Three experiments have incorporated stem wrapping treatments including black polythene, black woven polypropylene (Mypex) and even painting the stem with black bitumastic paint. This last treatment proved useless as well as being very expensive and unpleasant to apply. Wrapping very significantly reduces the number of epicormics while the treatment is in place but there is every sign that many new epicormics will emerge, perhaps even in greater numbers, once wrappings are removed. Wrapping does not seem an effective long-term solution to the epicormic problem and, of course, it is both expensive and very unsightly.

Figure 1. Effects of epicormic control treatments

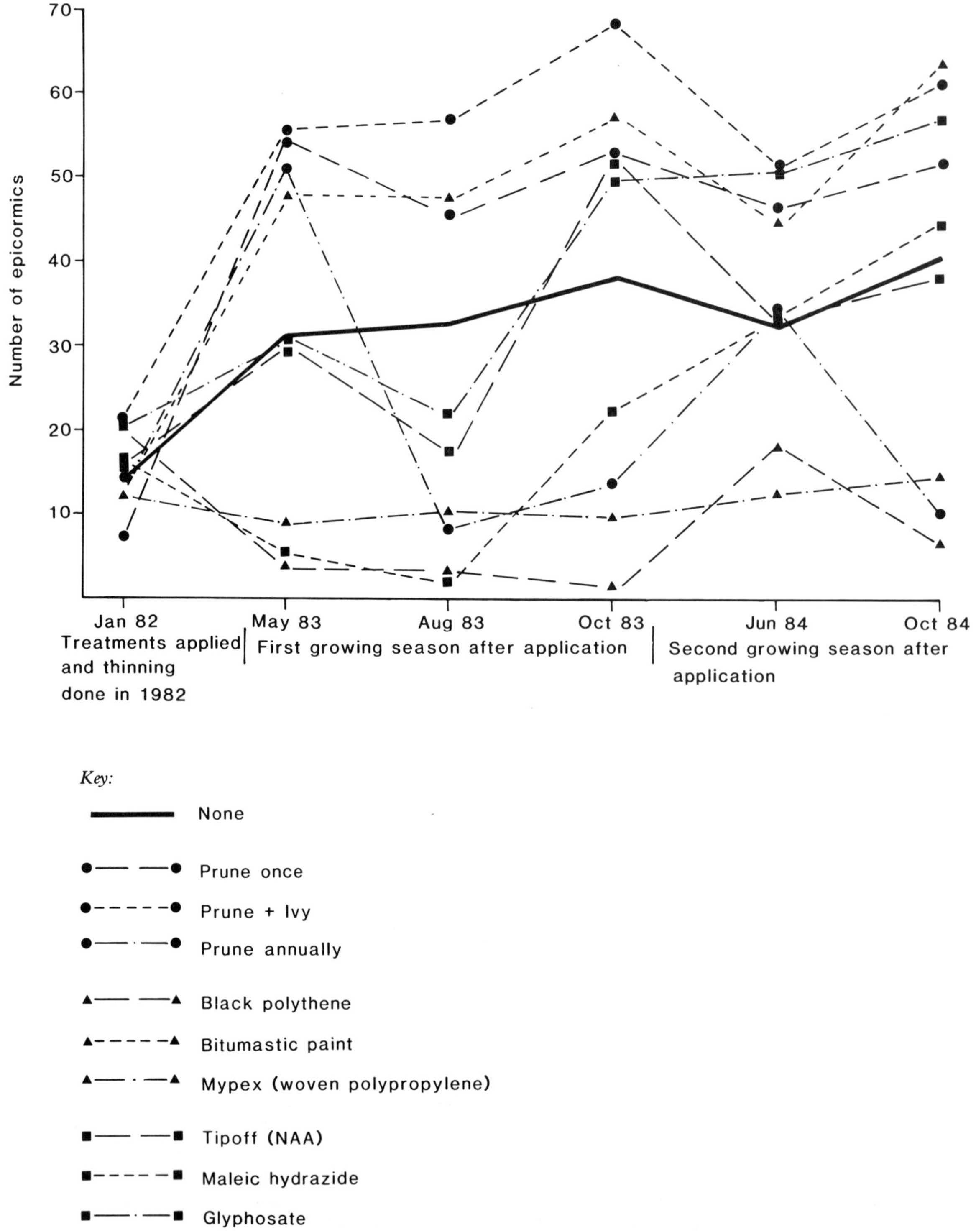

Table 3. Chemicals tested on young epicormic shoots of oak for immediate and long-term control

Chemical	Action
Daminozide (Alar)	Growth regulator
Dikegulac (Atrinal)	Growth regulator
Fosamine ammonium (Krenite)	Contact herbicide inhibiting new growth
Glufosinate	Contact herbicide only translocated in leaves
Glyphosate (Roundup)	Herbicide absorbed and translocated
Maleic hydrazide (Burtolin and Mazide)	Growth retardant
Mefluidide (Embark)	Growth regulator and herbicide
l-naphthylacetic acid (Tipoff)	Growth regulator
Paclobutrazol (PP 333)	Growth regulator
Copper EDTA	Defoliant

In addition to the above chemicals, maleic hydrazide has been used in formulation with Mixture B (a surfactant) to improve solubility and uptake.

Chemical control of epicormics

As long ago as 1960 2,4,5–T was applied to epicormic branches on oak as a possible means of controlling them (Holmes, 1963). This powerful herbicide killed the shoots but gave no long lasting suppression of new epicormic growth. Also, when applied to epicormics more than one year old the treatment left the branches as dead sticks which were slow to fall off. At about this time there was much interest in the growth retardant maleic hydrazide (Burtolin) for suppressing grass growth and hedge management. Use of this chemical for control of side branches was also recommended from this time by one of the manufacturers but no trials of it were conducted by the Forestry Commission.

In recent experiments maleic hydrazide and a large number of other chemicals have been evaluated in one or more trials – Table 3.

In the experiment, illustrated in Figure 1, glyphosate and naphthylacetic acid (Tipoff) killed young epicormic shoots but did not suppress emergence of new ones the following year. Maleic hydrazide achieved both control of existing and substantial suppression of new epicormics in the year after application. Of the other chemicals only fosamine ammonium (Krenite) and glufosinate have shown some promise but both appear less effective than maleic hydrazide.

Current trials are evaluating the optimum time and rate of application of maleic hydrazide and whether benefit is gained from addition of Mixture B, a surfactant to improve contact between the chemical and the tree. So far a late May application on very young shoots appears much the best time to treat trees and even using a weak solution of half manufacturer's (Diamond Shamrock) recommended rate (0.625 litre of Burtolin per 5 litres of water) is quite adequate. However, very strong solutions, in excess of 2.5 litres Burtolin per 5 litres water, if applied to the bark in February or March, will suppress epicormics throughout the year.

Maleic hydrazide is an effective means of controlling epicormics probably over a 2 year period, though the longer term suppression sought by foresters is not evident (Figure 1). However, of greater concern to foresters is that the diameter increment of the oak treated with maleic hydrazide has been significantly less than all other treatments. This, of course, may be of advantage in street tree management.

Conclusions

For amenity trees the best approach to control epicormic branches is to use species and varieties which are not prone to them. Epicormic-ridden established trees should be gradually replaced as part of a long-term management policy. Reasonable short-term control, lasting up to 2 years, can be expected from maleic hydrazide applied to young shoots in early summer.

Acknowledgements

Several foresters have been involved with the research programme on epicormics but I am particularly grateful to Mr. T.D. Cooper, Mr. D. Elgy, Mr. C.W. Shanks and Mr. R.E. Warn.

References

EVANS, J. (1982). Epicormic branches and their control, with a report of current research in Britain. *Proceedings of the Third Meeting National Hardwoods Programme*, Oxford, 5–11.

HOLMES, G.D. (1963). Trials of 2,4,5-T for removal of epicormic shoots on hardwoods. *Report on Forest Research 1962*, 156–166. HMSO. London.

PATCH, D., COUTTS, M.P. and EVANS, J. (1984). *Control of epicormic shoots on amenity trees*. Arboriculture Research Note 54/84/SILS. DOE Arboricultural Advisory and Information Service, Forestry Commission.

WIGNAL, T., personal communication. (Post graduate student undertaking research on contract to Forestry Commission at East Malling Research Station on anatomical and physiological aspects of epicormic shoots in oak).

Discussion

R. SKERRATT (Independent Consultant, Hartlepool).

As the cutting of epicormics increases their numbers is annual pruning effective?

J. EVANS The results suggest that the numbers decline after reaching a peak. More intensive treatment such as rubbing off buds is effective but very labour intensive. More information on the physiological factors at work is needed before better control methods will be found.

J.E.G. GOOD (Institute of Terrestrial Ecology, Bangor).

There may well be a possibility that strong apical dominance may be associated with the tendency not to produce epicormics and this might be a way of selecting at the nursery stage trees with low epicormic tendency.

J. EVANS There is also a risk that trees easily vegetatively propagated may well be those with a strong tendency to produce epicormics.

Trees and Buildings

P.G. Biddle
Tree Conservation Ltd., Wantage, Berks

Summary

During the past 7 years measurements have been made of the patterns of soil moisture changes near trees growing on clay soils. Sixty trees of nine different species are now under observation. The soil moisture content is monitored with a neutron probe in five access holes at varying distances from each tree. This demonstrates the seasonal build-up of desiccation, the patterns of maximum seasonal soil moisture deficit and the deficit which persists from year to year. Poplar shows dramatically greater effects than other species. These data should provide improved guidelines on the safe proximity of trees to buildings.

Introduction

Following the extreme drought conditions of 1975/76, there was extensive criticism of the adequacy of the existing guidelines (BSI, 1972; NHBC, 1974) concerning the proximity of trees to buildings founded on clay soils. However, it was also realised that these guidelines could not be modified without more detailed information on the effects of trees on clay soils, which could only be obtained by extensive field observations.

This has led to Tree Conservation Ltd. being commissioned to undertake three independent, although related, projects. The first instructions in 1978 were from Milton Keynes Development Corporation, who were concerned about their extensive tree planting programme and the possible future effects of these trees on the adjacent house foundations. The first year of this project evaluated various techniques for measuring the effects of trees; these included measurement of soil movement by rod gauges and monitoring seasonal changes in moisture content by neutron probe. This demonstrated the great potential of the neutron probe, which has been used for all subsequent work, although also correlating these results with measurement of soil movement by rod gauge.

The main field instrumentation at Milton Keynes was set up for the 1979 season, and readings have been taken every year since. It includes observations on a total of 12 trees, of three different species growing on two different soil types (Oxford and Boulder clay). This project was enlarged in 1981 on instructions from the National House-Building Council to include a further 24 trees. This extended the scope of species and the variety of soils. In 1983 instructions were received from the Department of Environment for a further 24 trees. Twelve of these are plane trees (*Platanus* × *hispanica*) and form part of an experiment on the effects of pruning. The remaining 12 further increased the range of species. Table 1 see overleaf summarises the species and clay types currently under investigation in these three projects, and shows the range of plasticity index (PI) recorded at 1.5 m depth.

All of the trees are growing either in parks or in the surrounds of playing fields. They are either individual trees or, if part of a row or group, they are clear of any possible interference from other species, and they are clear of other factors such as streams or ditches which might influence the moisture content.

Methods

The moisture content of the soil has been monitored in five boreholes in the proximity of each tree using a neutron probe. The boreholes (known as access holes) are carefully drilled with a combination of a manual screw auger and a 1¾ inch (44.5 mm) mild steel ream. The ream is then removed and an aluminium tube of the same diameter is hammered into the hole to make as tight a fit as possible. The bottom of the tube is sealed with a cap, and the top fitted with a vandal-proof cap or bung to prevent entry of water. The probe is lowered down these access holes whenever measurements are required.

Table 1. Summary of trees being investigated for their effects on clay soils

Tree species	London 32–65	Gault 31–56	Oxford 27–36	Boulder 16–29	Clay silt –	Total
	Clay type and PI range					
Plane	12	–	–	–	–	12
Horse chestnut	4	3	2	2	–	11
Poplar	3	2	2	2	1	10
Leyland cypress	2	2	–	2	–	6
Lime	1	–	2	2	–	5
Oak	2	–	–	2	–	4
Norway maple	2	–	–	2	–	4
Silver birch	2	1	–	1	–	4
Whitebeam		–	–	2	–	4
Total	30	8	6	15	1	60

The probe contains a small radioactive source emitting fast neutrons. Such neutrons are slowed down and reflected by certain atoms, in particular hydrogen, and the probe contains a counter for these slow returning neutrons. As virtually the sole source of hydrogen in the soil is in the form of water, the number of returning neutrons is proportional to moisture content. The readings are influenced by the soil type, but the particular value of the instrument is its ability to detect very small changes in moisture content.

The five access holes are located at varying distances from the tree, related to the height of the tree. In most cases they are at 0.2, 0.4, 0.6, 0.8 and 2.0 times the tree height (referred to as 0.2h, 0.4h, etc.), but for the oak (*Quercus* sp.) this was increased to 0.25h, 0.5h, 0.75h, 1.0h and 3.0h, and for the poplar (*Populus* sp.) to 0.25h, 0.5h, 1.0h, 1.5h and 3.0h. In most cases the access tubes are located in rough mown grass, and the most distant hole was in comparable conditions and was intended to show changes away from all possible tree root activity. The maximum tube length which it is practical to insert is 4 m, and this was usually used for the holes close to the poplars and oak. Shorter 3.6 or 3.0 m tubes were used in other positions.

Readings are taken at 10 cm intervals down each tube. Each reading records the moisture content of the immediate surrounding sphere of soil (sensitivity to moisture is related to the inverse square of the distance), and so the close intervals of the readings provide a very accurate definition of the moisture content profile. With a radioactive instrument there is some inevitable variation in the readings, but differences between successive readings greater than about 0.75 per cent are statistically significant at the 95 per cent probability level.

In addition to monitoring changes in moisture content, measurements have also been made on vertical soil swelling and shrinkage using rod gauges, but results are not included in this paper. Correlations are also being made with meteorological conditions to enable predictions to be made on the extent of drying in severe drought conditions.

Results

A brief summary paper of this sort cannot begin to present the results from all 60 trees for even a single season, let alone from the past 6 years. Only a brief selection of results are included here; additional results will be found in Biddle (1983) and further details will be published in the near future.

Seasonal development of moisture deficit

A selection of the access holes are measured at regular intervals through the summer to determine the build up of the zone of soil desiccation through the season, and the differences between the species and different clay types. These readings are usually from the access holes closest to the trees.

Figures 1 and 2 illustrate results. Figure 1 (0.2h from a lime (*Tilia* sp.) on boulder clay) shows a slight deficit developing to 1.1 m by 27 June 1984; by 1 August the soil was drier and the depth to which the deficit existed had increased to 1.7 m. The driest readings were taken on 31 August, and by 14 September the soil was starting to rewet to a depth of 1.4 m. At 0.25h from a poplar on boulder clay (Figure 2) there is a similar pattern of a progressive increase in the depth of the zone of desiccation and the extent of soil drying, although here some effects extend right to the base of the access hole at 3.6 m. Driest conditions were encountered on 13 September 1984, and slight rewetting had occurred by 12 October.

This seasonal development of the moisture deficit is also being correlated with the meteorological records of rainfall and soil moisture deficit. There are clear differences between species. Comparison between Figures 1 and 2 immediately shows the far greater effects of poplar compared with lime, and that modest rainfall can have a considerable influence on the deficit produced by the lime. Results for Horse chestnut (*Aesculus hippocastanum*) indicate that deficits tend to build up earlier in the

Figure 1. LIME TREE 2.

BOULDER CLAY

Seasonal changes of moisture content profile at 0.2h during 1984.

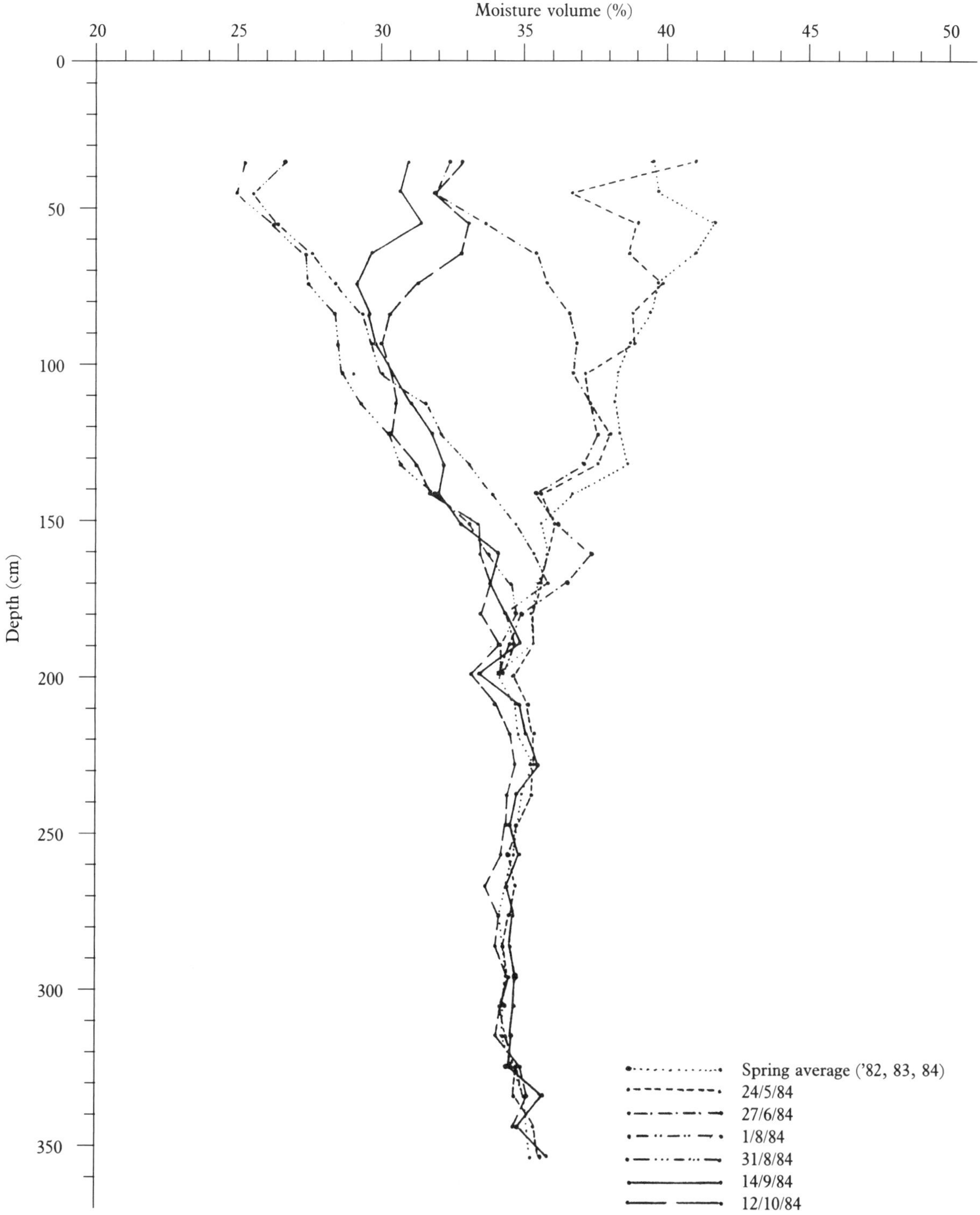

Figure 2. POPLAR TREE 4.

BOULDER CLAY

Seasonal changes of moisture content profile at 0.25h during 1984.

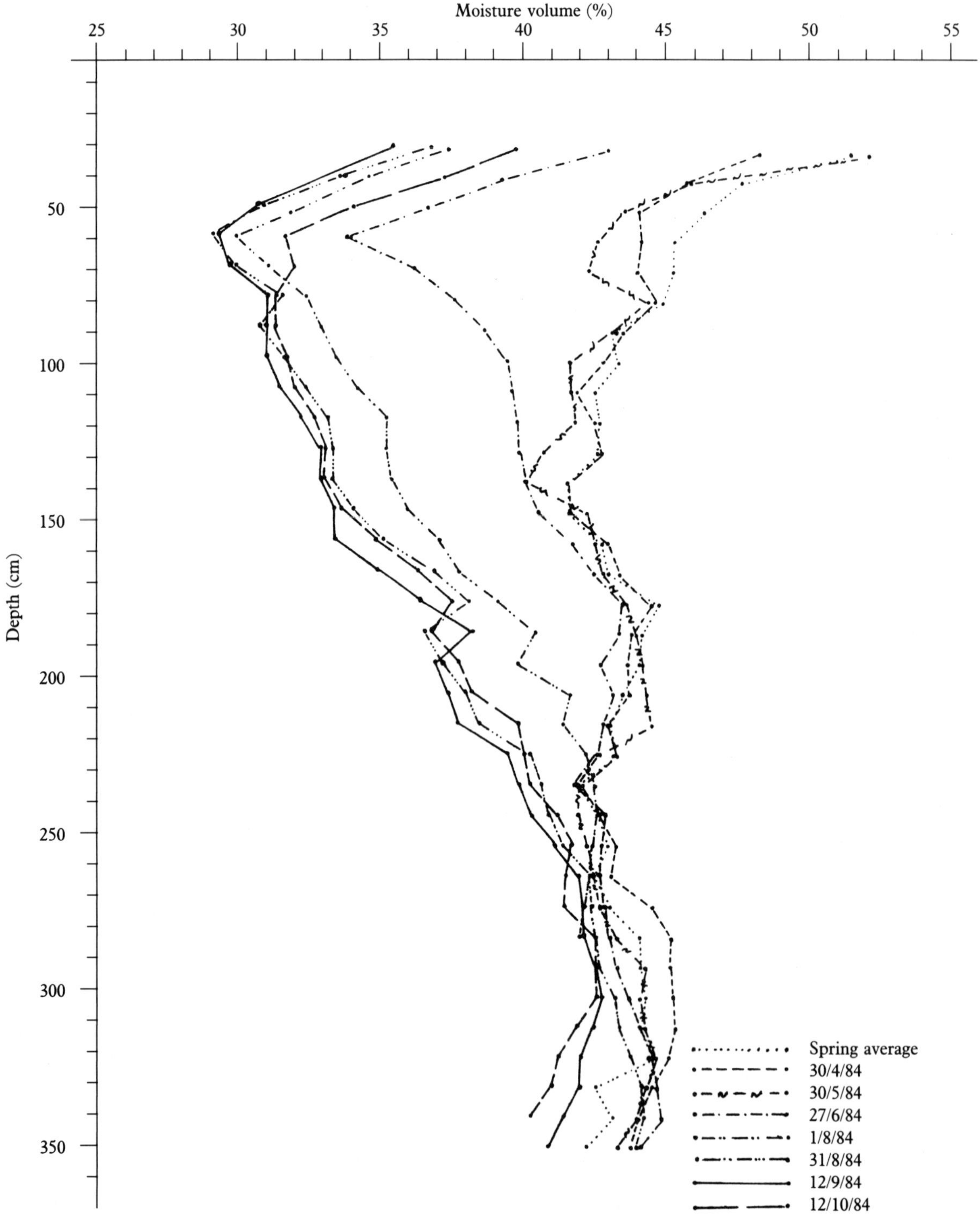

summer, with very little further development after early August, whereas the poplar continues to dry out the soil right up until leaf fall.

Seasonal patterns of moisture deficit of different species

Figures 1 and 2 illustrate how the greatest deficit occurs in early autumn, usually sometime during September. In order to determine the maximum seasonal change, readings are therefore taken on all of the access tubes during September of each year, and the soil moisture contents compared with readings taken in the spring (usually in late April or early May, when the soil is in its wettest condition).

For the presentation of these results the information from all five access holes on each tree are combined into a single diagram, with variation in density of shading used to indicate the amount of seasonal change in moisture content. Typical results for the different species are shown in Figures 3–7. These figures show the average result obtained for all of the individuals of each species (but excluding those trees where soil abnormalities produced anomalous results).

The results again demonstrate the greater effect of poplar compared with lime and Horse chestnut, and also show the insignificant effect of Silver birch (*Betula pendula*) and Leyland cypress (x *Cupressocyparis leylandii*).

Seasonal and permanent water deficits

Figures 3–7 show the seasonal changes which are occurring and which will cause a seasonal cycle of the soil shrinking each summer, and rehydrating and swelling each winter. However, it is known that in some situations there can be long-term progressive movements; evidence of this is seen in the swelling of soil and heave of foundations following the removal of trees. This will occur if there is a zone of permanently desiccated soil which persists through each winter.

The existence of a permanently desiccated zone of soil can be shown if it is assumed that the comparison hole provides a measure of the normal moisture content profile in the absence of tree root activity. Figure 9 shows the moisture content profiles at 0.8h and 1.8h from a poplar on London clay in both spring and autumn. It can be seen that seasonal variation at 0.8h extends to 1.6 m and with slight effects at the comparison hole to 1.3 m. In addition, it can be seen that the soil to the bottom of the 0.8h access holes is apparently drier; this suggests that there is a zone of permanent desiccation.

Figure 10 shows the same profiles for a poplar on boulder clay, but here the seasonal changes at 0.2h extend to 3.0 m, and to 1.1 m at the comparison hole. All the profiles are similar below 3.0 m, demonstrating no permanent deficit below this level, but the comparison hole is slightly wetter than the spring profile of the 0.2h hole down to 3.0 m. This shows that the winter rainfall manages to penetrate to the full depth of the desiccated soil but still not in sufficient quantity to entirely eliminate the persistent deficit.

This illustrates the importance of soil permeability. The highly plastic London clay tends to be very impermeable, and seasonal changes often do not extend below 1.5 m, whereas the more permeable boulder clay allows the deficit to be corrected more rapidly.

It is known that this approach of using the comparison hole as a measure of the normal soil conditions can produce misleading results (the soil is not necessarily sufficiently homogeneous), but the poplar and oak invariably demonstrate a permanent effect with all soil types under investigation. Figure 3 shows the average seasonal deficit for the poplars; this should be contrasted with Figure 8 which shows the permanent effects; at 0.25h these appear to extend beyond the depth of the access tubes.

None of the other species of tree produce similar clear cut effects. Some individual lime and Horse chestnut trees suggest a permanent deficit, but this could be the result of heterogeneity of the soil. These results therefore demonstrate the massive effect of poplar (and to a slightly lesser extent oak), and that problems of long-term heave must be expected when these species are removed. Other species are less likely to produce any permanent effect, particularly on the more permeable boulder clay.

Applications

This research has confirmed the previously held belief that poplar can cause very extensive desiccation of clay soils. Perhaps what was not sufficiently realised previously was just how much greater were the effects of poplar compared with most other species. Recent investigations of damage have also highlighted oak as a troublesome species, and this research has confirmed that its effects are almost as severe as poplar. This is of particular relevance as the existing guidelines (BSI, 1972; NHBC, 1974) are based on observations on the proximity of trees to damaged buildings where "most of the trees are poplars and elms, oak is fairly frequent and others are alder, willow, ash, scyamore, chestnut, birch and hawthorn" (Ward, 1953). It appears that the building industry may not have appreciated the significant differences between species, but for simplicity have tarred all trees with the same reputation as poplar.

New guidelines prepared by the National House-Building Council were published in 1985. These identify

Figure 3. POPLAR AVERAGE TREES 3,4,9,10,13,14,22,23 & 24

BOULDER/OXFORD/GAULT/LONDON CLAYS

Reduction in soil moisture content in Autumn 1983

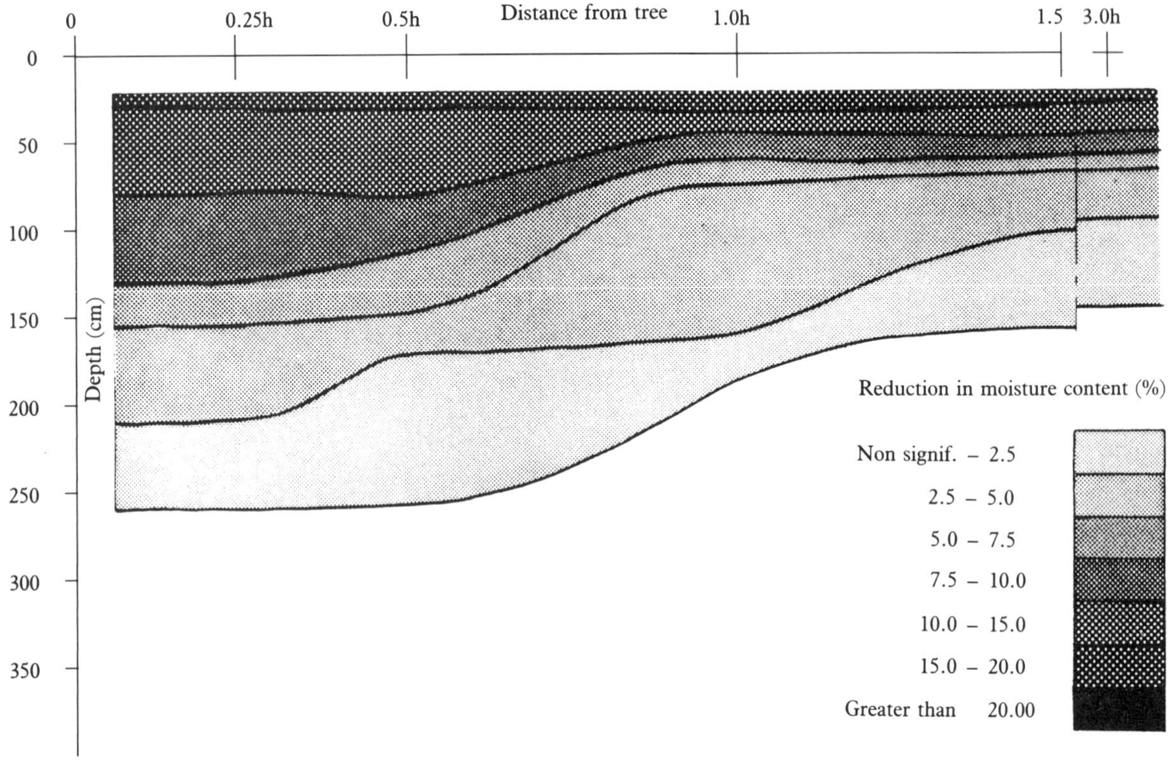

Figure 4. LIME AVERAGE TREES 1,2,7,8 & 29

BOULDER/OXFORD/LONDON CLAYS

Reduction in soil moisture content in Autumn 1983

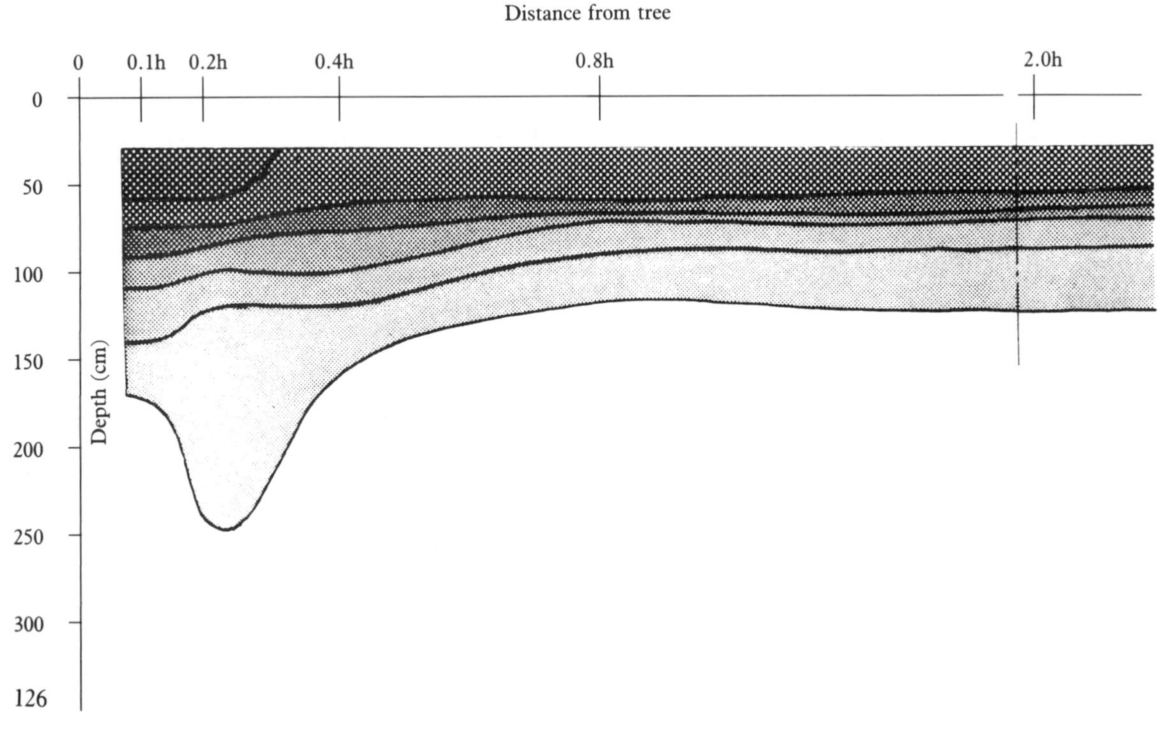

Figure 5. HORSE CHESTNUT AVERAGE TREES 5,6,11,12,16,17,25,26,27 & 28

BOULDER/OXFORD/GAULT/LONDON CLAYS

Reduction in soil moisture content in Autumn 1983

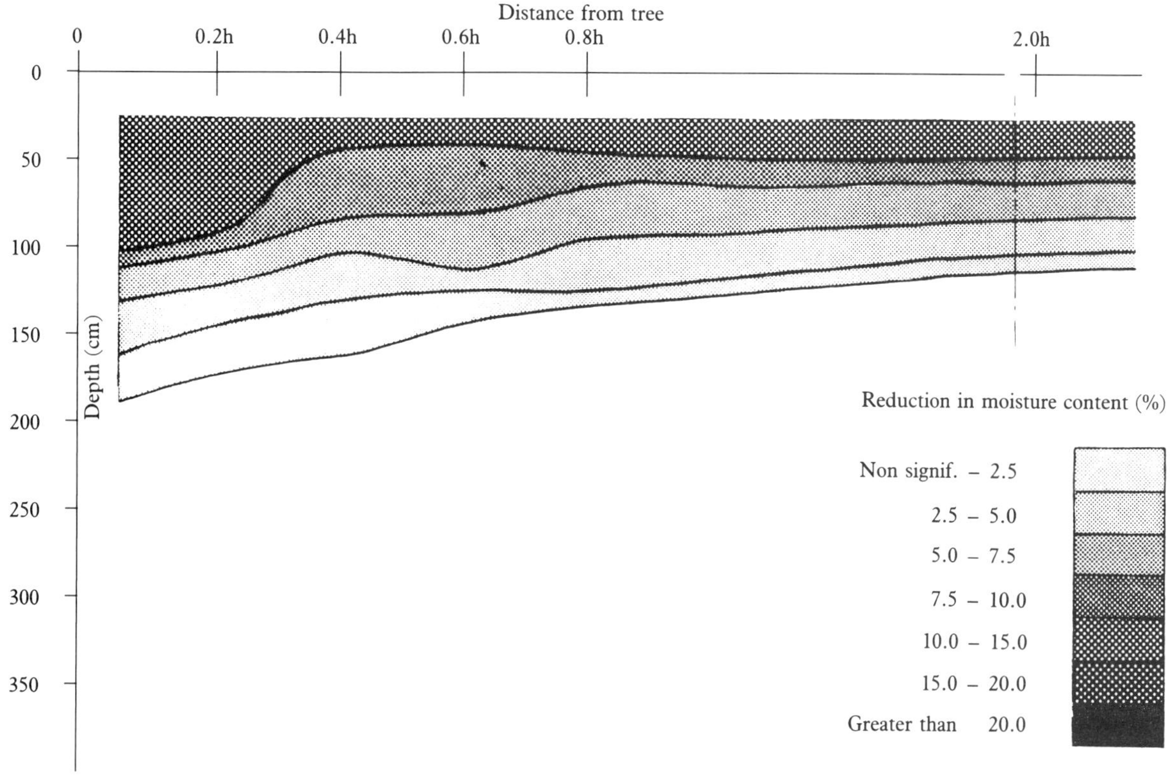

Figure 6. SILVER BIRCH AVERAGE TREES 31 & 34

BOULDER/LONDON CLAYS

Reduction in soil moisture content in Autumn 1983

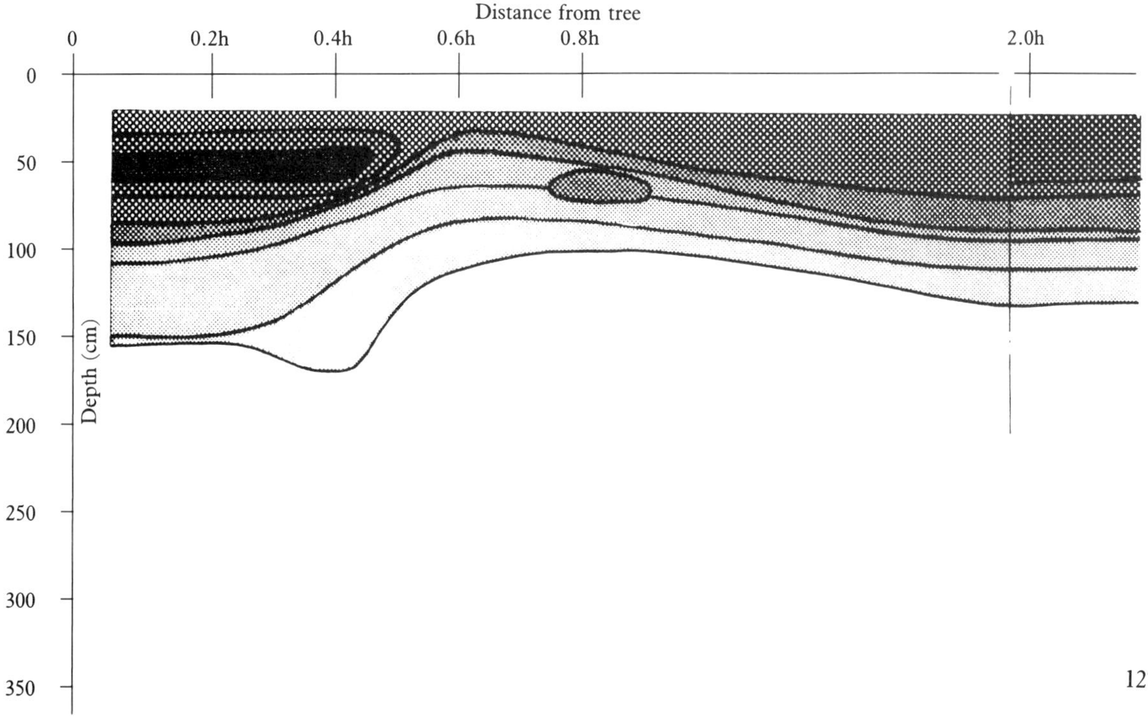

Figure 7. LEYLAND CYPRESS AVERAGE BOULDER/GAULT/LONDON CLAYS

Reduction in soil moisture content in Autumn 1983

TREES 20,21,32,33,35 & 36

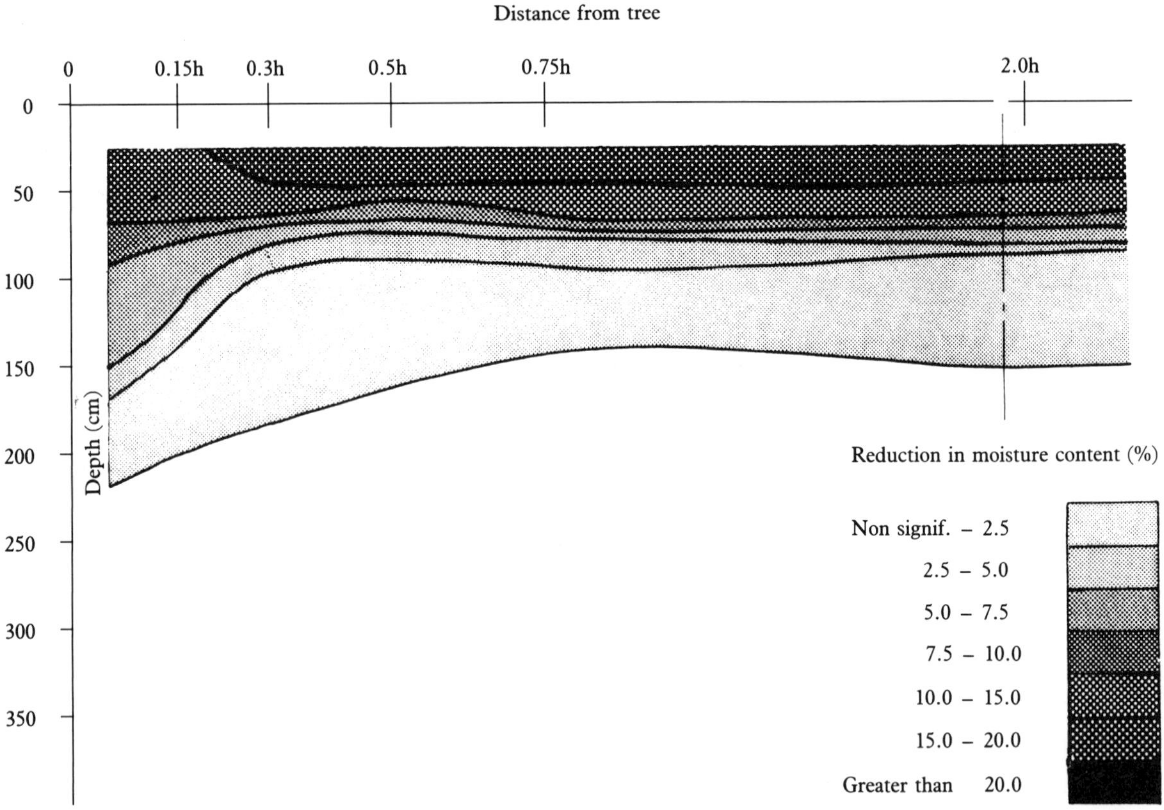

the required foundation depths dependent on the proximity of different species of tree on different clay soils. They subdivide trees into three categories of water demand, of which poplar, elm, willow and oak are identified as high demand. The required foundation depth with these species is marginally greater than previously. However, significant reductions in foundation depths are permissible with the majority of trees which fall in the moderate water-demand category, and even greater relaxation is permissible with those trees identified as low water-demand.

The new proposals also distinguish three categories of clay shrinkage potential, based on determination of the plasticity index of the clay. Reductions in foundation depth are permissible with clay of medium or low shrinkage potential.

It is hoped that these new guidelines will provide more detailed information on the anticipated effects of trees. This should help builders to minimise the risks of structural damage to buildings, without the necessity to incur additional expense by the use of unnecessarily deep foundations.

Acknowledgements

Tree Conservation Ltd. gratefully acknowledges the funding given by Milton Keynes Development Corporation, National House-Building Council and the Department of Environment, and the assistance of many local authorities and others in making sites available for this work.

Figure 8. POPLAR AVERAGE TREES 3,4,9,10,13,14,22,23 & 24

BOULDER/OXFORD/GAULT/LONDON CLAYS

Total reduction in soil moisture content (permanent + seasonal) in Autumn 1983

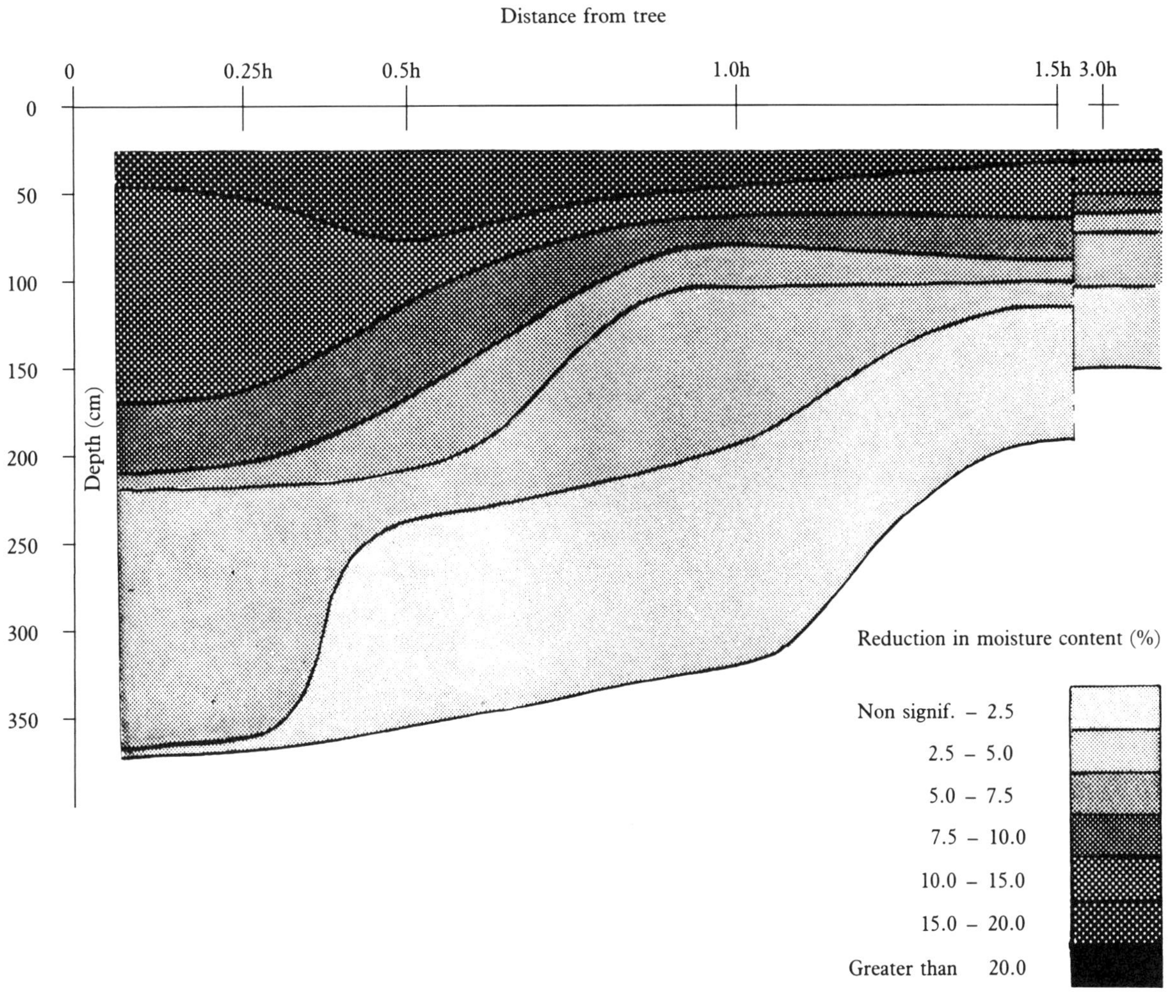

References

BIDDLE, P.G. (1983). Patterns of soil drying and moisture deficit in the vicinity of trees on clay soils. *Geotechnique* **33** (2), 107–126.

BSI (1972). *British Standards Institution Code of Practice for Foundations*. CP 2004:1972. British Standards Institution, London.

NHBC (1974). Root damage by trees – siting of dwellings and special precautions. *National House-Building Council Practice Note 3*. London.

WARD, W.H. (1953). Soil movement and weather. *Proceedings of the 3rd International Conference on Soil Mechanics, Zurich* **4**, 477–482.

Figure 9. POPLAR

LONDON CLAY

Soil moisture content profiles for access holes at 0.8h and 1.8h in autumn and spring

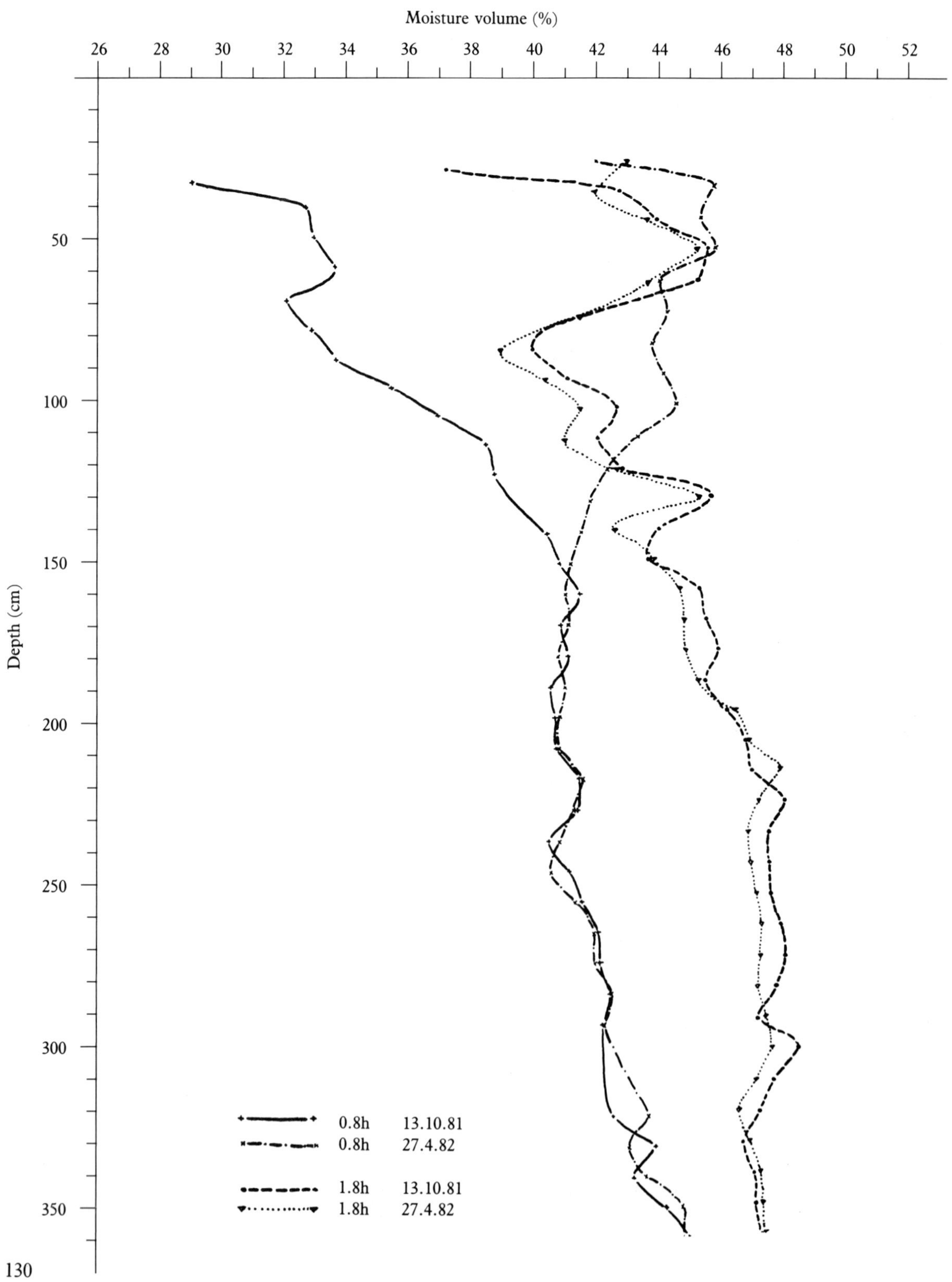

Figure 10. POPLAR

BOULDER CLAY

Soil moisture content profiles for access holes at 0.2h and 2.0h in autumn and spring

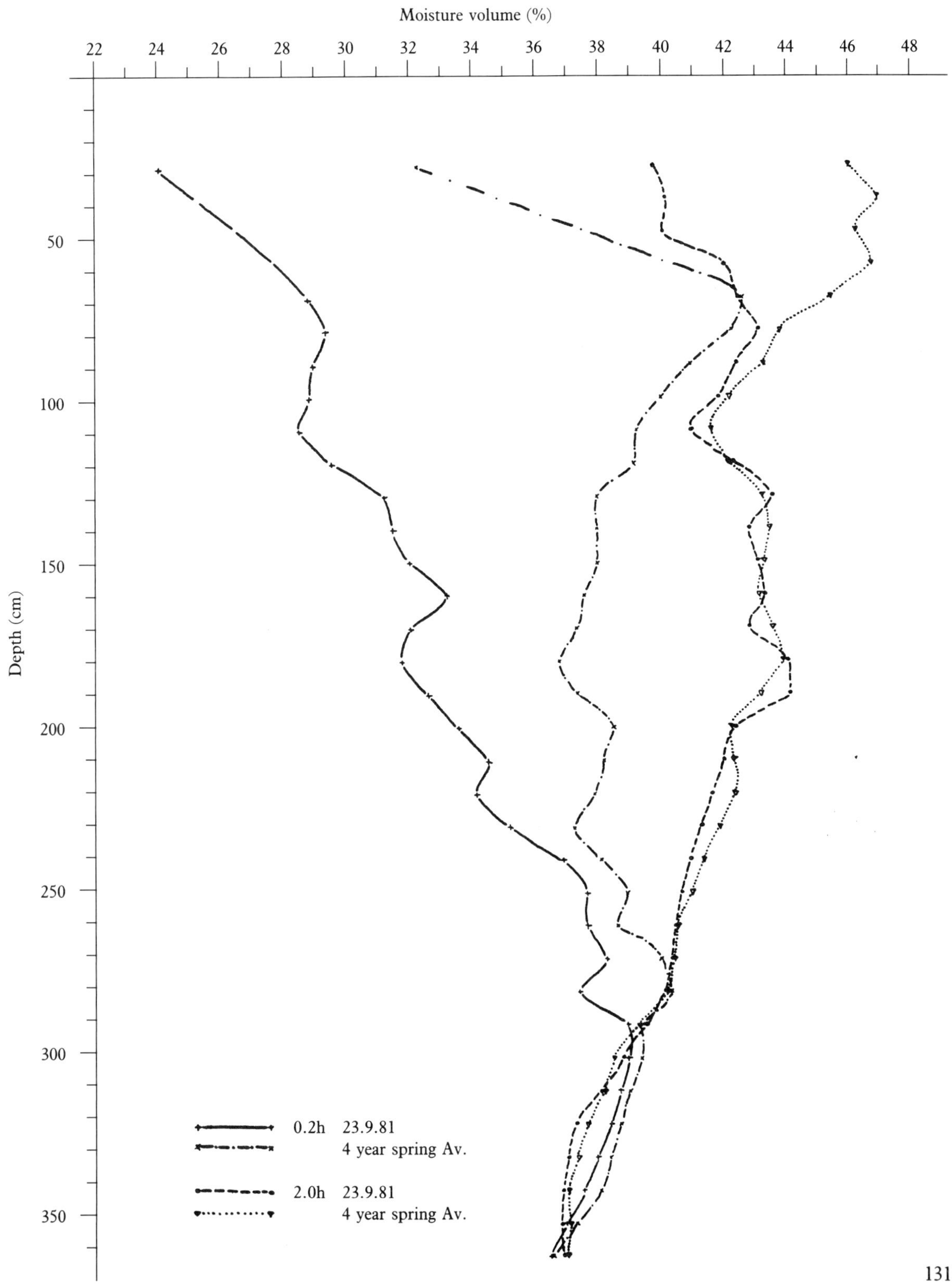

Discussion

R. TAYLOR — How long does the effect of pruning on soil moisture content last?

P.G. BIDDLE — This aspect will be followed up over the next few years and information on the necessary frequency of pruning obtained.

R. TAYLOR — Will regular pollarding and artificial shaping become more common?

P.G. BIDDLE — There must be a balance between the expensive pruning and the benefit in reducing water abstraction from clays.

(Property Services Agency)

Are there any data on how long 'heave' caused by re-wetting of clay continues?

P.G. BIDDLE — It can continue as long as 25 years, as shown by the example quoted by Samuel and Cheny of elm at Windsor on London clay. Clays vary in their permeability and some rewet quickly, e.g. boulder clay.

R. FINCH — (Roy Finch Tree Care Specialist)

Root barriers are being used more but there appears to be no research on their effectiveness. What is the effect of tree roots on soil moisture outside the rooting area?

P.G. BIDDLE — Root barriers are not effective if roots can find their way beneath or round them. The moisture content is not affected more than 30 cm beyond the extent of fine roots – clay has a low permeability.

L. ADAMS — Has watering been tested as a means of maintaining soil moisture content?

P.G. BIDDLE — No, because this is unlikely to be practical when most water is needed during drought.

L. ADAMS — Are you recommending pruning trees in parks?

P.G. BIDDLE — No, the work referred to was an experiment.

P. BARCLAY — Is the shrinking effect more exaggerated on shallow soils?

P.G. BIDDLE — No information on this but shallow soils may present better prospects for foundations. Certainly bands of sand in clay have dramatic effects because tree roots take up moisture from the sand preferentially.

T.P. MARSH — (Southampton City Council)

Could further advice be given on the question of insurance where trees under a Tree Preservation Order are involved.

P.G. BIDDLE — This is too big a question to be dealt with here.

T.J. CRAMB — (Durham County Council)

Is it feasible to have a code of practice defining the distances trees of different species may be planted from sewers?

C.G. BASHFORD — Department of Environment would have to seek advice on this.

P.G. BIDDLE — Tree roots could cause soil movement in some instances which may crack pipes. Modern pipe materials were flexible and this provided a solution. Generally, tree roots will only penetrate pipes if there are defects in the pipe.

Wildlife Research for Arboriculture

H.W. Pepper and J.J. Rowe
Forestry Commission

Summary

Relatively little research has been undertaken specifically on wildlife in arboriculture. This is largely because arboricultural wildlife problems are generally similar in all but scale to problems occurring in forestry, horticulture or agriculture and research recommendations aimed at problems in these fields have been adapted to provide arboricultural solutions. In addition, scattered locations and small scale encourage managers to treat problems as one-off situations. There has been no comprehensive survey of the frequency of occurrence which might suggest research priorities should be higher or changed in emphasis. Measurements of the field of wildlife enhancement for conservation have been neglected although small plantings are encouraged because of their potential wildlife value. Most advances have been made in the field of individual tree protection where most effort is now directed at reducing costs of materials and labour since satisfactory specifications for the commonest animal damage problems are available.

Introduction

Relatively little research on wildlife in relation to arboriculture has been undertaken directly. The usual response to a wildlife problem in arboriculture has been to use or adapt a solution originally researched to supply forestry, horticultural or agricultural needs. There may be two reasons for this. The first is that in the relatively small scale of arboriculture, problems often appear to managers to be one-off; each unique in some respect and therefore solved as they arise using whatever advice or expertise is available. The second reason is that relatively few wildlife problems have in fact been identified as peculiar only to arboriculture – and it has seemed adequate to assume that the same answer is appropriate for the common cause.

This paper reviews the main animal and bird problems so far identified, describes the damage and outlines the research done to produce solutions. In general, reduction of costs (labour or materials) is the main aim of research since the range of available solutions appears to provide reasonably satisfactory answers. This assumes of course, that the damaging animal is correctly identified, the appropriate choice of preventive measure is made and that protection techniques are correctly applied and maintained.

Wildlife Interactions with Arboriculture

The most common interactions are those of animals which damage ornamental trees and shrubs and, not unnaturally, much Forestry Commission (Wildlife) staff time is devoted to answering such problems. While there is a considerable amount of publicity for the benefits of arboricultural plantings to wildlife there has been extremely little actual measurement of before and after effects. The advantages of planting native trees and of particular weeding and other regimes are described qualitatively but not quantitatively, although there are some excellent guidelines such as the Greater London Council booklet on urban landscapes (GLC, 1985).

While no comprehensive survey has been carried out of the extent of wildlife damage problems in arboriculture, it is possible to use the range of enquiries which come into the Wildlife and Conservation Research Branch of the Forestry Commission to give some idea of the order of problems. Over a period of 12 months Grey squirrels, rabbits, Roe deer, other deer, domestic stock, voles, starlings and hares have produced up to 12 enquiries each relating to non-forest situations. Red deer, Red squirrels and Fallow deer have each produced a single enquiry. At other times, moles and rats have also figured in the list of enquiries.

Grey squirrels

Grey squirrels are frequently a problem in small groups of semi-mature trees. The most severe problem is bark stripping, which is solved most practically using trapping and poisoning of the group of squirrels doing the damage at the time that the damage is being done. The main constraints on this approach are the problems of carrying out such control by humane methods in a situation which is often subject to considerable public access (Rowe, 1985). Traps are liable to be disturbed by dogs and their owners. In some cases, building poison hoppers into the bottom of litter bins has ensured that the guilty squirrels are the most likely to suffer. Also the urban Grey squirrel, unlike its rural cousin, does not have to cope with fluctuating food supplies. The natural diet is supplemented by food from litter bins, bird tables and people providing food specifically for squirrels.

The ready availability of this additional food reduces the effectiveness of the control methods developed for forest protection which rely on enticing squirrels to artificially laid baits. Another difficulty associated with urban areas is that the squirrel is a pest to the person whose trees are being damaged and often a cuddly animal to those who are not affected. Where park trees or woodlands are being damaged and a greater proportion of the Grey squirrel population needs to be killed, a local public relations exercise may have to be devised to explain the problem and the reasons for the control measures. A local problem can be exacerbated by inaccurate and exaggerated press reports or letters suggesting that local Grey squirrel extermination rather than damage prevention is the objective.

A Grey squirrel problem of the future is likely to be bark-stripping of pole stage trees on motorways. In this case, poison hoppers may be the most appropriate control measure since, unlike the use of traps, daily visits are not essential.

Rabbits

The rabbit problem is a sporadic but widespread one and the damage can write-off groups of small trees. As with Grey squirrels the attitude has been to utilise control methods developed originally for agricultural damage prevention, but recently the development of tree shelters and individual tree guards has provided a more practical approach to damage prevention than attempts to control rabbit numbers, although one must not neglect the legal obligation to control rabbits (Pests Act, 1954). In addition work being done on rabbit fencing for agricultural and forest protection has some relevance to arboricultural situations although it has not been developed primarily for that purpose. Where warrens or burrows can be gassed then poisoning techniques using safe replacements for cyanide such as metal phosphides may equally be appropriate (Pepper, 1982).

Deer

Roe deer problems are most commonly found in parks and gardens especially where roses are being grown in association with small woodlands and shrubs. This is perhaps the only situation in which a chemical repellant may be more appropriate than control, which is often both difficult and potentially dangerous, or fencing which is difficult to keep secure if people and their dogs are allowed on to the site. Other deer such as Fallow deer and Muntjac can also provide an occasional nuisance but no research on these animals in arboricultural situations has been attempted. Exclusion, and the use of individual tree protection, appear, to be the most practical approaches (Pepper *et al.*, 1985). This also applies to domestic stock problems where ponies, goats and Jacob sheep have all figured in enquiries.

Small mammals

Field voles are one of the species which perhaps provide a more widespread general problem for arboriculture than do their occasional epidemics in forestry. They are widely distributed and common and where long grass is allowed to develop in plantings of small trees vole populations can quickly build up in numbers. This has become particularly common in relatively isolated situations such as motorways. In these habitats the use of warfarin poison and techniques which require fairly frequent visits to the site can be inhibiting – even for research. Once again the main emphasis in research has been on the development of individual tree guards. Weed control can inhibit voles from crossing bare ground to a tree. However, it is not known what effect this has on the incidence of damage other than it will have no benefit in conditions of lying snow.

Moles do not usually damage trees directly, though occasionally tunnels can cause problems to newly planted stock by removing soil from around the roots. Damage is generally associated with grass management where mole hills make grass cutting difficult, again adaption of traps and poisons available for agricultural control have been the only approaches utilised. Similarly with rats the wide range of poisons and traps available has been utilised where these animals cause problems by girdling or bark stripping, though this appears to be an extremely infrequent occurrence.

Birds

The only bird that has actually caused problems is the starling roosting in winter in considerable numbers in small blocks of trees and shrubs. Both the noise and the associated droppings are unpleasant especially in an urban situation. Such roosts can be dispersed with the use of bird-scaring cartridges and amplified distress calls (Currie *et al.*, 1977) but this technique can be as disturbing for local residents as for the birds!

In general, small bird management for conservation reasons, i.e. provision of nest boxes and food, is likely to be the main interaction of this group in arboriculture and, as already said, relatively little quantitative research has been undertaken. The use of nest boxes in more mature woodland is advocated as for the garden situation but the extent to which the species range can be modified by the use of suitable boxes and feeding has yet to be detailed.

The need to quantify the risk of damage

It is thus possible to outline generally the sorts of problems that exist. It is extremely difficult in the absence of survey information to say how much these sorts of problems cost in terms of time and capital expenditure or in terms of numbers of trees and shrubs potentially at risk. It is clear however that the number of constraints in arboricultural situations compared with forest, horticultural or agricultural situations can alter the approach to solving the problem. More account may have to be taken of local landscape considerations; the impact on other wildlife, domestic stock or pets; and on local human residents. There is another problem in that in many arboricultural situations the risk of vandalism not only to the trees but to any protective device is also a factor to be considered. In the absence of more definitive surveys on the extent of the various problems so far identified it is difficult to predict the degree of risk. Certainly for some animals the methods of prediction that are used in other situations are quite appropriate and it is possible that work currently being done for example on voles may make it practical to suggest when counter measures should be carried out. Research encompasses the use of an index method involving measuring the frequency of occurrence of pellets and grass clippings. Keeping an eye on the presence of predators, such as local concentrations of kestrels homing in on higher vole populations, should be used as an indicator of a potential problem.

Damage prevention research

In reviewing the main problems, we have also identified the main methods of approach to damage prevention. As in the other fields of wildlife research these have concerned the development of chemical repellants, fencing and individual tree protection. On the whole, improved control methods have less prominence in arboriculture than other fields. Where animals must be killed, it is perhaps more important to take the actual culprit rather than just to reduce population numbers generally.

Chemical repellants

The use of chemical repellants is potentially more appropriate for arboriculture than for other fields. However the only effective repellants found in the screening process so far are three contact repellants which must be sprayed on annually and which may be phytotoxic unless sprayed in winter on dormant trees. While the search continues and some 60 chemical repellants have been screened over the last 20 years, no systemic repellant whereby the chemical can be applied to roots and foliage and taken up by the tree has been found (Pepper, 1978): and although there are some smell repellants, they are only effective for a very short time – a few days at best.

Fencing

Fencing is not the cheapest method of protecting small areas of woodland. However there are some situations in which it is the only practical method and here the necessary length of life, mesh size, supports, possibility of re-use if it is a relatively short period and the extent to which it may or may not be vandal proof are all considerations (Pepper and Tee, 1986).

There are situations where fencing inappropriate for forestry, such as post and rail fencing or the use of larch lap, may be satisfactory in arboricultural situations.

Individual tree protection

It is in the field of individual tree protection that most work directly applicable to arboriculture has been undertaken (Pepper, *et al.*, 1985). Research on the design of guards, most appropriate size, most suitable type of material and their disadvantages and advantages compared with tree shelters has produced individual tree guards suitable for use against rabbits, Roe deer, Fallow and Red deer, hares and Field voles. Currently research is looking, as with fencing, at the development of cheaper and more effective tree guards involving less capital and less maintenance. We have already found that well designed tree guards can both be more vandal proof and produce less of a litter problem in some urban situations (Pepper and Williams, 1982).

Conclusion

Overall, it can be seen that the field of wildlife research for arboriculture is a rather patchy one, with most research being directed at methods of tree protection and relatively little at the ecological background in which the damaging interactions occur, or to the extent of the different problems in British arboriculture.

References

ANON. (1984). *The Pests Act 1954*. HMSO, London.

CURRIE, F.A., ELGY, D. and PETTY, S.J. (1977). *Starling roost dispersal from woodlands*. Forestry Commission Leaflet 69. HMSO, London.

GLC (1985). *Nature conservation guidelines for London*. Ecology Handbook No. 3. Greater London Council.

PEPPER, H.W. (1978). *Chemical repellants*. Forestry Commission Leaflet 73. HMSO, London.

PEPPER, H.W. (1982). *Rabbit control – phostoxin*. Arboriculture Research Note 48/32/WILD. Arboricultural Advisory and Information Service, Forestry Commission.

PEPPER, H.W., ROWE, J.J. and TEE, L.A. (1985). *Individual tree protection*. Arboricultural Leaflet 10. HMSO, London.

PEPPER, H.W. and TEE, L.A. (1986). *Forest fencing*. Forestry Commission Leaflet 87. HMSO, London.

PEPPER, H.W. and WILLIAMS, R.V. (1982). Plastic mesh for urban trees. *Arboricultural Journal* **6,** 211–215.

ROWE, J.J. (1985). Human control of squirrels in the urban scene. *Proceedings of UFAW Symposium*.

Discussion

P.A. HEMSLEY	(Askham Bryan College of Agriculture and Horticulture) What is the best time of year to remove Grey squirrel dreys?
H.W. PEPPER	There is no point in removing dreys as they will be quickly rebuilt. Removal of dreys will give some indication of size of population by revealing those dreys which are in use.
J. WETHERELL	(North Beds. Borough Council) Are weasels any threat to trees?
H.W. PEPPER	No. They are to be encouraged as they prey on many other small mammals which do damage trees.
F.R. CLAXON	(Epping Forest District Council) What steps need to be taken to manage a good ecological wildlife balance in an urban woodland?
H.W. PEPPER	Rarely any need to interfere with nature. Most small mammals adapt easily to a semi-urban environment. However, if, for example, foxes were to be encouraged this could be aided by encouraging field voles as a food species by the provision of large rough grass areas in the form of tracks or glades.
R.P. DENTON	(R.P. Denton & Co. Ltd., Colchester) What can be done to remove a large starling colony in a small conifer wood?
H.W. PEPPER	There is no need to exterminate. By using electronic equipment to play a recording of starling distress calls over a period of several evenings, the birds will become sufficiently disturbed to move on. Simultaneous noise from bird scaring cartridges is helpful.
R. ASKQUITH-ELLIS	(Bath City Council) What can be done about badgers causing a nuisance?
H.W. PEPPER	This is very difficult as so many people love badgers. Also they are protected under the Wildlife and Countryside Act. They can sometimes be persuaded to move from a sett by the use of smelly substances. For example, a creosote soaked rag pushed into the entrance of a sett at night may cause them to leave. They would probably move to another sett in the vicinity.
J.B.H. GARDINER	(Forestry Commission Research Station) Which are the best guards against voles?
H.W. PEPPER	Plastic tubing slit lengthwise and 'sprung' round the stem of a tree is effective and cheap.

Insects and Trees: Present Knowledge and Future Prospects

H.F Evans
Forestry Commission

Summary

Studies of insects on trees almost invariably reveal that the phytophagous insect faunas associated with them are very diverse particularly on the more widespread tree species. As an explanation of this association it is argued that trees, by having complex physical and chemical structures, are able to support a great variety of insects through the provision of many microhabitats. The theme of a 'normal' condition with insects in balance with trees is developed with examples. An insect outbreak is thought to be a result of factors upsetting this natural balance thus allowing insects to develop without normal restraints. Examples of current outbreaks in Britain are discussed and the necessity for greater research into insect communities and also for increased vigilance in preventing importation of exotic insects is emphasised.

Introduction

The main criterion used by most observers concerned with associations of insects and trees is whether the insects are having any significant impact on tree growth and hence can be regarded as pests. Where the impact is great, such as the extensive mortality of oak and other deciduous trees in N. America resulting from attack by the Gypsy moth (*Lymantria dispar*) (Doane and McManus, 1981), recognition of pest status is simple, and appropriate control measures can be taken. However, in the majority of cases, particularly involving deciduous trees, assessment of the role of plant feeding insects is difficult and the need for control measures uncertain.

One of the first, and perhaps most dramatic, conclusions from even a cursory study of insects on trees growing in Britain is the diversity of the fauna. Winter (1983) catalogued phytophagous insects and mites recorded on trees in Britain, although the list was not exhaustive. Table 1 summarises the total numbers of insect species in a number of the commoner insect orders and tree species including, for comparison, both deciduous and coniferous trees. In general insect species richness was greater on the former, especially since the conifers were composites of a number of tree species. There was also wide variation in phytophagous insect faunas amongst the deciduous trees cited, no doubt in part representing particular characteristics of each species.

As a group, phytophagous insects are typified by enormous species diversity. Indeed among living species (plants and animals) they are thought to represent 25 per cent of the recorded total and rank equally with the total of other invertebrate and all vertebrate species in this respect (Strong *et al.* 1984). Compounding wide species richness with the high potential reproductive rate of most insects provides us with an alarming prospect for wholesale destruction of our trees. Yet this happens only rarely. The infrequency of major pest outbreaks and the relatively minor damage observed in most woodlands provide ample evidence for the opposite view. There are exceptions of course, and losses of potential growth and tree death resulting from insect attack can occasionally be considerable.

Nevertheless, in comparison with the observed lack of impact of the majority of tree insects, the elevation of some to pest status must require a particular set of conditions that represents a considerable departure from the 'normal' state.

This Paper provides insight into those factors that define the 'normal' endemic state of most insect popu-

Table 1. Some phytophagous insect orders associated with British trees (Data from Winter, 1983)

		Number of species in insect order:					
		Coleoptera	Hemiptera	Hymenoptera	Lepidoptera	Others	Total
Deciduous	Alder	3	4	17	19	5	48
	Ash	4	7	2	6	3	22
	Beech	5	3	1	7	7	23
	Birch	6	10	30	43	4	93
	Elm	7	7	2	18	3	37
	Lime	2	3	0	5	9	19
	Oak	16	17	41	55	4	133
	Poplar	17	15	24	29	16	101
	Sycamore	0	2	1	4	4	11
	Willows	18	14	57	27	20	136
Conifers	Cypresses	1	1	0	2	0	4
	Douglas fir	1	1	1	5	0	8
	Silver fir	2	2	1	7	0	12
	Juniper	4	6	2	16	2	27
	Larches	4	7	8	23	1	43
	Spruces	10	18	12	46	4	90
	Pines	27	17	12	49	4	109

lations. The circumstances that may result in a pest outbreak are examined with British woodlands as the main area of concern.

Interactions of Phytophagous Insects and Trees – The 'Normal' Condition

Traditional concepts in the study of ecosystems have centred on distribution and abundance of plants and animals (Andewartha and Birch, 1954). These fundamental attributes apply equally well today but recent emphasis has shifted to studies of communities as a whole. Research into insect diversity and the questions of coevolution between insects and plants are basic issues providing lively debate among insect ecologists.

There is increasing evidence that insect diversity is associated with two main characteristics of their host plants, namely geographical range (including plant density) and complexity of plant structure.

Geographical range

The more widespread the geographical distribution of a plant the greater the variety of phytophagous insects associated with it. This concept is known as a species-area relationship and was first observed for insects on trees in Hawaii by Southwood (1960) and discussed as an ecological concept by Janzen (1968). In Britain there is a linear relationship between the number of species of phytophagous insects per tree and the number of genera of trees in 10 km squares (Strong *et al.*, 1984).

More specific examples include the number of leaf hoppers (Claridge and Wilson, 1981) and leaf miners (Claridge and Wilson, 1982) on British trees such that in the latter case a species-area relationship was obtained for 239 species of insect on 37 species of tree and large shrub. However, the amount of variation directly explained by species-area is often low and many other factors including how closely related the host plants are may be significant. Thus Claridge and Wilson (1982) were able to explain only 19 per cent of the scatter of their data directly by a species-area relationship while a further 23 per cent of variation was attributed to the relationships of the host plants studied. When species-area effects were considered for a single plant genus Opler (1974) showed that area accounted for 90 per cent of the correlation between leaf miners and oaks in California. Strong *et al.* (1984) summarised the possible mechanisms that could result in a link between host area and insect diversity. These included:

(i) *habitat-heterogeneity*. Larger areas inevitably incorporate a greater variety of habitats and hence may satisfy the needs of more insect species, although climate may have an overriding influence.

(ii) *encounter-frequency*. A greater density of plants is likely to be encountered more frequently by a given population of insects.

(iii) *island biogeography*. An ecological island exhibits a balance between immigration and emigration of insect species. Larger islands support a more diverse population of insects at equilibrium.

These theories are not mutually exclusive and there is evidence for each, but much observed variation remains unexplained and requires further study.

Complexity of plant structure

The second major determinant of insect diversity takes us to the host plant itself. In essence, the more complex the growth form (architecture), both in size and variety of structure, of the plant the greater the variety of insects associated with it (Lawton and Schroder, 1977).

Examples to support this hypothesis are relatively common and all point to the existence of a more varied insect fauna on trees than on other plants. Lawton and Schroder (1977) give an order of insect diversity related to plant types namely trees>woody shrubs>perennial dicotyledenous herbs>weeds and other annuals>monocotyledons (excluding grasses). Data by Strong and Levin (1979) suggest differences in interpretation for the smaller plant types but confirm the predominance of trees.

On *a priori* grounds it is not difficult to see why trees should have such a rich fauna. They tend to be larger, have a wide range of microhabitats, are relatively long lived thus providing continuity and, when mature, are the dominant feature of the ecosystem. Thus potential insect colonists are provided with a wide range of sites catering for many oviposition and feeding habits within a single plant. Inevitably some insects will find suitable conditions and establish a breeding population. Additional variation within host plant types is provided by seasonal changes resulting from expansion of buds, development of leaves and, for deciduous trees, loss of leaves in the autumn. These constitute changes in morphology but also variation in nutritional quality and secondary chemical constitution of the plant structures. This may then be reflected in a changing insect fauna with time so that early season populations may differ markedly compared with those later in the year.

Excellent examples of some 'plant architecture' effects come from various studies of phytophagous insects on beech trees (*Fagus* sp.). Beech has an impoverished insect fauna relative to its morphology and this may be partially attributable to seasonal changes in structure and plant chemistry. Nielsen and Ejlersen (1977) in Denmark showed that insects tended to specialise in certain areas of mature beech stands with leaf mining moths and adult and larval weevils (Beech leaf miner *Rhynchaenus fagi* and another weevil *Phyllobius argentatus* concentrating on the understorey foliage while adult Beech leaf miners were often found in the overstorey or on the fringes of the stand). A more recent study in Wytham Wood, Oxfordshire provided broad confirmation that most feeding took place in the lower canopy (mainly adult *P. argentatus*, larval Beech leaf miner and leaf mining moths) while adult Beech leaf miners fed mainly in the upper canopy (Phillipson and Thompson, 1983). it was also demonstrated that most of these attacks took place within the period of active leaf growth hence linking attacks to the presence of young leaves, implying that older leaves were less suitable for phytophagous insects.

Seasonal trends of this type were demonstrated for the Lepidopteran faunas of a number of deciduous trees in Finland (Niemela and Haukioja, 1982). Trees completing shoot growth during the early part of the season tended to have greatest species diversity in the spring. Oak (*Quercus robur*) was taken as the type example and it is true also in Britain that on oak the greatest variety of leaf feeding Lepidoptera is found on the new leaves (Feeny, 1970). Other tree species exhibited continuous growth of leaves through to the autumn (defined by the authors as *Populus* type) and the Finnish study indicated that late season diversity was greatest on these (birch, alder, aspen), reflecting a mix of young and old leaves present throughout.

Why are 'Normal' Insect Faunas not Universally Destructive?

A number of features of trees as ecosystems have been described that generally result in the expectation of a large variety of phytophagous insects all extracting nourishment from the tree. Diversity plus density should therefore lead to significant tree damage in many cases. There are several potential reasons why this is not so (Table 2).

Evidence for interspecific competition between plant feeding insects is scant since it is difficult to distinguish from the effects of the other factors in Table 2. Strong *et al.* (1984) ventured the opinion that as a mechanism *per se* it is rare and cited many examples to support this view including, in a forestry context, lack of density compensation when the variety of species of oak leaf miners was reduced (Faeth and Simberloff, 1981). However, despite

Table 2. Some factors preventing destructive build-up of resident populations of phytophagous insects on trees. (It is assumed that faunal diversity is near its optimal level and the factors act only on numbers of insects within species).

Potential factor	*Evidence?*
Competition for resources so that populations decline before food becomes limiting.	(a) Between species. Little evidence since different species tend to occupy different niches and these are generally not limiting. (b) Within species. Outbreaks of insects result in competition but only if other factors (below) have insufficient effect.
Natural enemies act in total to reduce populations in a density-dependent manner.	A large body of literature emphasises the role of natural enemies, although this is not universally accepted.
Direct feedback from the host plant by altered chemical or nutritional status-induced defences.	Experimental and natural defoliations have reduced performances in subsequent insect populations.

its apparent rarity, interspecific competition has been demonstrated but tends to be one sided so that one species has an adverse effect on another but not vice versa. An example includes the Hemlock scale (*Fiorinia externa*) competitively displacing another scale *Tsugaspidiotus tsugae* resulting in higher mortality of immatures and lower adult fecundity (McClure, 1981).

Intraspecific competition, on the other hand, results when insect abundance oustrips other regulatory influences and an outbreak occurs. Here the limitation to population growth is availability of food; complete defoliation is commonly observed before direct reduction of insect populations is apparent. Gypsy moth may completely defoliate oak and birch trees in N. America but suffers an immediate decline in numbers when it has consumed all available food (Doane and McManus, 1981). In Britain, Brown tail moth (*Euproctis chrysorrhoea*) may build up to damaging population levels on a variety of trees and shrubs, outstripping its food supply (Sterling, 1985). Dempster (1983), however, attributes the majority of major population changes in Lepidoptera to fluctuations in availability of resources and includes the forest pests Gypsy moth (oak and other deciduous trees), Green oak tortrix (*Tortrix viridana*) (oak), Winter moth (*Operophtera brumata*) (oak and other trees) and Fall webworm (*Hyphantria cunea*) (various deciduous trees) among his examples.

Dempster's view is challenged by Hassell (1985) who argues the case for regulation of populations mainly through the action of natural enemies (Table 2). Indeed the role of natural enemies (predators, parasitoids and diseases) as major mortality factors preventing catastrophic build-up of pests is the perceived wisdom of the majority of insect ecologists (Varley and Gradwell, 1970; Southwood and Comins, 1976; Strong *et al.*, 1984). The study of Winter moth populations over a 20 year period by Varley and Gradwell (1970) is probably the best documented appraisal of mortalities acting on a population of forest insects. They showed that a 'key factor' in population regulation was density-dependent reduction of pupae over winter by ground-dwelling beetles and small mammals. Other evidence for the contributory roles of natural enemies comes from successful biological control by introductions of predators, parasitoids or diseases. Notable among these was control of Winter moth in Nova Scotia by the parasitoid *Cyzenis albicans* introduced from Europe (Embree, 1971), despite its apparent lack of effect in Varley and Gradwell's study.

Diseases, notably viruses and bacteria, also have strong influences on populations of insects and there are many cases where baculoviruses were instrumental in reducing outbreaks in an apparently density-dependent manner (Entwistle and Evans, 1985). The regulation of Gypsy moth in North America is a case in point and has led to the development of a viral insecticide called Gypchek (Podgwaite, 1981). We are also using a virus in Britain and this is now commercially available as Virox for use against European pine sawfly (*Neodiprion sertifer*). There are also prospects for the use of a baculovirus against Brown tail moth.

By feeding on the host plant a population of insects is exerting some influence on that plant. This may be of little consequence but there is increasing evidence that the plants are able to respond to this pressure by induced metabolic changes that in turn influence subsequent populations of the insects. Thus both changes in nutrient content and production of secondary plant compounds may occur making the plant less suitable for insect growth or less palatable. Returning to Gypsy moth on birch as an example, it was shown that pupal weights decreased in proportion to the number of years of defoliation (Leonard, 1970). This in turn led to smaller females and fewer eggs, a classic example of negative feedback resulting from changes in the host plant.

Factors Leading to 'Outbreak' Conditions

Central to the theme of an endemic population is the dynamic balance between the phytophagous fauna, its host plant and natural enemies. Any upset of this balance may result in increases in population levels. If these disturbing factors are density-dependent the insect population will tend to regain stability and no outbreak will take place. On the other hand, if density-dependent regulation is removed the population will increase without restraint and will pass from endemic to epidemic status.

One way in which this can happen is removal of natural enemies. Good examples and salutory lessons in potential misuse of insecticides are outbreaks of Gypsy moth in both Europe (White *et al.*, 1981) and the United States (Doane, 1968) initiated or prolonged by application of DDT and tetrachlorvinphos (Gardona) respectively. A further example, perhaps more relevant to Britain, was the application of Dimilin to control the Green oak tortrix and *Zeiraphera insertana* on oak in Germany (Horstmann, 1982). Although virtually 100 per cent larval mortality was achieved following insecticide application, Green oak tortrix had reached 70 per cent of its original level within one year and a year later was at a higher level than in untreated plots. This was attributed to reduced disease and parasitism in the treated plot, although in this case the insecticide did not kill the monophagous parasitoids directly but reduced their populations through removal of available hosts.

Adverse weather may also remove natural enemies, either through direct mortality of predators and parasitoids or by creating asynchrony between them and their hosts. In either case the host population may rise to critical levels and outstrip any effects of natural enemies.

There may also be direct influences of the host plant on the reproductive success of phytophagous insects. For instance, if nutrient status changes rapidly to the advantage of plant feeders there may be a parallel increase in reproduction that is greater than the natural enemy complex can control. This may lead to an outbreak that will only decline when the natural enemies 'catch-up' with their hosts by delayed density-dependent compensation. Major components of this interaction are the proportions of nitrogen and water in the host plant. Scriber and Slansky (1981) reviewed this topic and emphasised that nitrogen and water contents act together to define the relative growth rates (RGR's) of Lepidopteran and Hymenopteran larvae. Interestingly RGR's on mature tree leaves were generally less than half those on herbaceous plants.

Environmental stress may often lead to increased nitrogen availability in trees, in turn leading to improved feeding and reproduction by phytophagous insects (Mattson, 1980). This is particularly true for aphids while the reverse may occasionally be the case for chewing insects. Thus in this way food consumption rises but only after the insect population has responded through increased reproduction thus inducing a delay in effect.

There may also be an increase in damage but not in insect numbers if either nitrogen or water content drops. This results from a compensatory increase in feeding rate, or by prolonged feeding, in order to maintain growth. In the same way it is only the larger leaf feeding insects that can consume and assimilate sufficient nitrogen-poor foliage for normal development. Thus many North American deciduous tree pests are macrolepidoptera (mature larvae > 40mm) that exploit late season foliage too poor nutritionally for microlepidoptera that tend to feed on young nitrogen-rich foliage in the spring (Mattson, 1980).

Prospects for Phytophagous Insect Outbreaks in Britain

In Britain there are currently outbreaks of two major insect pests on conifers namely Great spruce bark beetle (*Dendroctonus micans*) on spruce and Pine beauty moth (*Panolis flammea*) on Lodgepole pine (*Pinus contorta*). A review of insect records in Britain reveals that neither of these insects were pests before the mid to late 1970s. They provide two examples of insects that have escaped their natural constraints.

Great spruce bark beetle is an exotic insect that was accidentally imported to this country in the early 1970s. It is therefore here with an abundant source of food but without its major natural enemy, a predatory beetle called *Rhizophagus grandis*. An example of an introduced insect more appropriate to arboriculture is the Horse chestnut scale (*Pulvinaria regalis*). This scale insect was first discovered in 1964 and has spread over much of southern England and the Midlands. It colonises urban trees including lime (*Tilia* spp.), sycamore (*Acer pseudoplatanus*) and Horse chestnut (*Aesculus hippocastanum*) and is easily recognised by the white wax secretions of the adults in May. It was originally thought to have no natural enemies but recent research indicates that a predator and a parasitic wasp have some effect in reducing populations (Speight and Nicol, 1984). It seems, therefore, that Horse chestnut scale has succeeded in establishing itself over a wide area and illustrates the potential of exotic insects to multiply in the virtual absence of natural enemies. Pine beauty moth, on the other hand, is a native insect normally resident on, and in balance with, Scots pine (*Pinus sylvestris*) but which has switched to the exotic Lodgepole pine. Like many insects that exploit non-native trees it has benefited from the greater suitability of the new host, especially those growing in less than ideal conditions, to increase reproductive rate and oustrip its small complex of natural enemies.

Few major problems are encountered on our native deciduous trees. Occasional defoliation of oak by tortricid and geometrid moths takes place but these have little long term impact. Similarly beech (*Fagus sylvatica*) is attacked by a number of leaf feeders and by the well studied Beech coccus (*Cryptococcus fagisuga*), one of the predisposing factors of Beech bark disease (Lonsdale and Wainhouse, 1987), but these tend to be local in effect and only rarely result in tree mortality. However, we must not be complacent since the theme of natural balance implies that the appearance of an exotic pest may well result in substantial defoliation before balance is achieved. Thus, both Gypsy moth, and Fall webworm, major defoliators of deciduous trees in North America, would pose significant threats should they become established here. Similarly, a complex of bark beetles from Europe and North America could have devastating effects on conifer production in this country.

We must also look closely at the planting of exotic trees since they too will be devoid of their associated insect communities. *Nothofagus* sp., for instance, are receiving considerable attention as potentially valuable hardwoods yet they are already suffering a remarkable level of insect attack despite their relatively short period in this country (Welch, 1981). This has been attributed to their taxonomic similarity to our native beech and oak, especially the latter. A study by Connor *et al.* (1980) who looked at accumulation of phytophagous insects on introduced trees confirmed that a greater variety of insects was found on trees closely related to native trees.

Insects may also exploit host trees across large taxonomic boundaries as evidenced by the appearance of Winter moth, which normally feeds on oak, on Sitka spruce (*Picea sitchensis*) in Scotland. It is therefore conceivable that insects may switch host plants in a most unexpected way despite ecological studies that pinpoint taxonomic relatedness as an important factor in limiting insect accumulation on new hosts.

Conclusion

Our ability to cope with insect outbreaks on British trees will in part depend on our knowledge of the communities of insects on the trees. Here we require a much greater input to the study of those communities so that decisions on pest management can be made in the light of that increased knowledge. Secondly it is essential that we maintain out barriers to accidental importation of exotic insects since, by virtue of their imbalance with the native ecosystem, it is these species that pose the greatest threats to our trees, deciduous and coniferous alike.

References

ANDREWARTHA, H.G. and BIRCH, L.C. (1954). *The distribution and abundance of animals.* University of Chicago Press, Chicago.

CLARIDGE, M.F. and WILSON, M.R. (1981). Host plant associations, diversity and species-area relationships of mesophyll-feeding leafhoppers of trees and shrubs in Britain. *Ecological Entomology* **6**, 217–238.

CLARIDGE, M.F. and WILSON, M.R. (1982). Insect herbivore guilds and species-area relationships: leaf-miners on British trees. *Ecological Entomology* **7**, 19–30.

CONNOR, E.F., FAETH, S.H., SIMBERLOFF, D. and OPLER, P.A. (1980). Taxonomic isolation and the accumulation of herbivorous insects: a comparison of introduced and native trees. *Ecological Entomology* **5**, 205–211.

DEMPSTER, J.P. (1983). The natural control of populations of butterflies and moths. *Biological Review* **58**, 461–481.

DOANE, C.C. (1968). Changes in egg mass density, size and amount of parasitism after chemical treatment of a heavy population of the Gypsy moth. *Journal of Economic Entomology* **61**, 1288–1291.

DOANE, C.C. and McMANUS, M.L. (Eds.) (1981). *The Gyspy moth: research toward integrated pest management.* U.S. Forest Service, Technical Bulletin 1584.

EMBREE, D.G. (1971). The biological control of the winter moth in eastern Canada by introduced parasites. In, *Biological control*, ed. Huffaker, C.B., 217–226. Plenum Press, New York.

ENTWISTLE, P.F. and EVANS, H.F. (1985). Viral control. In, *Comprehensive insect physiology, biochemistry and pharmacology*, Vol. 12 Insect control, eds. Kerkut, G.A. and Gilbert, L.I. Pergamon Press, Oxford.

FAETH, S.H. and SIMBERLOFF, D. (1981). Experimental isolation of oak host plants: effects on mortality, survivorship, and abundances of leaf-mining insects. *Ecology* **62**, 625–635.

FEENY, P. (1970). Seasonal changes in oak leaf tannins and nutrients as a cause of spring feeding by Winter moth caterpillars. *Ecology* **51**, 565–581.

HASSELL, M.P. (1985). Insect natural enemies as regulating factors. *Journal of Animal Ecology* **54**, 323–334.

HORSTMANN, K. (1982). Auseirkung einer Bekämpfungsaktion mit Dimilin auf eine Eichenwickler – Population (Lepidoptera, Tortricidae) in Unterfranken. *Zeitschrigt für angewandte Entomologie* **94**, 490-497.

JANZEN, D.H. (1968). Host plants as islands in evolutionary and contemporary time. *American Naturalist* **102**, 592–595.

LAWTON, J,H, and SCHRÖDER, D. (1977). Effects of plant type, size of geographical range and taxonomic isolation on number of insect species associated with British plants. *Nature* (London) **265**, 137–140.

LEONARD, D.E. (1970). Intrinsic factors causing qualitative changes in populations of *Porthetria dispar* (Lepidoptera : Lymantriidae). *Canadian Entomologist* **102**, 239–249.

LONSDALE, D. and WAINHOUSE, D. (1987). *Beech bark disease*. Forestry Commission Bulletin 69. HMSO, London.

MATTSON, W.J. (1980). Herbivory in relation to plant nitrogen content. *Annual Review of Ecology and Systematics* **11**, 119–161.

McCLURE, M.S. (1981). Effects of voltinism, interspecific competition and parasitism on the population dynamics of the hemlock scales, *Fiorinia externa* and *Tsugaspidiotus tsugae* (Homoptera: Diaspididae). *Ecological Entomology* **6**, 47–54.

NIELSON, B.O. and EJLERSEN, A. (1977). The distribution pattern of herbivory in a beech canopy. *Ecological Entomology* **2**, 293–299.

NIEMELA, P. and HAUKIOJA, E. (1982). Seasonal patterns in species richness of herbivores : macrolepidopteran larvae on Finnish deciduous trees. *Ecological Entomology* **7**, 169–175.

OPLER, P.A. (1974). Oaks as evolutionary islands for leaf-mining insects. *American Scientist* **62**, 67-73.

PHILLIPSON, J. and THOMPSON, D.J. (1983). Phenology and intensity of phyllophage attack on *Fagus sylvatica* in Wytham Woods, Oxford. *Ecological Entomology* **8**, 315–330.

PODGWAITE, J.D. (1981). NPV production and quality control. In, *The Gypsy moth: research toward integrated pest management*, eds. Doane, C.C. and McManus, M.L., 461–464. U.S. Forest Service, Technical Bulletin 1584.

SCRIBER, J.M. and SLANSKY, F. (1981). The nutritional ecology of immature insects. *Annual Review of Entomology* **25**, 183–211.

SOUTHWOOD, T.R.E. (1960). The abundance of the Hawaiian trees and the number of their associated insect species. *Proceedings of the Hawaiian Entomological Society* **17**, 299–303.

SOUTHWOOD, T.R.E. and COMINS, H.N. (1976). A synoptic population model. *Journal of Animal Ecology* **45**, 949–965.

SPEIGHT, M. and NICHOL, M. (1984). Horse chestnut scale – a new urban menace? *New Scientist*, 5 April 1984, 40–42.

STERLING, P.H. (1985). *The Brown-tail moth*. Arboriculture Research Note 57/84/EXT. Arboricultural Advisory and Information Service, Forestry Commission.

STRONG, D.R., LAWTON, J.H. and SOUTHWOOD, Sir R. (1984). *Insects on plants. Community patterns and mechanisms*. Blackwell, Oxford.

STRONG, D.R. and LEVIN, D.A. (1979). Species richness of plant parasites and growth form of their hosts. *American Naturalist* **114**, 1–22.

WELCH, R.C. (1981). Insects on exotic broadleaved trees of the Fagaceae, namely *Quercus borealis* and species of *Nothofagus*. In, *Forest and woodland ecology*, Institute of Terrestrial Ecology Symposium No. 8, eds. Last, F.T. and Gardiner, A.S., 110–115.

WHITE, W.B., McLANE, W.H. and SCHNEEBERGER, N.F. (1981). Pesticides. In, *The Gypsy moth: research toward integrated pest management*, eds. Doane, C.C. McManus, M.L., 423-442. U.S. Forest Service, Technical Bulletin 1584.

WINTER, T.G. (1983). *A catalogue of phytophagous insects and mites on trees in Great Britain*. Forestry Commission Booklet 53. Forestry Commission, Edinburgh.

VARLEY, G.C. and GRADWELL, G.R. (1970). Recent advances in insect population dynamics. *Annual Review of Entomology* **15**, 1–24.

Discussion

N. MUIR (Chichester)

Is Beech bark disease increased by drought conditions?

D. LONSDALE (Forestry Commission Research Division)

Stress induced by drought favours the development of *Nectria* which usually follows Beech coccus. Therefore there is an apparent similarity between the effects of BBD and drought. However, there is no evidence of a direct relationship between drought and BBD.

E. HAMILTON (The Woodland Trust)

Are British oaks less susceptible to oak wilt?

H.F. EVANS Oak wilt is mainly associated with Red oaks. However, the pathogen may adapt to British trees and *Scolytus intricatus* appears to be a suitable vector for the spread of the disease. Research into the ecology of the disease is being undertaken for the Forestry Commission by the Institute of Terrestrial Ecology.

E. HAMILTON (The Woodland Trust)

Is the Channel Tunnel likely to pose a threat to tree health?

H.F. EVANS Increased flow of traffic from the continent is likely to increase the risk of introduced pests and pathogens. The main danger is from accidental or deliberate introductions by private individuals.

A.G. GORDON (EFG Nurseries)

In view of the reported importation of 65 million transplants from continental nurseries last year, what confidence is there in current control measures?

H.F. EVANS The quantity and quality of inspection and control of imported plant material are being increased. More time is being spent talking direct to the importers to emphasise the need for care. Pheromone traps at ports will assist in monitoring.

Treatment of Fresh Wound Parasites and of Cankers

D.R. Clifford and P. Gendle
Long Ashton Research Station, University of Bristol, Long Ashton, Bristol

Summary

Wounds on fruit and other tree species are susceptible to fresh wound parasites, e.g. fungi causing cankering and silver leaf disease, for up to 28 days and need to be treated against such infections. By contrast, they are vulnerable to mature wound pathogens, e.g. wood rotting fungi, for at least 2 years and no completely effective chemical treatment is available.

Easily prepared gel formulations containing Bayleton BM fungicide maintain an eradicant dose of fungicide for up to 6 months and can stimulate callus tissue formation. In practice, they have given very good protection against cankering and silver leaf disease but do not persist sufficiently long to give adequate protection against mature wound pathogens. They also eradicate *Nectria galligena* from within established cankers without the need for prior removal of decayed tissue.

Introduction

Trees are well equipped to combat invasion of their wood by fungi (Shain, 1979). The mechanisms by which they do so have been clearly described by Shigo (1984) and Shortle (1979) who have shown that the tree 'walls off' invading pathogens by a series of processes collectively termed 'compartmentalisation'.

Tree wounds are susceptible to attack by two types of fungi, the fresh and the mature wound pathogens. Organisms of the second type comprise the wood rotting fungi (e.g. *Coriolus* (syn. *Trametes*) *versicolor* and *Coniophora puteana*) and usually become active only after the wood has become predisposed to their development as a result of its colonisation by a series of other micro-organisms, not all of which are pathogenic. By contrast, the fresh wound pathogens attack and colonise the wood directly but can only develop in wounds up to 28 days old; examples are *Nectria galligena* and *Chondrostereum purpureum* which cause cankering and silver leaf disease respectively. They destroy the tree's vascular system and cause dieback in the crown.

Walling off infection involves creation of new tissue for which energy has to be diverted from other important physiological processes. Thus wounds are likely to be most vulnerable at times of great energy demand such as bud-burst, flowering and preparation for dormancy. Nevertheless, there is some unpublished evidence (Clifford and Gendle) that trees are better able to contain fungi introduced during the growing rather than the dormant season. Because attack by mature wood pathogens is a relatively slow process, the tree has time to divert energy to contain them but fresh wound pathogens colonise the wood quickly and directly and a suitable wound dressing is therefore very necessary.

Conventional wound treatments establish a seal over the wound surface, to guard against the entry of pathogens, and often contain a fungicide whose function is to eradicate existing and prevent further infection. Such seals, and also some fungicide components, promote callus tissue formation, although the importance of this in wound occlusion seems to have been over-rated (Lonsdale, 1984). However, few sealant preparations successfully eradicate existing infections and/or impart long term protection against future infection, although the mercuric oxide based Santar can be effective for up to 14 months, (Mercer *et al.*, 1983). It is also important to recognise that some preparations are designed for a specific purpose and not for use as a general wound paint; for example, those containing thiophanate-methyl can give excellent protection against canker but may even encourage development of wood-rotting fungi! It is equally important to realise that the failure of many wound treatments can be attributed to the often overlooked fact that spores of some fungi, e.g. *C. purpureum*, can be sucked into the wood to some considerable depth, even seconds after the wound is created (Brooks and Moore, 1923; Lonsdale, 1984). Yet with few sealant preparations does fungicide penetrate into the wood. Therefore, in place of sealants, we have concentrated on incorporating systemic fungicides into gel formulations with the objective of encouraging active ingredients to penetrate into and persist within the wood beneath the wound for sufficiently long to combat fresh and also, ideally, mature wound pathogens. This paper reports work with such preparations and with some commercially available sealants for comparison. Because our prime interest has been in protection of pruning wounds in apple, pear and plum trees, our targets have been to prevent cankering by *N. galligena* and damage to wood by *C. purpureum* and, to a lesser extent, by other wood-rotting fungi.

Composition Properties and Performance of Gel Formulations

Gel formulations are produced by creaming the required amount of commercial fungicide formulation(s) in water and adding the gel component (2.5% w/v) to the vigorously stirred mixture; this produces a material with the consistency of a gel paint, which is applied over and around the wound surface with a paint brush. Factors affecting performance are discussed below, using as headings the different criteria suggested previously for a commercially viable wound paint (Clifford and Gendle, 1984).

Intrinsic activity of fungicide components

Although wound treatments appear to contain large amounts of fungicide (2.5 – 10.0% w/v), the amount which is applied to a 4 cm diameter wound is likely to contain as little as 0.1 g of each active ingredient. Only a proportion of this will move into and become distributed within the wood to a depth of 1 cm below the wound surface, which is considered to be the effective target area. Thus the compounds used must be intrinsically very active against the target fungi or be converted *in situ*, e.g. by enzymes, into very active fungicides; the rate of such conversion can vary according to season but with triadimefon has been found to be adequate irrespective of time of application.

Lack of phytotoxicity

No active ingredient or formulant should damage existing tissue or have an adverse effect on production of callus tissue or on compartmentalisation. This precludes the use of additives such as dimethylformamide or dimethylsulphoxide, which aid both miscibility of fungicides with water and their penetration into wood, as we found in previous work (Gendle *et al.*, 1983; Clifford and Gendle, 1984). Penetration of some fungicides into, and their movement within wood can also be enhanced by converting them into salts by adding, for example, hypophosphorous acid but this can also lead to dieback of bark and cambium.

No gel component or fungicide tested so far has been phytotoxic and the carbendazim present in Bayleton BM sometimes promotes callus tissue formation.

Maintaining the normal microbial population of the wood

At any given time, there will be a balanced population of micro-organisms within the wood, most of which are beneficial. Entry of formulants or fungicides could upset this balance even to the extent of causing a previously unimportant group of micro-organisms to become dominant or even pathogenic; alternatively, they could promote the development of pathogenic fungi if present (Gendle *et al.*, 1983). This has not been the case with any gel formulation containing Bayleton BM (a wettable powder formulation containing carbendazim and triadimefon, sold by Bayer (UK) Ltd).

Fungicide mobility

Because infection may be present in the wood very soon after the wound is made, it is important for fungicide to move into the target area as quickly as possible; this requires very mobile fungicides. However, the gel component can also serve to improve or moderate their mobility since an extremely mobile fungicide could move into and away from the target area before the wood ceases to be vulnerable to infection. Thus sodium alginate (Manutex KPR) encourages mobility, itself moving into and through the wood, whereas Kelzan, a xanthan gum of bacterial origin, also moves into the wood but more slowly, moderating fungicide movement. By contrast, Manucol E/RK, an esterified alginic acid which is less water permeable, remains just below the wound surface where it also tends to retain the fungicides. On balance, Kelzan is perhaps the most generally useful formulant and also gives a preparation with optimum consistency for application. It should be stressed that systemic fungicides applied to wounds move predominantly by a process of diffusion, aided by water movement, rather than by normal translocation processes.

Time of application

Trees are often pruned during the dormant season, although fruit growers increasingly wait until after bud burst to check how best to shape their trees. However, physical damage can occur at any time of the year. Work with Bayleton BM in sodium alginate or in Kelzan has shown that penetration, movement and persistence of fungicides are satisfactory for treatment of fresh wound pathogens irrespective of time of year; nevertheless, movement is greater in the growing than in the dormant season with a corresponding decrease in persistence as will be discussed below. Treatment is probably more important in the dormant season for at least two reasons; first, this is the period when ascospores of *N. galligena* and basidiospores of *C. purpureum* are very prevalent and when conditions are very favourable for their development and second, although compartmentalisation can occur at this time, it does so less rapidly than in the growing season, as shown in experiments where *Coriolus versicolor* was introduced deliberately into wounds (Gendle *et al.*, 1983).

Persistence of fungicides

When Bayleton BM in sodium alginate is applied to a wound surface, the gel and fungicide components move into the wood over a period of 2–4 days. They leave behind on the surface a white coating consisting of the inert materials used to prepare the Bayleton BM wettable powder formulation. Kelzan formulations behave similarly but penetration of fungicides is slightly slower; in both cases, the surface of the wood does not dry out and crack because the gel components allow permeation of water. With Manucol E/RK the gel components and fungicides form a partial seal just below the wood surface and both persist within 0.5 cm of wood beneath the surface for at least 6 months, little fungicide moving further into the wood. With both sodium alginate and Kelzan formulations, an eradicant dose of fungicide persists beyond the 28 days for which wounds are vulnerable to fresh wound pathogens. It lasts for up to 6 months, remaining longer beneath flush than stub wounds with both preparations, and longer with Kelzan than with sodium alginate.

Stimulation of callus tissue formation

Gel formulations which contain carbendazim can stimulate callus tissue formation, particularly when applied during the growing season, and this is especially marked with formulations based on Manucol E/RK.

Are the components safe to use and readily available?

Bayleton BM is a readily available fungicide formulation used commercially for the control of fungal diseases in cereals. Its use in sodium alginate has received official provisional clearance for use on apple and pear trees and it is expected that this clearance will soon be extended to use on plums and to formulations incorporating Kelzan. The user mixes his own material as described above and formulations should contain a minimum of 2.5% w/v triadimefon (and therefore 5% w/v carbendazim). A leaflet explaining preparation of these gels is available from Bayer (UK) Ltd.

Performance in practice

Bayleton BM in alginate, Kelzan or manucol E/RK (5% w/v carbendazim + 2.5% w/v triadimefon; 2.5% w/v gel component) has given complete protection of wounds for up to 6 months in experiments carried out over 3 years in orchards with high canker or silver leaf disease inoculum potential. It is therefore a valid treatment for fresh wound pathogens but has insufficient persistence for a mature wound pathogen treatment.

Commercial Sealants

We have tested three sealants in conjunction with our work with gel formulations. Santar, mercuric oxide in a latex-type paint, has been widely used and has frequently given good results; Negal-extra (captan + carbendazim in an aqueous plastic dispersion) is used in Germany, mainly as a canker paint; Pancil-T contains the biocide octhilinone in an acrylic resin formulation. All produce good seals when applied to freshly made wounds; those with Santar and Pancil-T remained intact for up to 8 months, while the seals with Negal-extra were still intact one year after treatment. All three materials promoted callus tissue formation. With Pancil-T, there was little movement of fungicide into the wood beneath the wound but, contrary to previous work with Santar and especially with plums, mercuric oxide had moved – to a limited extent – from the seal into the wood and penetrated to at least 1 cm, where it persisted for up to one year. With Negal-extra only the carbendazim component was mobile but control persisted for at least 6 months. The observed mobility of mercuric oxide from deposits of Santar helps to explain the good performance of this material since the compound is intrinsically very active and has a wide spectrum of activity.

Canker Treatments

Although pruning wounds are an important means of entry of infection, cankers also result from penetration of inoculum through leaf and fruit scars and through holes made by insects or by physical damage. Such cankers, if left untreated, can girdle the branch, destroying the vascular system and eventually cause it to die. Conventional treatment is either to remove the cankered wood completely or to scrape or cut away the degraded wood over the canker and then apply a canker paint. With very young trees such treatment is very undesirable; with older trees it is labour intensive and time consuming. When Bayleton BM gels are applied to unscraped cankers, carbendazim moves through the degraded wood into the area where the fungus is actively growing and completely eradicates infection. In contrast to the work with wounds, Kelzan gives better penetration of fungicide through the degraded wood than sodium alginate. Santar and Pancil-T have also given good canker control without prior scraping.

Discussion

The work reported here has concentrated upon fruit trees where treatment against fresh wound pathogens is standard practice. However, Moore (1969) lists at least nine other tree species which are vulnerable to attack by *Nectria* spp. and no less than 22 which are susceptible to *C. purpureum*; these include forest, urban and amenity trees. Thus, treatment of wounds and cankers probably deserves more attention than at present. The easily prepared Bayleton BM gel formulations are very suitable for this purpose but do not maintain protection for sufficient time to recommend their use for treatment against mature wound pathogens. They are at least as effective as the commercially available sealants in controlling fresh wound pathogens and eradicating *N. galligena* in cankers and could provide a good alternative to mercury-based preparations.

Finally, greater attention needs to be paid to proper pruning practice, i.e. cutting above rather than below the branch bark ridge as emphasised by Shigo (1984), in order to encourage compartmentalisation and thus limit infection.

Acknowledgements

We gratefully acknowledge gifts of fungicides by many manufacturers, particularly Bayer (UK) Ltd., valuable discussions with colleagues in the Forestry Commission and the technical assistance of Mr N. Smith.

References

BROOKS, F.T. and MOORE, W.C. (1923). The invasion of wood tissues by wound parasites. *Transactions Cambridge Philosophical Society* (*Biol. Ser.*) **1.**

CLIFFORD, D.R. and GENDLE, P. (1984). Tree wounds and their treatment. *Arboricultural Journal* **8,** 109–114.

GENDLE, P., CLIFFORD, D.R., MERCER, P.C. and KIRK, S.A. (1983). Movement, persistence and performance of fungitoxicants applied as pruning wound treatments on apple trees. *Annals of Applied Biology* **102,** 281–291.

LONSDALE, D. (1984). Available treatments for tree wounds: an assessment of their value. *Arboricultural Journal* **8,** 99–107.

MERCER, P.C., KIRK, S.A., GENDLE, P. and CLIFFORD, D.R. (1983). Chemical treatments for control of decay in pruning wounds. *Annals of Applied Biology* **102,** 435–453.

MOORE, W.C. (1969). *British parasitic fungi*. Cambridge University Press, Cambridge. 430 pp.

SHAIN, L. (1979). Dynamic responses of differential sapwood to injury and infection. *Phytopathology* **69,** 1143–1147.

SHIGO, A.L. (1984). Compartmentalisation: a conceptual framework for understanding how trees grow and defend themselves. *Annual Review of Phytopathology* **22,** 189–214.

SHIGO, A.L. (1984). Tree decay and pruning. *Arboricultural Journal* **8,** 1–12.

SHORTLE, W.C. (1979). Mechanisms of compartmentalisation of decay in living trees. *Phytopathology* **69,** 1147–1151.

Discussion

N. HARRIS — Did you measure the effect of treatments at shorter intervals within 200 days?

D.R. CLIFFORD — Yes, at 5–10–15–60–150 days. These figures were not included in the tables because of the need to condense 5 years' figures into a 20 minute talk.

F.R. CLAXON — (Epping Forest)

Would Cuprinol be a suitable material for treating cut surfaces on large old pollards?

D.R. CLIFFORD — No, Cuprinol is a timber preservative and is phytotoxic and quite unsuitable for live tissue even if restricted to the centre of pruning wounds.

P.S. KEELING — (Livingstone)

Did you include alginate on its own in your experiments?

D.R. CLIFFORD — No, the experiments already involved destruction of many apple trees and it was not worthwhile extending the testing to include alginate alone as it would be very unlikely to give fungal control, though it might aid callus formation.

Prospects for Long Term Protection Against Decay in Trees

D. Lonsdale
Forestry Commission

Summary

The natural defences of trees provide the only proven form of long term protection against decay, but they are too often rendered ineffective by severe wounding and other avoidable damage.

A few fungicides and sealants, including mercuric oxide and polyvinyl acetate, have delayed fungal colonisation of wounds for several months, while the biological control agent *Trichoderma* sp. has given significant protection over 4 years. Proper evaluation of such treatment would require fundamental research and long term trials. For future plantings, genetic selection might help to reduce the incidence of decay, at least in clonally propagated species.

Possibilities for curative treatment remain conjectural, but use of existing knowledge could help to avoid making decay worse by unsuitable treatment.

Introduction

Some decay is present in most trees, but the affected wood is in most cases limited to the immediate vicinity of the wound, for example a dead branch, where the decay was initiated. Decay would develop much more extensively were it not for the physical and chemical barriers produced by the tree, which deter the growth of the decay-causing micro-organisms within the wood (Shigo and Marx, 1977). Even if these barriers are overwhelmed by microbial attack, the eventual occlusion of the infection court by callus growth can apparently help to retard the decay process by restricting gaseous exchange (Toole, 1965). Despite the effectiveness of these internal and external defences, it is possible for decay to develop so extensively that, after several – or perhaps many – years, the structural strength of a tree is seriously reduced. Only a small proportion of infections progress so far, but there are so many possible infection courts on a tree that decay remains a major problem in arboriculture and silviculture.

It was long believed that conventional wound dressings could prevent the entry of micro-organisms by either fungicidal or mechanical sealing effects, but it had become clear several years ago that such materials were at best able to delay this process slightly (Mercer, 1980). Against organisms dependent on fresh wounds for their attack, a short term protection is, however, of great value (Clifford and Gendle, 1987).

The prospects for long term protection will inevitably remain uncertain until research workers have carried out detailed and very lengthy experiments far beyond the scope of anything which has been attempted up until now. We can, however, gain some useful clues from some of the short and medium term data which are already available, and from examining the conditions, natural and artificial, which on circumstantial evidence seem to restrict or prevent decay.

Enhancement of the Tree's Natural Defences

The serious failure of the tree's natural defences against decay fungi is more the exception than the rule. It is therefore interesting to examine the possible factors which are responsible for the failures. If they can be identified it may be feasible to reduce their prevalence.

A few adverse factors are fairly obvious and their effects have been demonstrated experimentally in the last few years. Large wounds are an important example, since they are more likely to favour decay than small ones (Mercer, 1982), especially where many annual rings are injured. Thus, large branches should never be removed unless there is an inescapable need, and prevention of damage to stems by vehicles and other machinery should be a major priority. Unfortunately such injuries continue to be inflicted in situations where they could be avoided and this implies a lack of awareness of the risk. Greater awareness of the importance of large wounds might be just as useful as the formulation of a truly effective wound treatment. Dieback of large branches or roots may, like injury of these parts, give decay fungi the opportunity to overcome the tree's defences. This dieback is sometimes avoidable when it results from man-made disturbances such as soil compaction, alteration of drainage and de-icing salt injury. Here again, existing knowledge could be used to improve the prospects for the prevention of decay.

Perhaps less obvious than the dangers of large wounds and severe dieback is the effect of 'improper' pruning. On the basis of repeated observations, Shigo (1982a) has stated that flush cuts increase the risk of stain and decay, compared with cuts made distal to (i.e. 'outside') the 'branch bark ridge' and to the branch collar if one is present. Long stubs are also undesirable since they may become foodbases for decay fungi and since they tend to prevent wound closure. The advised pruning position is shown in Figure 1. We have traced the course of the vessels in branch junctions of several species, and in all of them found that the 'non-flush' cut avoided damage to main stem vessels above the branch. One of the methods used was to infuse the wood with particulate dyes which, like bacteria and fungal spores, could pass along vessels, but not from one vessel to another. This helped not only to demonstrate the course of vessels but also to locate the positions of vessel endings. There is some evidence that, in certain species, many of the vessels which pass down a branch into the main stem have their endings just below the branch junction. This could represent a natural 'safety valve' which is at risk of being breached by a flush cut.

Another reason for avoiding flush pruning is its effect on callus production. As indicated earlier, callusing does not prevent the initiation of decay. A completely closed callus seems usually to arrest or greatly retard its progression and callus thus forms part of the tree's defences against decay. Shigo (1982) has stated that callus production is impaired above and/or below flush wounds and is stimulated laterally. Preliminary results from our own replicated experiments confirm this observation and also indicate that the rapid callusing on the sides of a flush wound contributes surprisingly little to wound closure since much of this vigorous growth takes the form of an outward bulge of tissue. In any case, the flush pruning of a branch produces an unnecessarily large wound which cannot be quickly occluded even by a fast-growing callus. At the other extreme, stub wounds become occluded very slowly or, if the stub dies back, will never be occluded until after the stub is completely shed. In summary, although more data on callusing are needed, there is good reason to recommend the pruning position shown in Figure 1.

Whichever pruning position is adopted, callusing can be enhanced by use of a wound dressing, and by far the most effective which we have so far tested is one containing 6 per cent thiophanate methyl (Mercer, 1983; Lonsdale, unpublished). In a recent trial on beech wounds, the first season's callusing showed a mean stimulation of 144 per cent in vertical closure rate and 129 per cent in horizontal closure rate on thiophanate methyl-treated wounds cut as shown in Figure 1. Flush cuts treated in the same way showed an even greater benefit of treatment (they callused very little in the vertical direction if untreated). The same was true of stub wounds (which callused very little in either direction if untreated). Earlier trials showed that this beneficial effect extends beyond the first growing season after pruning (Mercer, 1983).

Another way in which trees can be helped to maintain their defences against decay is to safeguard their general health. There is some indication that the tree's ability to confine decay to a small volume of wood depends on its available energy reserves (Boddy and Thompson, 1983; Thompson and Boddy, 1983). These can be depleted by the general stress arising for example from root injury, environmental changes and severe defoliation (Wargo and Houston, 1974). In some cases dieback may occur and, as mentioned above, this can provide entry points for decay fungi. Reduction in photosynthetic capacity due to drastic pruning may also cause problems, while a less obvious threat is the excessive use of treatments which involve drilling holes into the stem (Shigo and Campana, 1977). Multiple holes, especially if deep and close together, can reduce the continuity and the volume of the living sapwood. There is a need to evaluate this risk more fully and, to this end, data are being gathered from elm trees which have been drilled during the course of injection or pellet implantation treatments against Dutch elm disease.

Preventive Treatment of Wounds

Although, in general, the long term value of protective treatment seems doubtful, (Mercer, 1980; Shigo and Shortle, 1983) it is worth considering those treatments which have had a delaying effect on the growth of decay organisms and to ask whether any of these is likely to be capable of improvement so as to provide a really long term benefit.

The most successful treatment so far in the medium term has been the application of the fungus *Trichoderma* to stem wounds in beech (*Fagus sylvatica*), and this limited the incidence of decay fungi to *c.*3 per cent as compared to *c.*14 per cent in controls over a 4 year trial (Mercer and Kirk, 1984). A trial on pruning wounds (as opposed to stem wounds) was also carried out using a commercial

Figure 1. Section of sycamore stem showing the relationship between branch base anatomy and a recommended pruning position. (*A10558*)

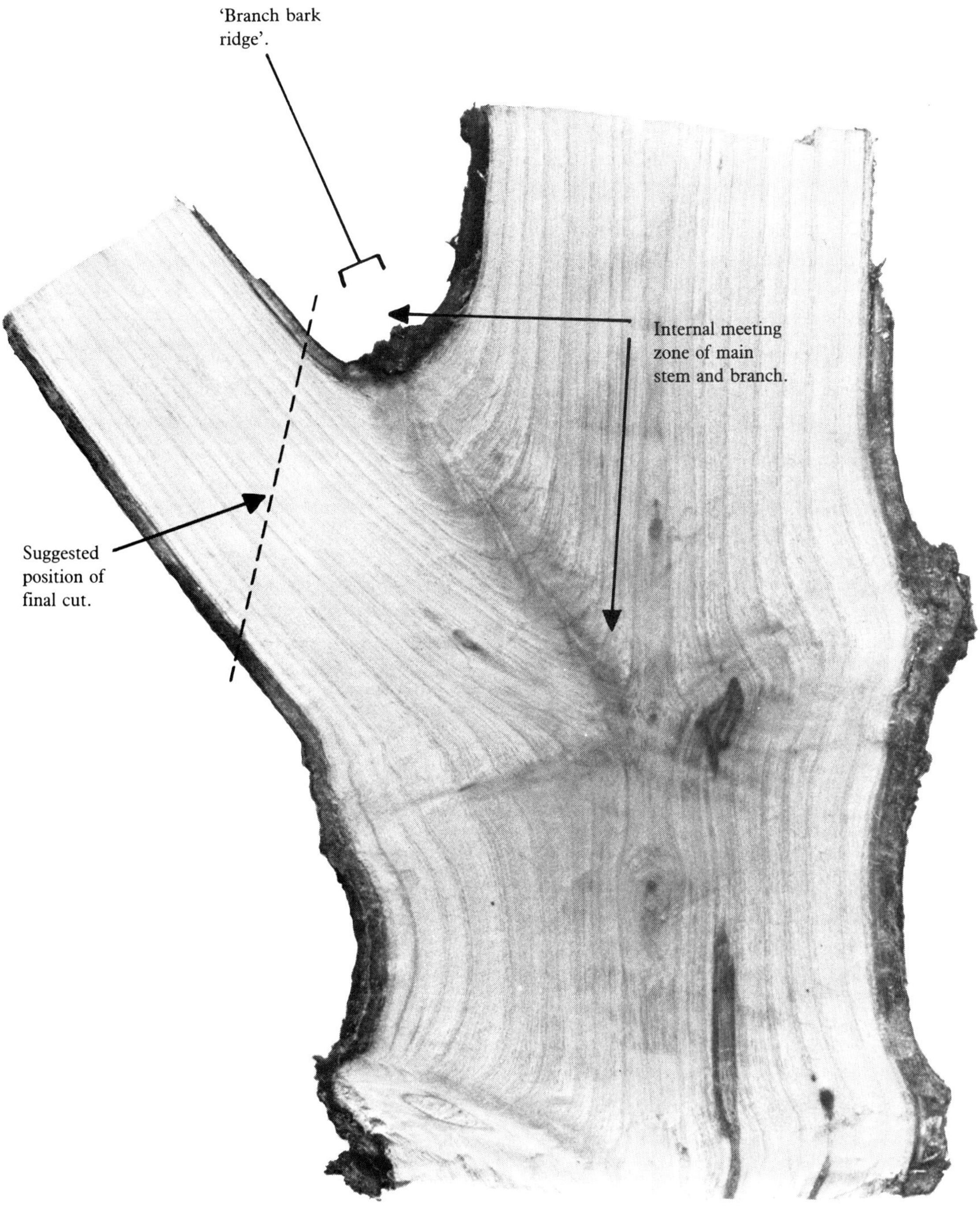

Table 1. Tests of sealant strength, using compressed air

Treatment:	*Control*	*'Sanseal'*	*'Novaril Rot'*	*'Lac Balsam'*	*'Arborseal' (Australian)*	*'Polybond'*
	no sealant	wound paint, composition unspecified, plus captafol	plastic emulsion wound paint	polyvinyl proprionate wound paint	polyvinyl acetate wound paint, plus captafol	general polyvinyl acetate adhesive
Mean leakage rate* (cm^3 per sec per cm^2 of wound cross-section	43.57	27.46	9.04	6.04	2.77	<0.02

* Lengths of *Fraxinus excelsior* stem, approx. 17 mm diameter and 55 mm long, were debarked and, while fresh, coated at one end with the test material. After drying and removal of surplus sealant each stem length was connected to a pressure chamber containing a known volume of air at 1.4 kg cm^{-2} pressure, and the pressure allowed to dissipate by air-flow along the xylem vessels.

preparation of *Trichoderma* spp. and gave encouraging results over a 2 year period (Mercer and Kirk, *op. cit.*). It seems from this work and from unpublished data that good establishment of *Trichoderma* requires the application of a non-fungicidal sealant on top of the treated surface. Current work includes the evaluation of *Trichoderma* on tree species other than beech, the initiation of longer-term trials and the use of *Trichoderma* in combination with chemicals which can selectively encourage it to grow in the wood and/or discourage the growth of decay fungi. The prospects for the successful use of biological control as a long term treatment remain uncertain, especially since there are fundamental reasons for suspecting that *Trichoderma* may eventually die out due to exhaustion of its foodbase in the wood. A trial on Red maple (*Acer rubrum*) in the U.S.A. in which *T. harzianum* showed a decline in survival within 3 years (Pottle *et al.*, 1977) lends weight to this suspicion. One way of obviating this problem might be to use a biocontrol agent which, unlike *Trichoderma*, can utilise lignocellulose and could thus persist for much longer. In the selection of such an agent it would be essential to establish that it has no ability to invade the tree much beyond the immediate vicinity of the wound. It must also be able to resist subsequent invasion by harmful decay fungi. A few organisms seem to have these attributes and we have begun some trials with them.

As far as fungicides are concerned, most products tested have been of only short term value if any. Mercuric oxide paint ('Santar') has given protection over several months (Mercer *et al.*, 1983), but is no longer available for general arboricultural use because of fears over its toxicity to vertebrates. Few other chemicals have even this persistence of effectiveness, and this can be related to a number of fundamental problems which include chemical or biological detoxification, leaching and weathering and the failure to provide an unbroken layer which bars every possible avenue of infection.

Although chemicals may fail to prevent the eventual establishment of decay fungi in wounds, they may be able to retard the decay process for at least a few years. Thus, several workers have found that certain proprietary products could significantly reduce the rate of extension of decay columns following stem injury in species such as Red maple (Houston, 1971), and Norway spruce (*Picea abies*) (Rohmeder, 1952; Bonnemann, 1981). However, Shigo and Shortle (1983) dismiss such effects as being too transitory, having themselves carried out trials lasting for up to 7 years.

The mechanical sealing of wounds is subject to some of the same limitations as the use of fungicides, namely imperfect coverage and breakdown or weathering. As far as sealants without added fungicides are concerned, the best results seem to have been obtained with polyvinyl acetate (PVA)-based materials (Dye and Wheeler, 1968). Mercer *et al.* (1983) have obtained several months' delay in fungal colonisation of beech pruning wounds with the PVA-containing product 'Australian Arborseal' which also contains the fungicide captafol. This delaying effect of PVA may be related to its ability to form a good mechanical seal on the moist wound surface. We have tested the mechanical strength of seals formed by various proprietary dressings and non-arboricultural adhesives including PVA products. The PVA-based materials have scored much better than any others (see Table 1). Although PVA gives only short term protection against fungal invasion, the physical and chemical properties which are responsible for its good adhesion to moist wood could, perhaps, be identified so as to help us formulate materials with similar desirable properties but with much greater long term sealing ability. Alternatively, PVA can be used as an adhesive for a 'wound cap' in the form of a durable membrane, and we have recently developed a method using butyl rubber sheeting which, in theory, could function as an 'artificial callus'. However, the assessment of its performance against decay fungi will be lengthy and, in any case, considerations of convenience and labour may demand a simple 'one coat' dressing.

Curative Treatments

The glimmerings of hope which can be seen in respect of preventive treatments are far less discernible where curative treatments are concerned. The risk of damaging the tree's natural defences in the process of cutting out decayed wood around cavities has been stressed by Mercer (1980) but it needs to be re-emphasised. In cases where decay appears not to have continued following such excisions, it is tempting to claim that the treatment has been successful, but it may well be that the decay had already been restricted by the natural barriers around the cavity.

These negative observations should not imply that curative treatment is impossible. Indeed, there are at least three possible methods which deserve further study. The first of these has already been discussed in relation to preventive treatment, namely the restriction of gaseous exchange. This happens naturally when a cavity is totally occluded by callusing, and it also happens in the lower part of a cavity when water is permanently present (perhaps another reason for not inserting drainage pipes, quite apart from the breaching of natural barriers in the wood). If filling materials can provide a durable and hermetic seal when pressed upon by the developing callus, it may be possible, in effect, to accelerate occlusion. The second method is removal of rotten wood which could, if allowed to remain, provide a foodbase for the decay fungus. In the pruning of dead branches or old stubs the final cut should be immediately outside the developing callus. In cavities, the removal of decayed wood carries with it the risk of breaching the natural barriers but, if confined to loose and obviously rotten material, it could be a worthwhile preliminary to filling. Finally, and most speculatively, there is the possibility of using a biological control agent which could destroy the decay fungi already present. Some agents can do this to some extent, the best known example being *Scytalidium lignicola*. We have also found that *Verticillium lecanii* can replace some decay fungi (e.g. *Inonotus hispidus*) in wood strips in the laboratory. Some decay fungi can replace other decay fungi (see Rayner, 1975), and they could perhaps be used so long as they lacked the ability to invade the tree beyond the original confines of the decay. It does, unfortunately, seem to be a general rule that these 'replacement' effects tend to occur only where the decay fungus is in a rather inactive state, i.e. not at an active front of invasion. This is, at least, the case with timber being invaded by *Lentinus lepideus* when *S. lignicola* is the intended biological control agent (P. Morris, personal communication.).

Genetic Selection

The ability to resist invasion by decay fungi varies not only between tree species, but also within species. This has been shown most convincingly in poplars, where replicated experiments have been made possible by the availability of genetically identical trees (Shigo *et al.*, 1977). Such vegetatively propagated individuals are not so readily available for most tree species but it might be possible, even in trees of unknown genetic constitution, to recognise anatomical or biochemical resistance factors which would be used to aid selection procedures for the development of strains with improved decay resistance. The resistance would need to be measured against a number of decay fungi and in relation to various modes of infection. It would also be necessary to maintain other desirable traits such as vigour, form (including good branching habit) and resistance to other diseases.

Some Questions for the Future

The development of decay is often a very slow process, and so the evaluation of any protective treatment is necessarily a long term task. This demands a major commitment of time and resources and it is therefore important that treatments under test should be chosen on the basis of a valid expectation that they might work. Some at least of the proprietary wound dressings might have been expected to work, judging by their fungicidal content or their apparently good covering ability, but research seems to show that they don't work well against decay fungi in the long term. This implies either that the use of sealants or fungicides is doomed to failure for fundamental reasons or that none of those yet marketed has the right characteristics. If such characteristics can exist, we could perhaps aim for a 'tailor-made' wound treatment, but this needs more than trial and error; we need to know more about the processes which determine success or failure. Too little is known also about the modes of action of biological control agents. The simple inhibition of decay fungi is only one possible mechanism in the case of *Trichoderma* spp. (Smith *et al.*, 1981). Given that some treatments are worthy of long term field evaluation, it is important that suitable methods are used in setting up and evaluating the experiments.

In particular, the type of artificial wounds used in experiments should be comparable with wounds which would be treated by the arboriculturist, and it is perhaps unfortunate that little work has yet been done using wounds of the 'recommended' type (Figure 1). Previous work has only involved stem wounds or pruning stubs, and these may provide an unsuitable model. As far as evaluation of results is concerned, our main criteria of the efficacy of a treatment must continue to be the presence or absence of decay fungi and the extent to which they have invaded the wood. We have little way of knowing whether the mere presence of a decay fungus at a certain depth after, say, 4 years' trial indicates that it would eventually have caused great damage to the tree or, conversely, never have progressed further. Depth of staining is an important marker of the boundaries of tissue reaction, although the precise significance of such data is open to doubt. In particular it is difficult to distinguish between staining which marks the position of an advancing decay front and 'protective' staining of the type which presumably develops, for example, in response to the use of *Trichoderma* as a biological control agent (Mercer and Kirk, 1984).

The progress of recent years has helped to dispel a number of misconceptions about decay and its prevention. In particular, we now realise the value of encouraging the natural defences of the tree. We also tend to accept that some sort of microbial colonisation of wounds is inevitable and that biological control is thus perhaps more realistic than the idea of keeping wounds 'sterile'. This new awareness is refreshing but it leaves many questions unanswered for the practising aboriculturist. There is, however, another awareness that needs to be fostered from now on; the awareness that we have reached a stage where theory needs to catch up to some extent with practice. This does not imply a need for academic research for its own sake, but it does emphasise that good 'applied' research is not simply a matter of conducting 'trials'. We must first define the questions that need to be asked, and this exercise depends very much on co-operation between the research worker and the practitioner.

Acknowledgements

I thank Drs. J.N. Gibbs, D.R. Clifford and P.C. Mercer for useful discussions and Miss S.E. Chuter for assistance with fieldwork. I gratefully acknowledge the receipt of test materials from ICI-Midox, Hortichem, Transatlantic Plastics and Aagrunol-Stahler. Permission for the use of trees was kindly granted by staff of the Forestry Commission and of Hillier's Arboretum, Hampshire. The research project is financed under a contract from the Department of the Environment.

References

BODDY, L. and THOMPSON, W. (1983). Decomposition of suppressed oak trees in even-aged plantations. I. Stand characteristics and decay of aerial parts. *New Phytologist* **93**, 261–276.

BONNEMANN, I. (1979). Untersuchungen über die Entstehung und Verhütung von 'Wundfaulen' bie der Fichte. *Dissertation zur Erlangung des Doktorgrades der Forstlichen Fakultat der Georg-August-Universitat zu Göttingen*. 173 pp., 53 appendix tables.

CLIFFORD, D.R. and GENDLE, P. (1987). Treatment of fresh wound parasites and of cankers. In, *Advances in practical arboriculture*, ed. D. Patch. Forestry Commission Bulletin 65, 145–147. HMSO, London.

DYE, M.H. and WHEELER, P.J. (1968). Wound dressings for the prevention of silver-leaf in fruit trees caused by *Stereum purpureum* (Pers.) Fr. *New Zealand Journal of Agricultural Research* **11**, 874–882.

HOUSTON, D.R. (1971). Discoloration and decay in Red maple and Yellow birch: reduction through wound treatment. *Forest Science* **17** (4), 402–406.

MERCER, P.C. (1980). Decay – its detection and treatment. In, *Research for practical arboriculture*. Forestry Commission Occasional Paper 10, 10–15. Forestry Commission, Edinburgh.

MERCER, P.C. (1982). Basidiomycete decay of standing trees. In, *Decomposer basidiomycetes*, eds. Franklin, Hedger and Swift. British Mycological Society Symposium, Cambridge University Press, Ch. 8, 143–160.

MERCER, P.C. (1983). Callus growth and the effect of wound dressings. *Annals of Applied Biology* **103**, 527–540.

MERCER, P.C. and KIRK, S.A. (1984). Biological treatments for the control of decay in tree wounds. II. Field tests. *Annals of Applied Biology* **104**, 221–229.

MERCER, P.C., KIRK, S.A., GENDLE, P. and CLIFFORD, D.R. (1983). Chemical treatments for control of decay in pruning wounds. *Annals of Applied Biology* **102**, 435–453.

POTTLE, H.W., SHIGO, A.L. and BLANCHARD, R.O. (1977). Biological control of wound hymenomycetes by *Trichoderma viride*. *Plant Disease Reporter* **61** (8), 687–690.

RAYNER, A.D.M. (1975). *Fungal colonization of hardwood tree stumps*. Ph.D. Thesis, University of Cambridge.

ROHMEDER, E. (1953). Wundschutz an verletzten Fichten. *Forstwissenshaftliche Zentralblatt* **72**, 321–335.

SHIGO, A.L. (1982a). A pictorial primer for proper pruning. *Forest Notes* (Society for Protection of New Hampshire Forests), Spring issue.

SHIGO, A.L. and CAMPANA, R. (1977). Discolored and decayed wood associated with injection wounds in American elm. *Journal of Arboriculture* **3**, 230–237.

SHIGO, A.L. and MARX, H.G. (1977). *Compartmentalization of decay in trees*. USDA Forest Service, Agriculture Information Bulletin 405. 73pp.

SHIGO, A.L. and SHORTLE, W.C. (1983). Wound dressings: results of studies over 13 years. *Journal of Arboriculture* **9**, 317–329.

SHIGO, A.L., SHORTLE, W.C. and GARRETT, P.W. (1977). Genetic control suggested in compartmentalization of discolored wood associated with tree wounds. *Forest Science* **23** (2), 179–182.

SMITH, K.T., BLANCHARD, R.O. and SHORTLE, W.C. (1981). Postulated mechanism of biological control of decay fungi in Red maple wounds treated with *Trichoderma harzianum*. *Phytopathology* **71** (5), 496-498.

THOMPSON, W. and BODDY, L. (1983). Decomposition of suppressed oak trees in even-aged plantations. II. Colonization of tree roots by cord- and rhizomorph-producing basidiomycetes. *New Phytologist* **93**, 277–291.

TOOLE, E.R. (1965). Inoculation of bottom-land Red oaks with *Poria ambigua*, *Polyporus fissilis* and *Polyporus hispidus*. *Plant Disease Reporter* **49** (1), 81–83.

WARGO, P.M. and HOUSTON, D.R. (1974). Infection of Sugar maple trees by *Armillaria mellea*. *Phytopathology* **64**, 817–822.

Discussion

R. FINCH (Roy Finch Tree Care Specialist)

You said that we should avoid making large wounds wherever possible but some species, like the Sweet chestnut in one of your slides, can receive large wounds without much risk of decay. Others, like poplars, have very little resistance.

D. LONSDALE There are great differences between species, as you say, but large wounds always represent a greater risk than small ones. I would like to be able to advise on the maximum permissible wound size for a given species but it is very hard to do this on experimental evidence. Perhaps we can build up our knowledge from case histories in our advisory work.

C. YARROW (Chris Yarrow and Associates)

Is a vigorously growing tree more resistant to decay and does it respond to wound treatment?

D. LONSDALE Healthy trees appear to be more resistant to decay than stressed trees but the evidence is somewhat tenuous.

R. SKERRATT (Independent Consultant)

Do wound treatments which initially benefit callus formation lose effectiveness?

D. LONSDALE Untreated wounds callus more rapidly after 5 years and the callus growth effect of treatment is temporary.

Ash Decline

R.G. Pawsey
Oxford Forestry Institute

Summary

The history of ash decline (or dieback) in Britain and the results of a survey of the disease in the east-central region of England are discussed. The survey indicated that the disease is a chronic condition related to stresses imposed on the trees by the effects of modern agricultural practices linked with other adverse physical and biological factors. The need for government-funded research to identify the factors associated with the development of the disease is emphasised.

Introduction

This paper summarises the extent of our knowledge of a condition more usually called ash dieback, but the term 'decline' somehow more aptly captures the atmosphere of impotence or apathy which surrounds a major problem of one of the foremost hardwood trees in Britain.

The Forestry Commission's census of woodlands and trees 1979–82 (1983) confirms the importance of ash (*Fraxinus excelsior* L.) in this country, not only as a woodland tree, but especially as a dominant element in the non-woodland environment. In England, ash is the most common broadleaved tree in non-woodland situations (being third and fourth most common in Wales and Scotland respectively). In the amenity context particularly, its pre-eminence in England is underlined by the revelation that ash is the most common non-woodland broadleaved tree in more than half of the individual English counties. The importance of ash relative to other tree species in the English landscape will no doubt increase as elm disease losses increase in the north, west and east of the country.

The first significant reference to ash dieback was made by T. R. Peace who founded the Pathology Branch at the Forestry Commission's Research Station at Alice Holt, in his comprehensive textbook published in 1962. He suggested that dieback was due to moisture stress associated with major variations in soil water content, particularly on heavy soils. Forestry Commission records reveal Peace's serious concern about the extent and severity of ash dieback arising from his examination of trees in Northamptonshire during the building of the M1 motorway in the late 1950s. He recorded that although it was worse on older trees, smaller trees were also affected, and it occurred on trees growing on loam as well as on heavy clay soils.

Despite Peace's concern, and the growing general awareness of the widespread nature of the condition (Pawsey, 1973), ash dieback failed to attract the serious attention of foresters, arboriculturists or research specialists until it became the subject of recent surveys carried out from the Department of Forestry at Oxford University. However, lack of detailed study rarely suppresses conjecture (more often, the reverse), and apart from the effects of water stress, other factors cited as possible causes of ash dieback included pollution, fungal, bacterial and virus infections, and insect activity. To some extent, these suggestions were generated by the results of studies of ash dieback in the north-eastern U.S.A. (Ross, 1966; Hibben and Silverborg, 1978). However, on American ash species (notably *Fraxinus americana* L. and *F. pennsylvanica* March) the disease symptoms appear to be quite different from those on *F. excelsior* in Britain and no close parallel should be drawn between the factors causing ash dieback in Britain and North America. In America, early symptoms of ash dieback include chlorosis and reduction in the size of leaves, whereas in Britain leaf abnormalities are not characteristic of the disease.

During their survey of virus infections of hedgerow ash in Britain, Cooper and Edwards (1981) made numerical assessments of ash dieback in many different localities. Although they recorded widespread infection of hedgerow ash by Arabis mosaic virus (AMV), they concluded that dieback was not directly associated with the presence of virus infection. Over Britain as a whole (in the localities they surveyed), the incidence of ash dieback varied from 29 to 84 per cent, with the highest incidence occurring in the east-central region of England.

1983 Ash Dieback Survey

During the summer of 1983, a more detailed survey of ash dieback was made in the east-central region of England, on funds donated by relevant county and district councils (Pawsey, 1983). Preliminary observations indicated that saplings and young coppice growth were virtually free of dieback, and that the condition was absent or of very low incidence in woodlands and copses.

All the numerical data in the survey was based on observations made from a slowly moving car. The criterion for recording dieback on a tree was the same as that used by Cooper and Edwards (1981), i.e. 10 per cent or more of the branches without leaves, or with leaves only present at the shoot tips. Not least of the problem in working on ash dieback is the difficulty of defining the symptoms.

In the survey area (a broad band of country about 80 km wide, stretching from just north of York to just south of Aylesbury) the incidence of dieback was recorded in a series of east-west transects, each transect separated by a distance of 10 km. The incidence of healthy and dieback-affected trees (recorded on tally counters) was noted after travelling along each 10 km (nominal) section of the traverse (i.e. across each 10 km grid square of the 1:50 000 O.S. map), although the length of road between each recording point often exceeded this considerably. Wherever possible, the route of the east-west transects was along secondary roads and in order to minimise variation in observations, all assessments were made by the author of the present paper. The survey route was over 1700 miles (2600 km) long.

The survey, which included ash trees close to the roadside and in the area on both sides of the road up to a distance of about 150 to 200 m was confined to single-stemmed trees over 6 m (about 20 feet) in height and clearly distinguishable from trees of other species. In addition to roadside and hedgerow trees, the survey included trees in gardens, open parkland, recreation grounds and other amenity areas.

The survey confirmed the widespread occurrence of ash dieback. The incidence and severity of symptoms varied considerably in different areas, often between adjacent sections of the traverses. The average incidence of dieback over the whole survey area was just over 22 per cent with the highest incidence occurring in Northamptonshire and adjacent areas of neighbouring counties, where 40 per cent incidence (or higher) was commonly recorded. The highest incidence recorded in a single survey section, was 67 per cent, near Newport Pagnell. Full details of numbers of healthy and affected trees, as well as the percentage incidence recorded in each transect section, are given in the published report of the survey (Pawsey, 1983).

The Factors Involved

Soon after the start of the main survey, a strong impression was gained a) that the incidence of ash dieback was lower in non-agricultural locations (e.g. towns, suburbs and villages) than in the intervening areas of agricultural countryside, and b) that the highest incidence of dieback often occurred in field-side hedgerows in areas of intense arable farming. As the survey proceeded, a direct relationship between the incidence of dieback and the intensity of arable farming was fairly consistently (but not invariably) observed.

In order to compare the incidence of ash dieback in the agricultural countryside with that in non-agricultural situations, two supplementary surveys were made along routes selected to pass through a number of villages and towns clearly separated by agricultural countryside. Records were made of the incidence of dieback in the separate 'agricultural' and 'non-agricultural' sections along both routes. The routes of the two special surveys were:

a) from Halford (Warwickshire) to Newport Pagnell (Bucks); 75 miles
and
b) from Princethorpe (Warwickshire) to Raunds (Northants); 72 miles.

The incidence of dieback in towns and villages as compared with that in the agricultural countryside was: Halford to Newport Pagnell, 6% : 46%, and Princethorpe to Raunds, 5% : 46%.

The results of the survey suggested strongly that modern farming practices are directly associated with the development of ash dieback in hedgerow trees. No attempt was made to gain more detailed information on which practice or practices (acting separately or cumulatively) might be particularly involved, but these could include regular deep cultivation close to hedgerows, drainage operations, spraying with pesticides and herbicides, regular applications of large quantities of fertiliser, and stubble-burning. The effect of any of these factors would clearly be influenced by general adverse soil and climatic conditions, or might exacerbate stresses in the tree generated by unfavourable edaphic and climatic factors.

The likelihood of death of ash trees resulting directly from a long period of dieback is not known. The appearance of a considerable proportion of affected ash trees (15–20 per cent in certain areas) suggests that dieback can occur continuously over a period of several years, with long-dead branches accounting for 30 per cent or more of the general crown volume. Some of these trees are of course extremely old and their condition largely due to senescence, but it seems likely that death may ensue in a proportion of younger trees affected by dieback over a prolonged period.

Atmospheric pollution has been described by American workers as a significant contributory factor in the development of ash dieback in the north-east of the United States. Although atmospheric pollution has often been described as an important factor associated with ash dieback in Britain, there was no evidence of this during the 1983 survey, even though the survey route passed through areas of high industrial activity and close to steelworks, fertiliser and other chemical plants, and also through south Bedfordshire with its high concentration of brickworks.

An impression was gained at various times during the survey that the incidence of ash dieback might be directly related to the degree of exposure to which the trees were subject. However, this relationship was not consistent, but it would seem reasonable to assume that increased rates of water loss from foliage associated with exposure would enhance the effect of moisture stress caused by other factors.

Examination of many ash trees, both young and old, during and after the survey suggests that flowering behaviour of older trees might exert considerable influence on the pattern of development of vegetative shoots, which could be of importance in relation to the development of dieback. Regular annual production of flowers on ash does not usually occur until the trees are more than 15 years old, and thereafter a high proportion of lateral buds produced on shoots in the previous year develop only to form inflorescences, with vegetative extension in the spring being confined to growth from apical buds. Factors which lead to the death of buds, particularly apical buds, could have a profound effect on the extent of vegetative growth and foliage production during the following season, and on overall shoot development and crown form. The possible adverse effects of some agricultural practices, in relation to the development of dieback, might be linked directly or indirectly with their influence on bud survival.

In recent years, Forestry Commission pathologists have observed considerable damage to, and death of, buds on ash trees caused by the Ash bud moth (*Prays curtisellus* Dup.) (R. G. Strouts and S. C. Gregory, personal communications), and such damage was commonly seen during and following the 1983 survey. Previously in Britain (Chrystal, 1937; Gent, 1955) damage by *P. curtisellus* was regarded as being significant only on young ash. More recent continental accounts (Novak, 1976; Anon., 1983) indicate that the moth commonly attacks older trees, and it seems possible that killing of buds and terminal parts of shoots on older trees in Britain is extremely common, and could be an important factor contributing to the development of ash dieback.

Forestry Commission pathologists (R. G. Strouts and D. Rose, personal communication), have also observed that bud and shoot damage by Ash bud moth is commonly followed by infection of the shoots by the fungus *Nectria galligena* f. sp. *fraxini* (Flack and Swinburne, 1977). *N. galligena* Bres. has long been known as the cause of cankering on ash (and on a wide range of other trees, notably apple and pear) but considerable uncertainty exists as to the importance of infection by this fungus because of the similarity of *Nectria* cankers with those of the more prevalent canker-forming bacterium, *Pseudomonas savastanoi* (Smith) Stevens f. sp. *fraxini* Dowson. Boa (1981) described differences between the two types of canker on ash, but there is little information on the significance of more diffuse infection of ash shoots by *N. galligena*, either generally or in particular relation to dieback.

The Need for Research Support

As a result of the information provided by the 1983 survey, and the interest (and financial response) of local authorities and other organisations which facilitated the survey and a limited amount of research work, an application was made by the Department of Forestry, Oxford University, to the Agricultural and Food Research Council for a grant to fund a 3 year research programme on ash dieback. The object of the proposed research programme was to determine the relative importance of different agricultural practices and associated soil, climatic and pathological factors in the development of the disease, incorporating a) detailed studies on trees in selected areas influenced over varying periods of years by the effects of different farming techniques in a range of different site circumstances, b) comparative studies of the growth, development and reproduction of healthy and dieback-affected ash trees, and c) investigation of the insects, fungi and other pathogens associated with dieback.

The AFRC referred the application to the Natural Environment Research Council, but after due consideration the latter felt unable to support the proposed research.

One cannot help but ponder the criteria used by grant-giving authorities when weighing the priorities for the funding of research projects. Ash dieback may not be a catastrophic infectious condition like elm disease (which itself enjoyed the serious attention of an estimated one tenth of one pathologist in 1967, after the beginning of the present epidemic (Pawsey, 1973)), but there is no doubt about its prevalence in the ash population of Britain or about the potential importance of its effect on the rural landscape, to say nothing of broader and more fundamental ecological issues. The concern of local authorities and other organisations is evident from their generous response to the University of Oxford's Ash Dieback Survey and Research Appeals, but without central government support no really significant progress will be made in our understanding of this complex chronic problem. At the time of writing, research on ash dieback in Britain is limited to a part-time M.Sc. project at Birkbeck College, London, to be followed in the summer of 1985 by specialised surveys in selected areas financed from funds donated by County Councils and other bodies and deposited with the Department of Forestry at Oxford. Commencing in 1986, research has been funded by the Department of the Environment.

Acknowledgement

The author and Department of Forestry, Oxford University, gratefully acknowledge the financial assistance given by local authorities and other organisations which made possible this preliminary survey and other work on ash dieback.

References

ANON. (1983). Waldshutz: Schäden an Eschen und anderen Laubhölzern in Niedersachsen. [Forest protection: damage to ash and other broadleaved trees in Lower Saxony]. *Der Forst-und Holzwirt* **38** (15), 398.

BOA, E.R. (1981). *Ash canker disease*. PhD Thesis, University of Leeds.

CHRYSTAL, R.N. (1937). *Insects of the British woodlands*. Frederick Warne.

COOPER, J.I. and EDWARDS, M.L. (1981). Viruses of trees. In, *Forest and woodland ecology*, I.T.E. Symposium No. 8, eds. F.T. Last and A.S. Gardiner.

FLACK, N.J. and SWINBURNE, T.R. (1977). Host range of *Nectria galligena* Bres. and pathogenicity of some Northern Ireland isolates. *Transactions British Mycological Society* **68** (2), 185–192.

FORESTRY COMMISSION (1983). *Census of woodlands and trees 1979–1982*. Various reports. Forestry Commission, Edinburgh.

GENT, J. (1955). The cultivation of ash in relation to *Prays curtisellus*, the Ash bud moth. *Journal of the Oxford Forestry Society* **4** (3), 9–13.

HIBBEN, C.R. and SILVERBORG, S.B. (1978). Severity and cause of ash dieback. *Journal of Arboriculture* **4** (12), 274–279.

NOVAK, V. (1976). *Atlas of insects harmful to forest trees*. Vol. 1. Elsevier.

PAWSEY, R.G. (1973). Tree diseases and the British environment. Chairman's address. Forestry sub-section, British Association for the Advancement of Science, Canterbury. *Commonwealth Forestry Review* **154,** 325–334.

PAWSEY, R.G. (1983). *Ash dieback survey, summer 1983*. C.F.I. Occasional Paper 24.

PEACE, T.R. (1962). *Pathology of trees and shrubs*. Oxford University Press.

ROSS, E.W. (1966). *Ash dieback: etiological and developmental studies*. State University College of Forestry. Syracuse University, New York. Technical Publication 88.

Discussion

J.C. PETERS (Department of the Environment, Bristol)

With reference to fears of Ash decline being related in some instances to damage caused by stubble burning; national census records obtained from satellite at Farnborough are available, which record areas of stubble burning over several years.

R.G. PAWSEY It is more important to look at the trees to see what effect stubble burning, as a possible factor in Ash decline, actually has on ash trees. Research into the ecology of ash in this country is urgently needed.

Dutch Elm Disease: The Vectors

C.P. Fairhurst and P.M. Atkins
Department of Biological Sciences,
University of Salford

Summary

The distribution of elm bark beetle species in Britain is reviewed and the factors which may be limiting their role in transmitting Dutch elm disease are discussed. Consideration is given to the application of trap and brood tree techniques to controlling the disease vectors.

Introduction

It is inevitable that Dutch elm disease, probably the greatest landscape tree disease of the Northern hemisphere, has produced fascinating research problems and necessitated the fundamental biological work upon which control strategies are based. Many of the recent laboratory and field experiments in Europe have been summarised (Burdekin, 1983). In Britain, the main research on the pathogen has been carried out by the Forestry Commission at its Research Station, Alice Holt Lodge. The ecology, distribution and effects of the bark beetle carriers of Dutch elm disease have been studied by a research team at Salford University during the past 10 years and this paper is a brief summary of progress over this period. Much of the research has still to be published in academic journals and investigations continue, for there are many areas where local authorities have successfully controlled Dutch elm disease and a few areas with many elms, which have still to experience this disease and its cost in terms of landscape, shelter and finance. For most of southern Britain, with the notable exception of East Sussex, Brighton and Hove, Dutch elm disease is a historical fact with recurrent problems which in turn become practical problems when the elm seedling and sucker growths are old enough to be re-infected.

The Elm Bark Beetle Species

Apart from root graft infection, spread of Dutch elm disease is effected by human carriage of infected material

Table 1. Bark beetle vectors of Dutch elm disease

Scolytus species	Length (mm)	Number of eggs per maternal gallery	Number of generations per year	Development time day degrees (°C)
S. scolytus	3.5–5.5	40	2(3)	930
S. laevis	3.5–4.5	30	?2	?
S. multistriatus	2.0–3.5	50	1(2)	1010
*S. pygmaeus**	1.5–2.5	55	2(3)	1005

*not recorded in Britain

and transmission by bark beetles of the genus *Scolytus*. Such insects infest healthy trees by 'crotch feeding' and the spores of the disease are introduced into the vascular tissue at this time. The beetles reproduce under the bark of elm trees and the young emerge carrying the disease which can be introduced into other elms. The insects are able to find other elms initially by tree odours enhanced by attractant substances produced by the beetles themselves.

In Europe, many species of bark beetles are recognised as potential vectors but in Britain there are only three species (Table 1). *Scolytus scolytus*, averaging 4–5 mm is present throughout Britain. Its large size and consequent spore carrying capacity and flying ability lead to it being considered as the most effective vector of the disease in this country (Webber and Brasier, 1984). *S. laevis*, thought of as the northern European vector of Dutch elm disease (Yde-Anderson, 1983), is a recent discovery in Britain, (Atkins *et al.*, 1981) although it has probably been present for a considerable time. The ability of this species to create pupal chambers in the sapwood (Kirby and Fairhurst, 1983) poses particular problems for control strategies based on debarking logs. *S. laevis* is now confirmed as a vector from experiments in which this species was collected emerging from diseased logs and caged in mesh screens on healthy English elms, 0.5 m tall. The trees contracted the disease, whilst a similar number of elms caged with *S. laevis* which emerged from disease-free logs remained healthy (Fairhurst, unpublished data).

There is relatively little overlap in geographical range between *S. laevis* and the smaller European bark beetle, *S. multistriatus*, which although too small to carry many spores of the pathogen (Brasier and Webber, *op. cit.*) are abundant in the south of Britain (Figure 1). Competition in the smaller branches where both these smaller beetle species can exist, may have ensured that the northward spread of *S. multistriatus* in the wake of the epidemic has been slowed. There may also be other species able to transmit Dutch elm disease, such as *S. pygmaeus*, which is common in the low countries of north-west Europe, but they have not yet been recorded in Britain.

Figure 1. Approximate geographical range of three elm bark beetle species.

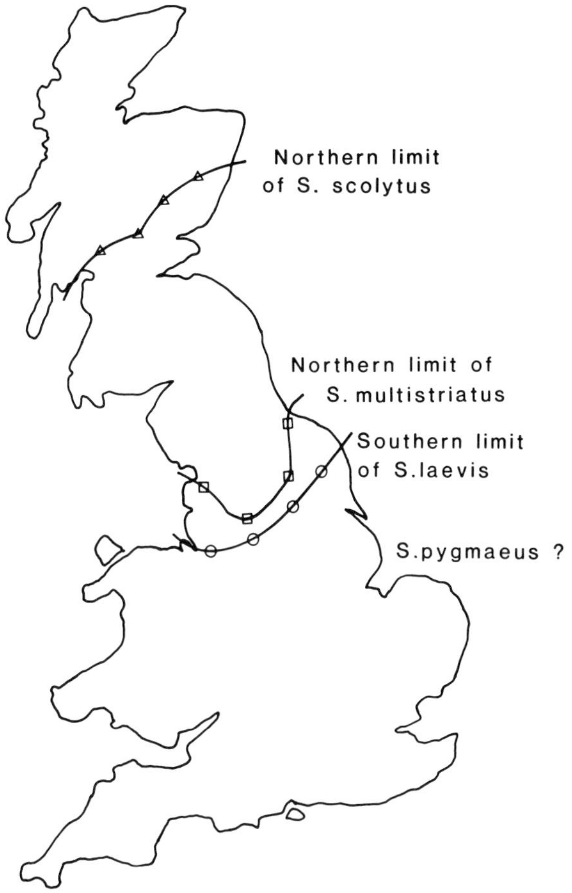

Elm Tree Preference

The beetles have preferences for varieties of elm and probably parts of the trees for breeding purposes. In general, English and Wheatley elms (*Ulmus procera* and *U. carpinifolia*) are preferred to Wych elm (*U. glabra*), and this, in part, explains differences in the incidence of the disease. Over several years of a sanitation programme in Blackpool during which about 12 per cent of the original total elm population succumbed to Dutch elm disease, 15 per cent of the original English elm population became diseased while the corresponding figures for Wheatley and Wych elm are 13 per cent and 7 per cent. Differences are even more marked if the 'brood trees' are considered. These are trees which became extensively colonised by larvae. Overall in Blackpool, only 1 in 18 diseased trees became brood trees, whilst the figures for English and Wych elms were 1 in 6 and 1 in 23 respectively. Indeed, it is rare to find *S. multistriatus* breeding in Wych elm in the wild, although this elm species can be used successfully to breed *S. multistriatus* in the laboratory. A major factor in restricting beetle breeding in Wych elm is believed to be the fungus *Phomopsis oblonga* which is antagonistic to beetle breeding (Webber, 1981) and is more common in the north and in Wych elms.

Figure 2. Annual totals of number of flight days (⩾22°C).

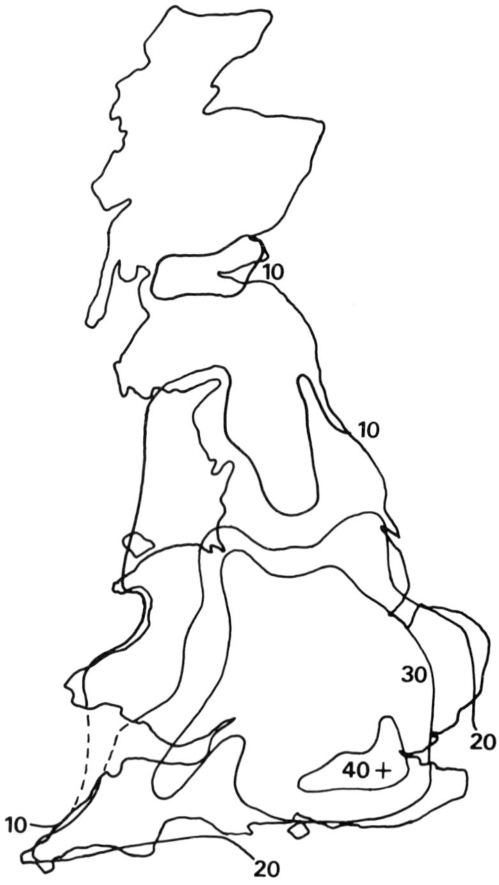

Beetle Dispersal

Beetle productivity and dispersal is dependent on climatic factors. In Britain winter cold temperatures are rarely severe enough to exert large mortality. This is due to the Atlantic climate in Britain and the insulation properties of bark. However, the effect of temperature on larval development determines how many generations are produced per year and there are indications that in the high summer temperatures of 1982, 1983 and 1984 partial extra generations were completed. It is known that beetle flight can occur above 18°C but in the field the majority of flying beetles are caught when the daily maximum temperature exceeds 21°C, particularly when this coincides with an emergence period (Fairhurst and King, 1983). Figure 2 indicates the number of these flight days taken from 25 year meteorological averages, and Figure 3 shows how this can alter between years. The distribution of temperature within the year is also important as indicated by 1982, which did not have a particularly high number of 'flight days' but such conditions coincided with emergence periods for the insects.

Normally beetle emergence begins in late May/early June and extends to the first week of September. However, with the unusually warm conditions of 1984, beetle flight was recorded at the end of April. Normally symptoms of new aerial infections of the disease can be noticed 3 weeks after the first major flight. Although such infections may occur some 8 km from known brood sources (Birch *et al.*, 1981) most attacks are confined to less than 500 m away from where the beetles emerge.

Practical Application

The question is how can this information aid control programmes? Certainly the fundamental basis of a regional policy is for sanitation of affected timber. It is when the expertise, manpower and finance is limited that Dutch elm disease attains epidemic proportions. Felling or pruning trees in the private sector or in woodlands is always a problem, and despite legislation, the willingness of owners or local authorities to provide payment and ensure prompt removal of diseased elms is usually lacking. Current research has suggested ways in which local authorities can pinpoint and concentrate on removal of the relatively few brood trees and log stacks, leaving the other diseased trees for later attention, as long as the risk of root graft is low. It is admitted that often the smaller beetle species may be breeding in the crowns of such trees, but a felling priority can be established. Sticky traps with attractants, while not within their own right a method of controlling beetle numbers, enable staff to know where and when to look for infections or beetle broods. Considerable research has gone into the development of such attractants or baits, and the result of one experiment is given in Table 2.

White plastic sticky traps were fixed to telegraph poles or trees other than elms at 2–3 m height and about 30 m apart. Traps were in place from 1 May to mid-October, throughout the flight season, and were deployed in triplets. In each triplet, one trap was left unbaited (blank), one was baited with the full multilure bait (4-methyl-3 heptanol (H), (−)-2-multistriatin (M) and (−)-2-cubebene (C)), and one with the two pheromone components only (H+C). All pheromone components were released from Conrel bellow fibre dispensers (Albany International, Mass., USA).

The trap catches confirmed the results of Grove (1983) that low release rates of 2-multistriatin (M) enhanced the responses of *S. multistriatus* to baits containing 4-methyl-3 heptanol (H) plus the host tree produced compound

Figure 3. Number of flight days (daily maximum >22°C) in Manchester and Blackpool 1961–1984.

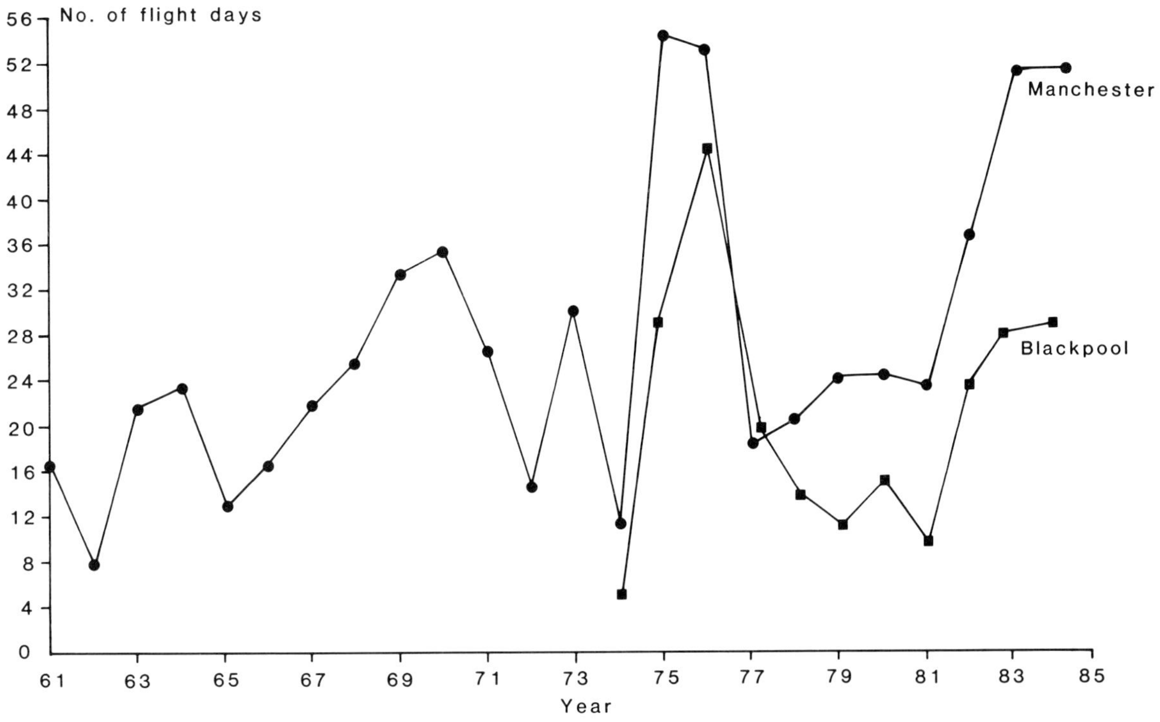

Table 2. Pheromone trap catches in Blackpool, Crosby, Wirral and Southport in 1984

BAIT (n = number of trap replicates)	Mean catch of *S. scolytus* (total catch)	Mean catch of *S. multistriatus* (total catch)	Mean catch of *S. laevis* (total catch)
*BLANK (n=15)	2.7 (41)	1.9 (28)	0.07 (1)
H + C (n=27)	25.0 (686)	22.0 (593)	0.2 (6)
MULTILURE (n=27)	21.3 (575)	99.0 (1436)	0.04 (1)

Key:
H – 4-methyl-3-heptanol.
C – (−)-d-cubebene (host produced)
M – (−)-d-multistriatin
MULTILURE – H+C+M

*1. Blank traps not used in Blackpool (12 replicates of { H+C, Multilure

2. Crosby, Wirral, Southport (5 replicates in each area { 5 × Blank, 5 × H+C, 5 × Multilure

2-cubebene (C). Although Grove (*op. cit.*) suggests that the release rate of M from the Multilure baits may not be optimal for *S. multistriatus*, the Multilure bait caught significantly more beetles than H + C alone. The addition of M to baits containing H does not have a significant effect on the number of *S. scolytus* caught (Blight *et al.*, 1978) and high release rates of M appear to reduce the catch (Blight *et al.*, 1983). This effect is confirmed by the 1984 trap results shown in Table 2. One further aspect which emerged from these results is that no single bait has yet been devised to attract both *S. scolytus* and *S. multistriatus*. Furthermore, no information about the pheromone of *S. laevis* is available. Only very small numbers of *S. laevis* were trapped in the 1984 experiment (Table 2).

Trials of the 'trap tree' technique, successful in North America, have not met with similar success in Britain. The idea is to attract large numbers of beetles to a few trees by injecting cacodylic acid into selected elms (O'Callaghan and Fairhurst, 1983; O'Callaghan *et al.*, 1984); once infested the treated trees can be safely disposed of. Unfortunately the influence of a relatively unpredictable climate and the presence of the fungus *Phomopsis oblonga* means that beetles frequently find the trees treated in this way unsuitable for boring and thus the method is only appropriate for trees which, by reason of ownership or access, cannot be felled promptly, to stop them producing further generations of beetles.

Northern Spread of Dutch Elm Disease

It is now clear that the spread of Dutch elm disease in northern Britain is a function, not only of active control programmes in some areas, but also of certain more natural phenomena. Amongst these is the greater proportion of Wych elms, less attractive to beetles and producing less beetle brood, the greater occurrence of *Phomopsis oblonga*, lower average temperatures for beetle development, fewer days suitable for beetle flight and in general the provision of more information about the disease, its incidence and carriage. Despite a general public attitude that Dutch elm disease is part of history, several local authorities are still successfully controlling the disease and therefore spreading felling and replacement costs over a longer period.

Future Control

Apart from individual tree treatments, which, like injection with Ceratotect are proven but costly, other methods such as *Pseudomonas syringae* and *Trichoderma viride* show promise and are the subject of continuing research. We are always interested in observing whether such treatments lead to fewer trees eventually becoming brood trees, and we hope to be able to report on this in the near future. Biological control methods may also be applicable to the beetles themselves, although the difficulties of following a life cycle under bark and obtaining acceptable replication makes testing basic control agents difficult.

Artificial agar media layered between wood and glass are currently being developed, so that larvae can be maintained under controlled conditions and direct observations can be made. With this procedure chemical and biological pesticides can be evaluated with no more background mortality than experienced in logs. Initial results from nearly 1000 larvae raised in this way and tested with 12 micro-organisms, show that *Trichoderma* spp. and *Bacillus thuringiensis* cause significant mortality (Yassim, 1984). This screening work continues and attention can be paid to ascertain if such biological control agents can be successfully applied to standing or cut timber.

Acknowledgements

Finance for further work has been provided by Salford University, CAVE fund, the Forestry Commission and the local authorities of Merseyside, Tyne and Wear County and District Councils, Blackpool, Colwyn Bay, Greater Manchester and Cheshire. Such funds have given employment to the research staff; Dr S.G. Kirby, Dr D.P. O'Callaghan, Mr P.M. Atkins, Dr J.F. Walsh, Mr E.M. Collins, Mr A.W. Wright, Mrs L. Adams and Mr A. Cartwright. The enthusiasm and assistance of the staff of the University, local authorities and the Forestry Commission has been an essential part of the programme. Material support and encouragement has also been provided by Dr J. Ricard of Bioinnovation.

References

ATKINS P.M., O'CALLAGHAN, D.P. and KIRBY, S.G. (1981). *Scolytus laevis* (Chapins) (Coleoptera: Scolytidae) new to Britain. *Entomologist's Gazette* **32**, 280.

BIRCH, M.C., PAINE, T.D. and MILLER, J.C. (1981). Effectiveness of pheromone mass trapping of the smaller European elm bark beetle. *California Agriculture*, Jan.–Feb., 6–7.

BLIGHT, M.M., FIELDING, N.J., KING, C.J., OTTRIDGE, A.P., WADHAMS, L.J. and WENHAM, M.J. (1983). Field response of the Dutch elm disease vectors, *Scolytus multistriatus* (Marsham) and *S. scolytus* (F.) (*Coleoptera:Scolytidae*) to 4-methyl-3 heptanol baits containing α−, β−, γ− or δ-multistriatin. *Journal of Chemical Ecology* **9**, 67–84.

BLIGHT, M.M., KING, C.J., WADHAMS, L.J. and WENHAM, M.J. (1978). Attraction of *Scolytus scolytus* (F.) to the components of Multilure, the aggregation pheromone of *S. multistriatus* (Marsham) (*Coleoptera: Scolytidae*). *Experentia* **34**, 1119–1120.

BURDEKIN, D.A. (ed.) (1983). *Research on Dutch elm disease in Europe*. Forestry Commission Bulletin 60. HMSO, London.

FAIRHURST, C.P. and KING, C.J. (1983). The effect of climatic factors on the dispersal of elm bark beetles. In, *Research on Dutch elm disease in Europe*, ed. D.A. Burdekin. Forestry Commission Bulletin 60, 40–46. HMSO, London.

GROVE, J.F. (1983). Biochemical investigations related to Dutch elm disease carried out at the Agricultural Research Council Unit of Invertebrate Chemistry and Physiology, University of Sussex, 1973–1982. In, *Research on Dutch elm disease in Europe*, ed. D.A. Burdekin. Forestry Commission Bulletin 60, 59–66. HMSO, London.

KIRBY, S.G. and FAIRHURST, C.P. (1983). The ecology of elm bark beetles in northern Britain. In, *Research on Dutch elm disease in Europe*, ed. D.A. Burdekin, Forestry Commission Bulletin 60, 29–39. HMSO, London.

O'CALLAGHAN, D.P. and FAIRHURST, C.P. (1983). Evaluation of the trap tree technique for the control of Dutch elm disease in northwest England. In, *Research on Dutch elm disease in Europe*, ed. D.A. Burdekin. Forestry Commission Bulletin 60, 23–38. HMSO, London.

O'CALLAGHAN, D.P., ATKINS, P.M. and FAIRHURST, C.P. (1984). Behavioural responses of elm bark beetles to baited and unbaited elms killed by cacodylic acid. *Journal of Chemical Ecology* **10**, 1623–1634.

WEBBER, J.F. (1981). A natural biological control of Dutch elm disease. *Nature*, London **292**, 449–451.

WEBBER, J.F. and BRASIER, C.M. (1984). The transmission of Dutch elm disease: a study of the processes involved. In, *Invertebrate-microbial interactions* (British Mycological Society Symposium 6), 271–306, eds. J. Anderson, A.D.M. Rayner and D. Walton. Cambridge University Press.

YASSIM, H.K. (1984). *Insect pathology: effects of biological control agents on Scolytus spp*. Unpublished MSc Thesis, University of Salford.

Discussion

J.C. PETERS (Department of the Environment)

Has there been any work on the photoperiodism of *Scolytus* beetles and if so what is its effect on the life cycle of the beetle?

C.P. FAIRHURST Unaware of any work on this.

Recent Advances in Dutch Elm Disease Research: Host, Pathogen and Vector

C.M. Brasier and J.F. Webber
Forestry Commission

Summary

Recent developments in 'short term' disease control such as the treatment of elms with *Pseudomonas*, *Trichoderma* and 'Ceratotect' fungicide are summarised. Developments in longer-term research are reviewed with emphasis on recent discoveries providing new insights into the processes of the disease or suggesting possible control approaches. These include the virus-like agent ('d-factor') of the fungus; the effect of the fungus *Phomopsis* on beetle breeding; research on the efficacy of disease transmission; recent spread of the strains/races of the pathogen; and the likely behaviour of the pathogen in the post-epidemic period in relation to the future of young elms.

Introduction

Dutch elm disease involves three different organisms – elm tree host, fungal pathogen and beetle vector – plus many accessory organisms ranging from viruses to man. Since this paper therefore encompasses several very wide fields of research we shall have to be selective in what we consider to be 'recent advances'. Fortunately Fairhurst and Atkins (1987) have covered recent advances in knowledge of the biology of the beetles so we shall consider research on the beetle *only* from the point of view of its behaviour as a vector of the disease.

Regrettably, the Dutch elm disease problem has not been solved. We are not, therefore, going to be able to present practitioners with a list of new disease management recommendations. What we shall attempt to do, in reviewing research progress, is to highlight some recent discoveries in what may be called 'longer-term' research which are producing a deeper understanding of how the disease 'works', and show where some of these could provide new insights into how the disease might be controlled in future. At the same time we will try to draw out any implications for current disease management where appropriate.

Some of the progress being made in research is exciting even if its rewards in control terms will not be immediate. If the slowness of progress in control seems disappointing to some, it must be remembered that some of the recent quantum leaps in the sister sciences of medicine and agriculture have come about mainly through heavy investment in painstaking 'in depth' research, laying great emphasis on the physiology and genetics and, most recently on the molecular biology of the organisms concerned.

Short Term Research

Protection or cure of high value amenity elms

Before moving on to longer-term research, it is important to briefly summarize progress in a vital area of short term research, that of the attempted control of the disease among high value amenity elms through some form of injection or inoculation. Three methods of 'individual tree' treatment are currently under investigation: injection with the bacterium *Pseudomonas* sp., inoculation with the fungus *Trichoderma* sp., and injection with the fungicide TBZ or 'Ceratotect'. The Forestry Commission has been involved together with others in trials of all three systems.

Table 1. Results of *Pseudomonas* experiments

1. Preventive injection[#] of English elm (Shi and Brasier, 1986)

Pseudomonas strains	*No. of trees used*	*Disease (mean % defoliation) after 13 weeks*	
Nine different isolates	98	95.1–100%	No significant difference in all nine treatments from control.
Control (*Pseudomonas* absent)	9	98.1%	

2. Preventive injection[#] of Commelin elm (Scheffer, 1983, 1984)

1982: *Pseudomonas strains*	*No. of trees used*	*Mean disease index (0–6 scale) after 9 weeks*	
WCS 085	9	1.8	All treatments significantly different from control.
WCS 361	9	1.0	
WCS 374	9	0.7	
M27+	9	0.7	
Control (*Pseudomonas* absent)	9	4.1	

1983: *Pseudomonas strains*	*No. of trees used*	*Mean disease index (0–6 scale) after 11 weeks*	
WCS 374 I (injected at 2.5 cm intervals)	10	2.3	Only treatment WCS 374 II differed significantly from Controls A & B. All other treatments and controls not significantly different.
WCS 374 II (injected at 5.0 cm intervals)	10	1.7	
Control A (no treatment)	10	3.4	
Control B (liquid medium only)	10	3.2	
Control C (liquid medium only)	10	2.8	

[#]Tree injected first with *Pseudomonas* then after an interval challenged with an inoculum of *O. ulmi* spores.

1. *Pseudomonas injection*

The bacterium *Pseudomonas* produces a strong antibiotic against fungi in culture. It was suggested by Strobel and Lanier (1981) that if injected into elms *Pseudomonas* might continue to live in the tree and produce *in situ* an antibiotic against the Dutch elm disease fungus *Ophiostoma (Ceratocystis) ulmi*. A strong wave of interest in *Pseudomonas* followed in the early 1980s, particularly in North America, including an International Symposium on Dutch elm disease in 1981 at which *Pseudomonas* was strongly highlighted (Kondo *et al.*, 1982).

During 1982–1983 Forestry Commission staff preventively injected over 300 clonal English elm (*Ulmus procera*) with *Pseudomonas* after first screening bacteria for antimycotic activity (Shi and Brasier, 1986). Typical results are shown in Table 1. As can be seen, no difference was obtained between preventively injected trees and control trees, and it was concluded that *Pseudomonas* did not offer any useful preventive control potential on English elm. Scheffer (1984) working in Holland has had similar poor results with the Smooth-leaved elm (*U. carpinifolia*). However, in additional experiments which included the preventive injection of 'Commelin' hybrid elms with the *Pseuduomonas* bacterium, he observed a reduction in disease symptoms (Scheffer, 1983; 1984). Table 1 summarizes some of his best results with Commelin which indicate a striking suppression of disease level (1982 results), and some of his worst results (again with Commelin) which are more equivocal (1983). Scheffer has

Table 2. Results of preventive experiments with *Trichoderma*

(Fairhurst, Atkins & colleagues and Greig & Hickman: unpublished data)#

1. Artificial inoculation with *O. ulmi**

	Trichoderma treated trees		Control trees		
	Total number treated	Mean disease levels (% defoliation)	Total number	Mean disease levels (% defoliation)	
Farnham (1983-4)					
English elm	10	93%	10	94%	NS
Huntingdon elm	10	31%	10	31%	NS

2. Natural infection with *O. ulmi*

	Trichoderma treated trees		Control trees		
	Total number treated	Number of trees developing disease (% of total)	Total number	Number of trees developing disease (% of total)	
Heavy disease area					
Bexhill (1984)	100	26%	100	27%	NS
Light disease areas					
Blackpool (1984)	967	2.1%	990	4.1%	S
Tyne and Wear (1984)	203	8.4%	211	12.8%	NS

Data presented by kind permission of these authors.

* Tree inoculated first with *Trichoderma* pellets, then after an interval challenged with an inoculum of *O. ulmi* spores.

NS = treatment not significantly different from control; S = treatment significantly different from control ($p < 0.5$)

also obtained what appears to be a more dramatic suppression of disease levels in the hybrid elm 'Belgica' although the results were not statistically analysed (Scheffer, 1984). Thus *Pseudomonas* has so far only shown a significant effect in elms such as 'Belgica' and 'Commelin' that contain an element of *U. glabra* in their ancestry.

To summarize therefore, there is no evidence of any practical use of *Pseudomonas* injection on English elm or on the Smooth-leaved elm (and this probably includes Wheatley elm), but definite evidence of some form of disease suppression on 'Commelin' and 'Belgica', although the effect is not consistent enough to make this method anything more than of further research interest at present. The intense interest in the field injection of *Pseudomonas* for disease control in North America appears, so far as we can tell, to have declined.

2. *Trichoderma*

Several species of *Trichoderma* produce antibiotics against other fungi, especially in agar culture. Following suggestions about the possible effectiveness of *Trichoderma*

Table 3. Curative injection with TBZ (Greig and Coxwell, 1983)

1. Treatment of trees with new infections (≤ 5% of crown diseased) 1978–1981, Hove:

Number of TBZ injected trees	% Recovered from disease	Number of non-injected control trees	% Recovered from disease
37	83.8%	23	8.7%

2. Bioassay results; mean % inhibition of *O. ulmi* growth by twig samples after TBZ injection:

Weeks after injection (1979)	% inhibition of *O. ulmi* growth	Number of twig samples
1	87.5%	56
9	78.5%	56
42	55.4%	112
70	52.5%	80
104	13.7%	80

against Dutch elm disease (Ricard, 1983; and in commercial literature), the Salford University group and the Forestry Commission have implemented field trials. The results have been assessed co-operatively between the two groups, the Forestry Commission visiting the Salford trials and Salford being involved with the Forestry Commission trials. Table 2 summarizes the main data obtained from preventive inoculation experiments. Those at the top of the Table are from experiments involving artificial inoculation with the fungus, comparable to the *Pseudomonas* trials carried out in Britain and Holland. The remaining three sets of data are from field trials in which the trees were exposed to natural infection.

No suppression of disease was obtained on either English elm or on the hybrid Huntingdon elm in the experiments involving artificial inoculation. Similarly, in the field trial at Bexhill, a heavily diseased area, *Trichoderma* inoculation had no effect on disease levels. There is, however, some evidence of an effect by *Trichoderma* in the Blackpool trial, a very lightly diseased area where disease was around the 4 per cent level in 1984.

In the *Pseudomonas* experiments there was little evidence, perhaps hardly surprisingly, for any significant upward movement of the bacterium in the tree from the injection point, although it did appear to reach the roots in some cases (Shi and Brasier, 1986). *Trichoderma* also shows little or no movement away from the point of introduction. Furthermore, *Trichoderma* appears to have exerted some effect only in areas where disease levels are low (as in Blackpool), while *Pseudomonas* tended to suppress disease symptoms in elms that were already moderately resistant to the Dutch elm disease pathogen. One is left, wondering therefore, whether any disease reduction caused by inoculating an antibiotic producing bacterium or fungus into a tree may be due as much to the triggering and intensification of the tree's internal resistance mechanisms as to the presence of an alien organism and its antibiotic effect on *Ophiostoma ulmi*. There may be interesting avenues here for longer term research.

3. *TBZ fungicide*

After many years of experimentation and development, often with rather disappointing results, a systemic fungicide has been found which has acquitted itself well in field trials. This is the fungicide TBZ (thiabendazole) or 'Ceratotect' (Stennes, 1981; Greig and Coxwell, 1983).

Table 3 shows the results of Forestry Commission trials of *curative* injection with the fungicide on trees naturally infected with the disease. The data are consistent, and show that this fungicide can be used curatively with a high probability of success on new infections of ≤ 5 per cent disease in the crown (Greig and Coxwell, 1983). Two North American workers, Stennes and French, have carried out similar carefully controlled experiments on American elm (*U. americana*) with TBZ. They have stated that "when properly applied the chemical will thoroughly protect American elms from artificial inoculation with an aggressive strain of *O. ulmi* for at least 13 months" (Stennes, 1981). Unfortunately, the data on which this statement is based have yet to be published. However, the suggestion that TBZ may give a degree of protection for a second season, is supported by Forestry Commission bioassay data on levels of residual fungicide in trees 12 or more months after injection (see Table 3).

The attraction of TBZ injection is that, although not cheap and requiring specialised equipment, if used curatively it need not be applied to large numbers of trees, but simply to treat new infections. It could therefore be a useful part of an integrated control programme in a large urban area, or suitable for a country landowner with a small number of important specimen elms. For recommendations of use see Greig and Coxwell (1983).

4. *Fenpropimorph*

Since this article was written in mid 1985, the popular press has reported on research carried out in the Netherlands which suggests that the antifungal chemical, 'fenpropimorph', may be effective in the curative injection of quite heavily diseased elms. This discovery arose from the search for a chemical that interfered with the ability of *O. ulmi* to switch from its specialised yeast-like mode of growth to its more usual mycelial growth mode. We understand that this research is still at the developmental stages; trials on the efficacy, persistence and phytotoxicity of fenpropimorph have yet to be completed. The Forestry Commission hope to conduct field trials with this 'new' chemical in 1987.

Risks of short term approaches

In an epidemic the like of which we are experiencing with Dutch elm disease, the sort of short term research approaches summarized above are a vital part of an overall research strategy. However, falling within the short term context there have been suggestions for a number of miracle cures for Dutch elm disease over the past three decades, particularly in North America. Some rather optimistic claims have been made for their efficacy sometimes in the press and sometimes in advertising literature. Many of these cures come and go rather like modern-day clothes fashions, but they can bring attendant problems for which arboriculturists should be on guard.

(i) Firstly there is a risk that, however unproven a short term remedy may be, a local authority or other organisation may be tempted to opt for it and in so doing divert its resources away from measures of initially greater expense but otherwise of proven effectiveness such as sanitation control.

(ii) Another concern is that short term methods are most often directed to control of the disease amongst what are relatively small numbers of urban elms, whilst the enormous problem of disease control in the vast mass of the elm population in the countryside tends to be largely neglected.

(iii) A third consequence of short term remedies – to the scientist at least – is that if they become temporarily fashionable, pressures often tend to build up that can result in the diversion of finite research resources away from in-depth research aimed at understanding the underlying causes of the disease, and into the short term effort. If resources are already stretched this can be quite damaging to the balance of a research programme particularly since, as mentioned above, it is through the effort of painstaking longer term research that a basis is most likely to be found from which lasting disease control may spring. At present both in North America and Europe in-depth research on the biology of host, pathogen and vector is at a very low ebb.

Longer-term Research

We shall now examine some recent developments in the longer term research context to show how our understanding of Dutch elm disease is changing, although of necessity restricting ourselves to certain key topics.

Spread of the disease: the new strains and races of the pathogen

One fact which has emerged with increasing force and poignancy from the research of the past decade is that the epidemic in Britain is not an isolated event, but a small part of a much larger disaster extending across the Northern Hemisphere. A summary of what we have learned so far is:

(i) The aggressive and non-aggressive strains of the fungus were identified over a decade ago (Gibbs and Brasier, 1973; Brasier and Gibbs, 1973). Dutch elm disease probably only arrived in Europe and North America during the early part of this century. The non-aggressive strain is now believed responsible for the first epidemics of the disease in Europe, North America and south-west Asia during the 1920s–40s.

(ii) The aggressive strain is responsible for the present second epidemic of the disease. It has also recently been discovered that there are two distinct forms or races of the aggressive strain: the Eurasian (EAN) race and the North Amer-

Figure 1. Summary of the known distribution of the EAN and NAN races of the aggressive strain of *Ophiostoma ulmi* in Europe and south-west Asia in 1983. Based on over 1500 samples. From Brasier (1987).

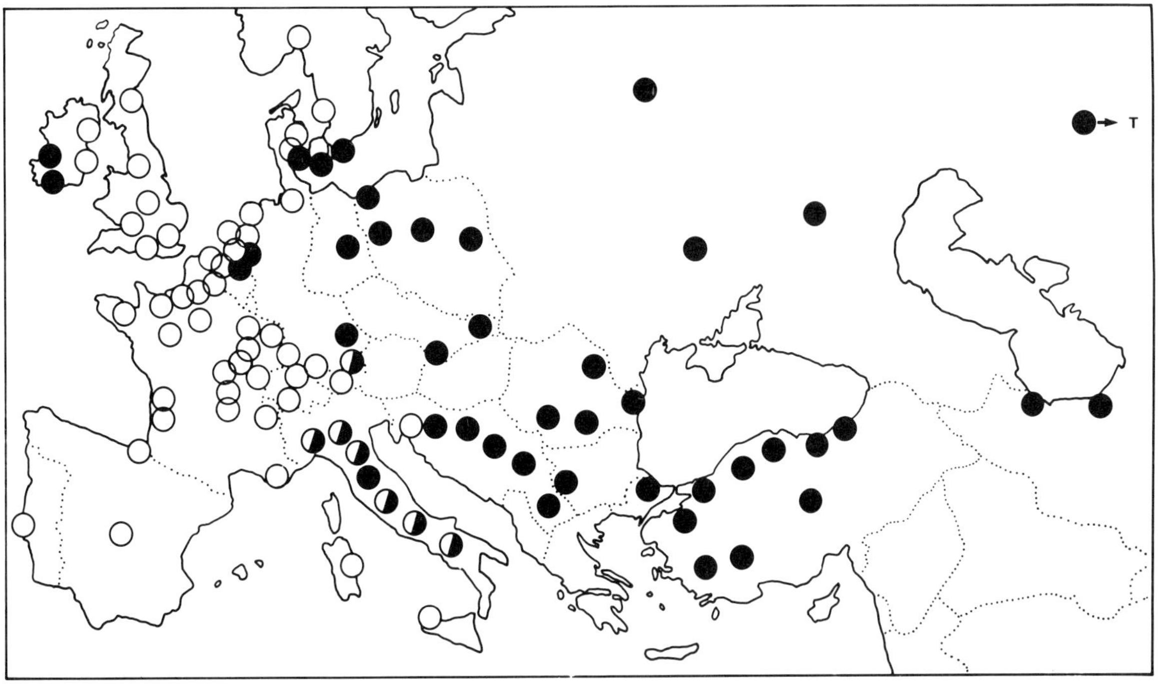

T=Tashkent.

ican (NAN) race. Their distribution is shown in Figure 1. The NAN aggressive was probably introduced into Britain from North America on diseased elm timber (Brasier and Gibbs, 1973) and has since spread into neighbouring parts of Europe (Brasier, 1979). The EAN aggressive has for some decades been migrating overland from central Europe or from further east (Brasier, 1979).

Thus Europe is currently experiencing two epidemics, one originating from the west and one from the east. Whereas these epidemics have been under way in parts of North America, in Britain and Romania for many years now, they are only just building up momentum in countries such as Spain and Sweden (Brasier, 1982, 1983*a*, 1986*b*, 1987). Britain has so far suffered an estimated loss of 20–25 million elms as a result.

It is clear Britain is only on the edge of a much larger event, which may have its origins in the east. The geographical centre of origin of the disease before its arrival in Europe in the early 1900s, and hence the possible source of the three different strains and races of the pathogen, is generally supposed to be in eastern Asia. This hypothesis is based largely on the fact that Asia's elms are more resistant to the disease (Heybroek, 1976) and also on evidence that China appears to be the main centre of elm diversity (Heybroek, 1966). However although the epidemic caused by the EAN aggressive race extends at least as far as Iran and Tashkent (Brasier and Afsharpour, 1979; Brasier, 1982) the status of the disease in central and eastern Asia is almost unknown. It is important to try to identify the centre of origin of the disease for several reasons, including the possibility that there may be further strains of the fungus there. Interestingly, a single isolate of *O. ulmi* from the Himalayas does not conform to the non-aggressive, EAN or NAN aggressive strains (Brasier, 1983*b*).

A point to emerge from recent studies is that, if Britain had not had an epidemic caused by importation of the NAN agressive race from America, it is very likely that the EAN aggressive race would have reached our shores from Belgium, Holland, or southern Ireland. Indeed, a sad aspect of the present situation is that the EAN aggressive race – discovered only in 1979 – has been rolling westwards across Europe causing a massive new epidemic since the 1940s, yet this event received virtually no attention in the European forestry literature.

Future behaviour of the disease and the future of young elms

There is now a complex situation in Europe: three forms of the pathogen – the 'old' non-aggressive strain, the NAN aggressive and the EAN aggressive strain all thrown together in the same ecological niche. An important question for the future of the elm is what form or forms of the fungus will emerge victorious from this melting pot? In particular, what is the future for the large numbers of elm suckers and seedlings now appearing in the older epidemic areas where the mature elms have died? In Britain this question is particularly pertinent to the future of English elm root suckers, which at present are growing up in large numbers. Will the aggressive strain (NAN) die out as the large elms are killed, perhaps to be replaced again by the non-aggressive, or will it survive to return and kill the next generations of elm saplings and suckers?

This problem has been addressed both through laboratory research on the genetics of the pathogen, and through field monitoring of changes in the *O. ulmi* population at a number of sites in Britain and Europe (Brasier, 1982, 1983*a*, 1986*b*, 1987). The results of this research has led to the following qualified prognosis, summarized in Figure 2:

(i) the old non-aggressive strain will die out in competition with the aggressive strain, which will replace it (Figure 2 C–E).

(ii) future generations of young elms will continue to be attacked by some form of the aggressive strain once they are large enough to support a breeding population of beetles (Figure 2 G).

(iii) unless the aggressive strain attenuates in some way, i.e. it becomes less pathogenic, or the beetle is hit by a disease of its own, the elm in the shorter term is likely to be reduced to a scrub or marginal population with few trees, if any, reaching maturity (Figure 2H). Most surviving mature trees are likely to be escapes in woodlands, on islands and in upland valleys. Indeed, most large surviving single elms or groups of elms are best assumed to be escapes *unless* shown to be otherwise by experimentally controlled inoculations with the pathogen.

This is a bleak prospect. Any hope for the future lies in a balance being achieved between the genotype of the elm and that of the pathogen. At present our native elms are much too susceptible – or, the pathogen is much too aggressive. In this disastrous situation, for which man probably holds no small amount of responsibility, can we try to bring about a return to a better balance of elm and pathogen? One way is to consider breeding resistant elms.

Elm breeding

Breeding elms for resistance to the disease, is a long term research process which aims to raise the host's genetic level of resistance to the pathogen. Elm breeding is presently going on in at least seven centres: Wisconsin, Ohio and Washington DC, USA; Manitoba, Canada; Wageningen, the Netherlands; Florence, Italy and Volgograd, USSR. The intensity of interest in elm breeding (indeed the breeding programme in Florence has only recently begun) reflects a remarkably sustained interest in the elm in spite of Dutch elm disease, and is a tribute to the properties of the elm as a shelter and landscape tree which combines its beauty with toughness in terms of tolerance of wind, drought, cold, and salt spray. Some of these programmes are of long standing – the Dutch breeding programme, for example, dates back to the 1930s and has thrived in spite of a number of traumas, including the discovery of the aggressive strain of the disease.

Good progress is now being achieved in breeding for moderate to high levels of resistance to the aggressive strain through the incorporation of more resistant Asiatic parent elm material. Some very promising and interesting material is being released, and some of the product names are familiar: 'Sapporo Autumn Gold', 'Regal Elm', 'Jacan Elm', 'Lobel', 'Plantijn' and others (Heybroek, 1982, 1983; Lester and Smalley, 1972; Smalley, 1984).

These products are intended mainly for local consumption, e.g. as shade for the urban streets of the USA, for the hot climate of southern European cities, for wind and salt-tolerant shelter belts in the Dutch polders and for shelter belts in the hot arid areas of central USSR. The tolerance of many of these products to British climatic conditions has yet to be proven. A number of clones from the Dutch elm breeding programme and more recently some American material are undergoing trials in Britain as part of a wider EEC scheme initiated in 1979–83 (Heybroek, 1983). None of the programmes are presently committed to producing the equivalent of a disease resistant 'English elm' and one should not expect the elms produced to look like 'English elm'. Rather than having a dense globular form as in English elm, many of them have a rather open crown, are very fast growing and larger leaved, and have been said to look rather like poplars!

It is vital that while elm breeding is going on, behaviour of the fungal population is monitored so that elms are screened for resistance to any new variants that may arise, and so that breeders may know just what it is they are breeding for resistance to. The very latest research indicates that EAN/NAN aggressive hybrids are emerging in of all places Holland! (C.M. Brasier, unpublished). They are likely to arise in other areas where both NAN and EAN are present, and may be a pointer to different behaviour of the pathogen in future (Brasier, 1986*b*).

Figure 2. Changes in the size of the elm, fungus and beetle populations during the present epidemics of Dutch elm disease. Probable course of events from the pre-epidemic period (A), throughout the epidemics (B–F) to a possible outcome in the post-epidemic period (G–H). At A, host, pathogen and vector populations are in reasonable balance. At C, the non-aggressive strain population explodes following the arrival of the aggressive strain. At E, the non-aggressive strain is replaced by the aggressive, leaving future generations of young elms to be attacked by the aggressive strain, G–H. From Brasier (1987).

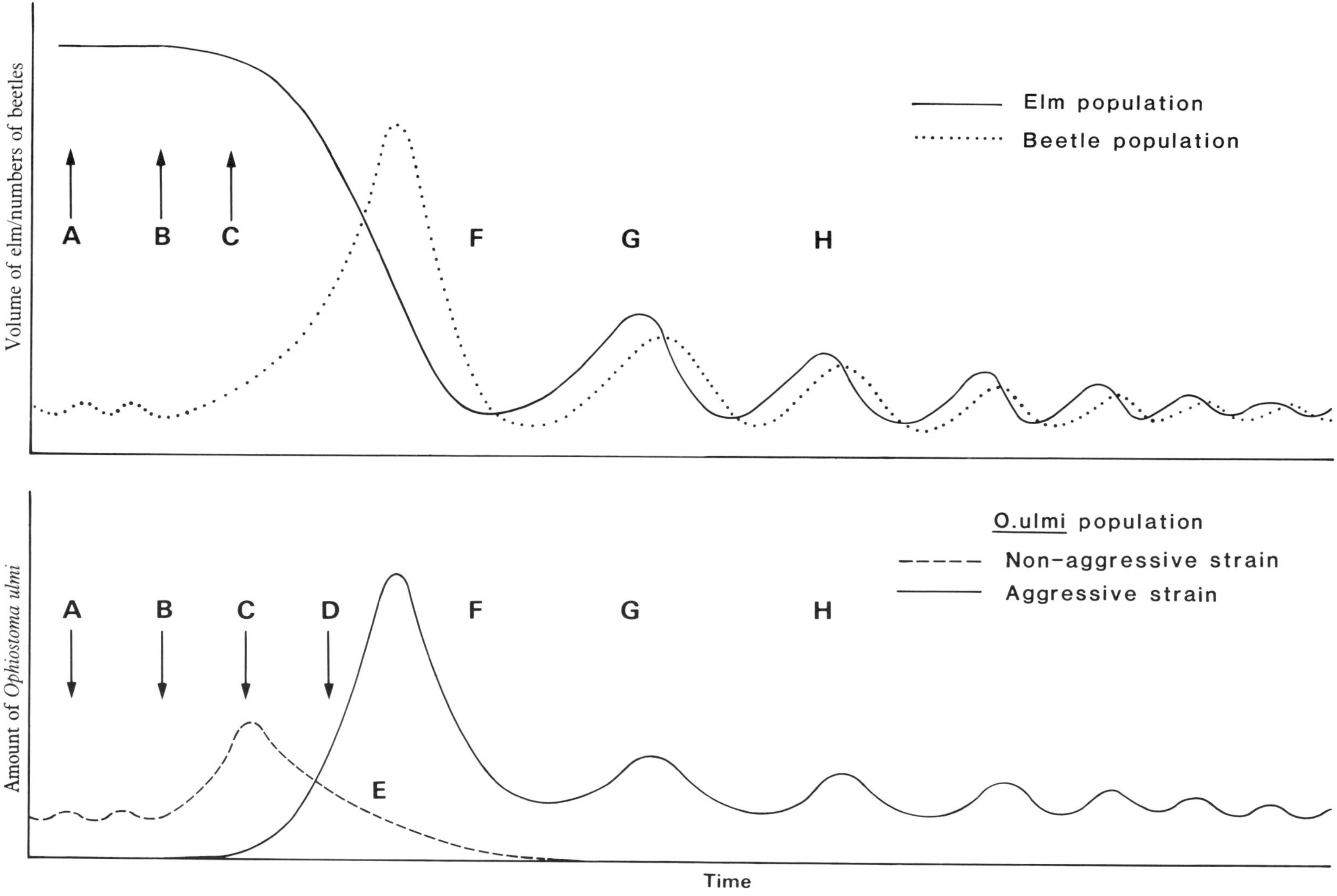

Clonally propagated elms are unlikely to be planted in large numbers outside urban areas or shelter belts, so the problem still remains of the low level of resistance to the aggressive strain in the bulk of the elm population, which consists of large numbers of sucker and sapling elms coming up where mature elms have died in the countryside. It is in the countryside that the real battle of Dutch elm disease control, that for a natural balance between host and pathogen, will be fought. Consideration could be given to ways of raising the base-line of disease resistance in native European field elms, for example by encouraging the spread of exotic disease resistant species with similar arboricultural properties to the native elms, such as forms of *U. japonica*. This might be achieved either by scattering seed, or in areas where European elms are fertile, through the release of pollen.

Another risk is that other forms of the fungus may be lurking in the east (e.g. China or the Himalayas) that could pose a threat to our resistant elms and all the hard work that goes into producing them. Fortunately there is increasing evidence that the genetic basis of resistance in the elm is a complex polygenic character (Lester and Smalley, 1972), which should prove fairly durable providing the level of aggressiveness of the pathogen does not show a gradual quantitative increase as a consequence of the presence of larger numbers of highly resistant elms. The latter is another question of great interest which may be illuminated, at least in part, by the current genetical research into the pathogen's population behaviour (Brasier, 1986*b*, 1987).

The beetle as vector: beetle feeding preference – Wych versus English elm

During the earlier stages of the present epidemic in Britain, Forestry Commission surveys showed that the Wych elm was surviving rather better than the English elm in the field. Indeed, the incidence of disease in English elm was significantly higher than in both Wych and in Smooth-leaved elm (Gibbs and Howell, 1972, 1974). It therefore became popularly assumed that Wych elm was more resistant to the aggressive strain of the fungus than was the English elm. However, when young English and Wych elm were inoculated with the pathogen, Wych elm was shown to be more, not less susceptible to the fungus than English elm (Brasier, 1977). It was therefore suggested that the better field performance of Wych elm in the present epidemic was probably due to its relative unattractiveness to the beetles rather than to its intrinsic resistance to the fungus (Brasier, 1977).

To examine this possibility, beetle feeding preference experiments were carried out by Webber and Kirby (1983) in which beetles were released in fruit cages containing randomised blocks of young English elm and Wych elm

Figure 3. Results of a beetle feeding preference experiment, showing the marked preference of the larger European elm bark beetle, *S. scolytus*, for feeding on English elm. From Webber and Kirby (1983).

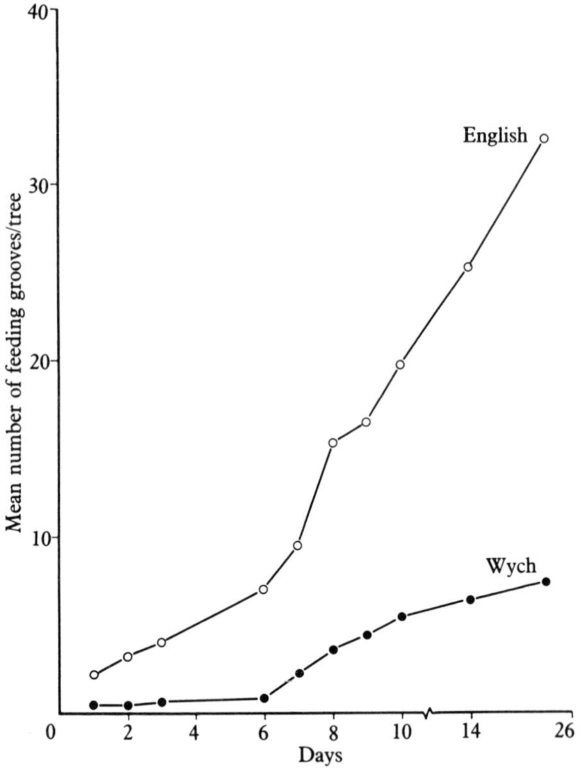

trees. The experiments showed very clearly that the larger European elm bark beetle, (*S. scolytus*), had a strong preference for feeding on English elm, as shown in Figure 3.

It seems very likely, therefore, that the better early field performance of Wych elm was due to preferential feeding by the beetles on English elm. In many areas, however, after the big English elms had gone, the Wych elms too have been killed, presumably as the beetles were forced to turn their attention in that direction. The same has probably occurred where there have been mixtures of English with either Smooth-leaved or Huntingdon elms. The better field performance of *U. laevis* compared with *U. carpinifolia* in Europe (Maaslov, 1970) might also be explained in terms of feeding preference.

The occurrence of beetle feeding preferences is valuable knowledge so far as elm breeding is concerned, since it offers the possibility of breeding for resistance to the beetle as well as for resistance to the fungus. None of the present elm breeding programmes involve any screening for resistance to the beetle. It is clearly important, in future research, to attempt to identify those factors which make elms unattractive to beetles for feeding, so they can be incorporated into the products of elm breeding programmes. In addition, for those involved in urban disease control, an awareness of the greater risk of scolytid feeding on say English as opposed to Wych elm, could well be helpful in strategic planning.

Natural biological control of beetle breeding by *Phomopsis*

Another recent discovery regarding the field performance of the beetle vector is that the naturally occurring fungus *Phomopsis oblonga* can exert a detrimental effect on beetle breeding (Webber, 1981). The fungus invades the nutrient-rich inner bark of elms dying from Dutch elm disease. In a series of experiments it was shown that, when given a choice, beetles refused to breed in *Phomopsis* invaded trees. When forced to breed in such trees, breeding was almost completely unsuccessful, and any larvae in the bark failed to pupate successfully and yield another generation of beetles. Webber (1981) concluded that the main reason for this effect was that the fungus reduced both nutrient and moisture levels in the bark, so that the bark became both unattractive to the breeding beetles and deleterious to their larval development.

Phomopsis is particularly common in the bark of Wych elm, and is especially common on bark of elms in the north and west of Britain (Webber and Gibbs, 1984). It is probably a major factor in reducing beetle numbers in potential brood trees (Fairhurst and Atkins, 1987) and combined with the effects of colder climate, has almost certainly been responsible for the significantly slower rate of spread of the disease in the north of England and in Scotland. Hence *Phomopsis* is an important ally of sanitation control programmes in these areas. The effect of *Phomopsis* is also likely to become particularly critical to the beetles in the post-epidemic period when their population will be much smaller as a result of the destruction of most mature elms (Figure 2 G & H).

Spore loads of the beetle vectors

Healthy elms become infected from spores carried on the surfaces of the vector beetles. A new generation of beetles leave the bark of elms in the spring and fly to twig crotches at the tops of healthy elms to feed. Any spores transferred from the beetles to the feeding grooves in the twig crotches may germinate and infect the xylem or sap-stream of the tree. Surprisingly, this crucial process, the weakest link in the disease cycle, has received very little research since the 1930s. The Forestry Commission has initiated a project in this area, which is supported by the Pilkington Trust. This research has shown:

Figure 4. Comparative *O. ulmi* spore loads on a sample of the smaller European elm bark beetle, *S. multistriatus* (above) and the larger European elm bark beetle, *S. scolytus* (below). Spore loads of ≥ 10 000 spores may often be required to achieve infection. Beetles without any spore inoculum indicated as unshaded squares. From Webber and Brasier (1984).

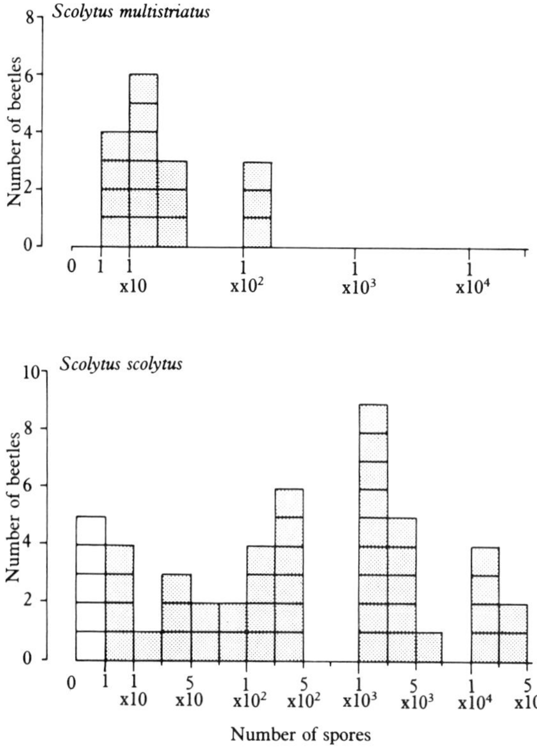

(i) the larger elm bark beetle (*S. scolytus*) carries many more spores than the smaller European elm bark beetle (*S. multistriatus*) (Figure 4). Since as many as 10 000 spores may be needed on a beetle leaving the bark to secure infection on English elm, *S. scolytus* is likely to be a much more effective vector than *S. multistriatus* (Webber and Brasier, 1984).

(ii) 60–90 per cent of beetles leave the bark contaminated with *O. ulmi* but only 10–50 per cent of beetles arrive at feeding grooves still contaminated. About 30 per cent of feeding grooves become contaminated with *O. ulmi*, and only 3–5 per cent of all feeding grooves lead to infection (Webber and Brasier, 1984).

Figure 5. The transmission of a d-factor from a slow-growing severely d-infected aggressive strain (below, white arrow) to a colony of a healthy isolate (above). Growth at the margin of the healthy colony has been arrested (black arrows) as a result of the transfer and spread of the d-factor.

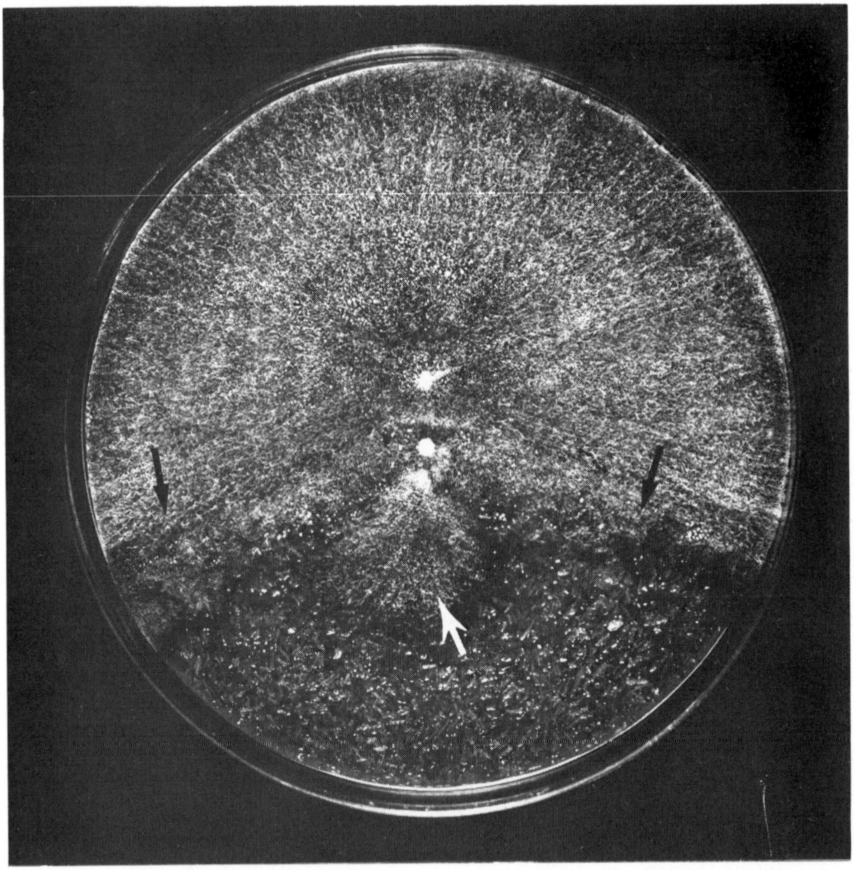

These latter observations indicate that during beetle flight and beetle feeding there is a very large fall-off in numbers of viable spores on the beetles. The pathogen is clearly very vulnerable at this time. Recent experiments suggest the main cause of this fall-off is desiccation of the spores (J.F. Webber, unpublished). It is also apparent that in control terms the longer beetle feeding is delayed the better, since more spores will die. There are two ways in which breeding elms for resistance could help here. Resistance to beetle feeding could delay eventual feeding of those beetles making a 'wrong' first choice of host, leading to death of their spore inoculum. Breeding for resistance to the fungus should raise the threshold number of spores required for infection from say 10 000 (the estimate for English elm) to some unknown figure – say 50 000 spores.

These results also suggest that when carrying out sanitation control, it is important to dispose of diseased elm material carrying broods as far away as possible from healthy elms. The further the beetles have to fly, the more likely it is that any inoculum of the pathogen they are carrying will fail to survive.

A disease of the pathogen

One exciting disovery to emerge in the past few years is that the fungus itself has a disease (Figure 5).

This disease or 'd-factor' is a virus-like agent which spreads from infected to healthy forms of *O. ulmi* across fusions between the hyphae or mycelial threads of the fungus. Infected isolates of *O. ulmi* are much less vigorous in growth, their spore germination is severely impaired, and their production of fruit bodies is reduced (Brasier, 1983*c*, 1986*a*). Present evidence indicates that the d-factor probably exerts its most deleterious effects on the fungus at two points in the fungal cycle:

Figure 6. Decline of the non-aggressive strain as a proportion of the *O. ulmi* population in Holland from 1974–1980. Adapted from Brasier (1983*a*).

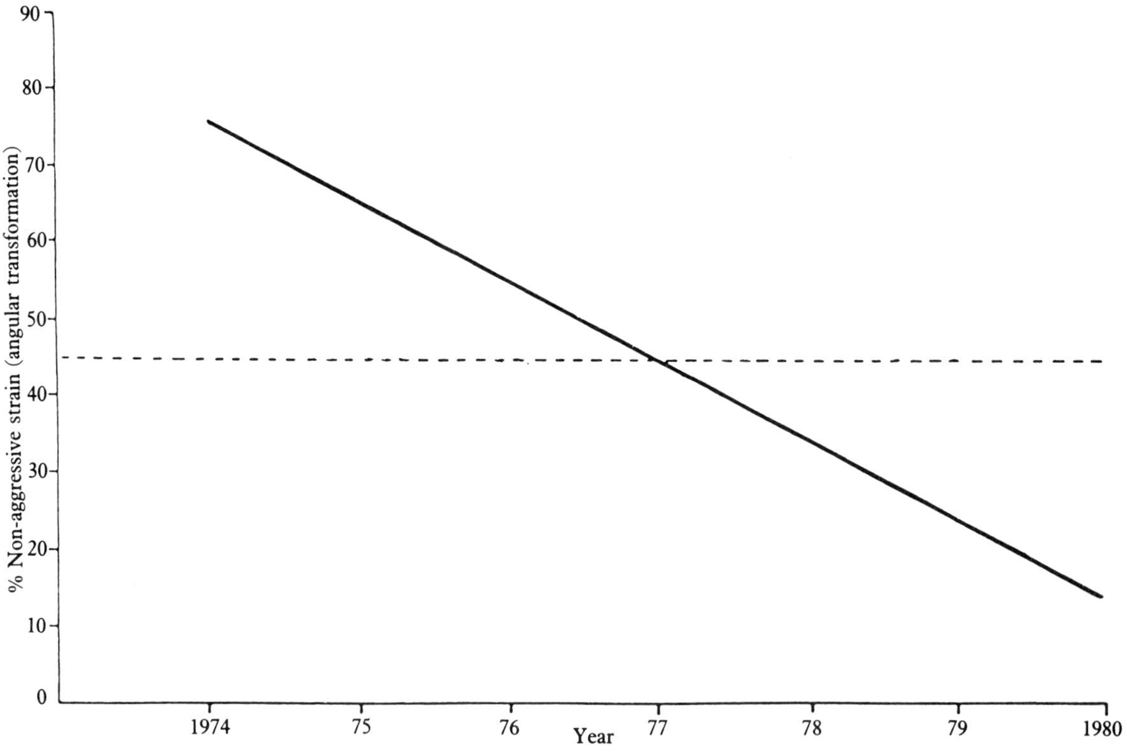

(i) during the long overwintering saprophytic phase of the fungus in diseased elm bark (Webber and Brasier, 1984); this is probably also when most spread of the d-factor occurs.

(ii) when the fungal spores are on the beetle surface – during beetle flight and feeding.

Two interesting questions arise from these observations. Firstly, could the d-factor cause a significant attenuation in the pathogenicity of the aggressive strain during the post-epidemic period (Figure 2 G & H) and in doing so restore the balance of host and pathogen by exerting a natural biological control on the fungus? A similar virus-like disease of the Chestnut blight fungus *Endothia parasitica* appears to have allowed recovery of the Chestnut in northern Italy from the Chestnut blight epidemics of the early 1950s (Grente and Sauret, 1969; Elliston, 1984).

Secondly, as a means of artificial biological control, could we deliberately breed and release beetles carrying spores of the pathogen infected with rather 'nasty' d-factors in order to spread the d-infection in the aggresive strain population. Such an approach might be used locally as part of an integrated disease control programme, or more likely as part of a geographically wider scheme aimed at reducing or attenuating the overall level of pathogenicity of the aggressive strain.

Much more research is needed both in the laboratory and in the field before the answers to these questions become clear, including work on the molecular basis of d-factors (Rogers *et al.*, 1986) and on ways in which the fungus may resist the effects and spread of d-factors (Brasier, 1984, 1986*a*).

Breeding the pathogen?

An interesting fact about the present epidemic revealed in recent research is the rapidity with which the 'old' non-aggressive strain is being replaced by one or other form of the aggressive strain (Figure 6). At the sites in Britain which have been researched in detail the non-aggressive appears to be declining at about 10 per cent per annum, and the same appears to be true for other sites in Europe and North America. As already mentioned, the non-aggressive strain may be heading for extinction.

How does this replacement occur? Although the different virulences of the aggressive and non-aggressive strains are probably a major factor, there is evidence of a variety of other contributory factors including possibly even direct antagonism between different individuals of the fungus (Brasier, 1986*b*). Certainly, the rapid demise of the non-aggressive strain raises an intriguing question. Could the same thing be done to the aggressive strain through some form of genetic manipulation? Would it be possible using some of the more recently acquired knowledge of the genetics and ecology of the fungus to attempt a radical new approach to control by taking a leaf out of the elm breeders book and, as well as breeding the elm, try to breed a new less harmful form of the fungus that would compete with and eventually replace the aggressive strain.

It certainly looks on paper as if it might be possible to breed such a 'model fungus' using the pathogen's own breeding system. It would need to be a weak pathogen, producing an acceptable level of disease much like the non-aggressive strain, and it would need a whole range of other attributes that would enable it to survive where it seems the non-aggressive has not. As knowledge of the biology of *O. ulmi* increases, the chances of producing a suitable 'model-fungus' through existing techniques and through the application of genetic engineering are likely to increase further (for a theoretical discussion of the possibility of breeding *O. ulmi*, see Brasier, 1986*b*). There would certainly not be any problem in spreading such a 'model fungus' to compete with the aggressive strain, since, as with the d-factor, the beetle vector is an ideal delivery system to convey it to its target. The biggest problem, apart from achieving success in the breeding programme itself, would be whether mother nature would smile on the model fungus, or indeed whether the aggressive strain would smile at it!

Conclusion

It seems likely that the post-epidemic period may actually present a better opportunity for attacking the aggressive strain of Dutch elm disease on its own terms than during the epidemic. In the epidemic, both the beetle and fungus populations are extremely large; in the post-epidemic period they are much smaller because of the lack of large trees available for both beetle and fungus to utilise. Under these conditions pathogen and vector must operate under more severe constraints: the relative detrimental effects of other competing organisms, the d-factor and climate are likely to increase. In essence the fungus and its vector beetle are likely to come under severe selection pressure. The post-epidemic period may therefore, in one sense, offer a better opportunity for control of the disease. Moreover, even if most of the mature elms are lost we still need to plan strategically for future elm generations including the elm suckers already appearing in profusion, and which in some areas of southern Britain have already reached a height of 5 m or more. This then is perhaps a time to be at least sustaining rather than decreasing the research effort on Dutch elm disease.

References

BRASIER, C.M. (1977). Inheritance of pathogenicity and cultural characters in *C. ulmi*. Hybridisation of protoperithecial and non-aggressive strains. *Transactions of the British Mycological Society* **68**, 45–52.

BRASIER, C.M. (1979). Dual origin of recent Dutch elm disease outbreaks in Europe. *Nature* **218**, 78–79.

BRASIER, C.M. (1982). Occurrence of three sub-groups within *Ceratocystis ulmi*. In, *Proceedings of the Dutch elm disease symposium and workshop*, Winnipeg, Manitoba, October 5–9, 1981, eds. S. Kondo, Y. Hiratsuka and W.B.G. Denyer, 298–321. Manitoba Department of Natural Resources, Manitoba, Canada.

BRASIER, C.M. (1983*a*). The future of Dutch elm disease in Europe. In, *Research on Dutch elm disease in Europe*, ed. D.A. Burdekin. Forestry Commission Bulletin 60, 96–104. HMSO, London.

BRASIER, C.M. (1983*b*). The origin of Dutch elm disease. In, *Report on forest research 1983*, 32. HMSO, London.

BRASIER, C.M. (1983*c*). A cytoplasmically transmitted disease of *Ceratocystis ulmi*. *Nature* **305**, 220–223.

BRASIER, C.M. (1984). Inter-mycelial recognition systems in *Ceratocystis ulmi*: their physiological properties and ecological importance. In, *The ecology and physiology of the fungal mycelium*, ed. D. Jennings and A.D.M. Rayner, 451–497. Cambridge University Press.

BRASIER, C.M. (1986*a*). The d-factor in *Ceratocystis ulmi*: its biological characteristics and implications for Dutch elm disease. In, *Fungal virology*, ed. K.W. Buck, 177–220. C.R.C. Press, Florida.

BRASIER, C.M. (1986*b*). The population biology of Dutch elm disease: its principal features and some implications for other host-pathogen systems. *Advances in Plant Pathology* **5**, 53–118, eds. D.S. Ingram and P.H. Williams. Academic Press, London and New York.

BRASIER, C.M. (1987). Recent genetic changes in the *Ceratocystis ulmi* population: the threat to the future of the elm. In, *Populations of plant pathogens*, eds. M.S. Wolfe and C.E. Caten, 213–226. Blackwells, Oxford.

BRASIER, C.M. and GIBBS, J.N. (1973). Origin of the Dutch elm disease epidemic in Britain. *Nature* **242**, 607–609.

ELLISTON, J.E. (1984). Hypovirulence. *Advances in Plant Pathology* **1**, 1–34.

FAIRHURST, C.P. and ATKINS, P.M. (1987). Dutch elm disease: the vectors. In, *Advances in practical arboriculture*, ed. D. Patch. Forestry Commission Bulletin 65, 160–165. HMSO, London.

GIBBS, J.N. and BRASIER, C.M. (1973). Correlation between cultural characters and pathogenicity in *Ceratocystis ulmi* from Europe and North America. *Nature* **241**, 381–383.

GIBBS, J.N. and HOWELL, R. (1972). *Dutch elm disease survey 1971*. Forestry Commission Forest Record 82. HMSO, London.

GIBBS, J.N. and HOWELL, R. (1974). *Dutch elm disease survey 1972–1973*. Forestry Commission Forest Record 100. HMSO, London.

GREIG, B.J.W. and COXWELL, R.A.G. (1983). Experiments with Thiabendazole (TBZ) for control of Dutch elm disease. *Arboricultural Journal* **7**, 119–126.

GRENTE, J. and SAURET, S. (1969). L''hypovirulence exclusive', est-elle controlée par des determinants cytoplasmiques? *Comptes Rendus Hebdomadaire, Academie des Seances, Paris, Serie D* **268**, 3173–3176.

HEYBROEK, H.M. (1966). Dutch elm disease abroad. *American Forests* **72**, 26–29.

HEYBROEK, H.M. (1976). Chapters in the genetic improvement of elms. In, *Better trees for metropolitan landscapes*, Symposium Proceedings, USDA Forest Service, General Technical Report NE22, 201–213.

HEYBROEK, H.M. (1982). The Japanese elm species and their value for the Dutch elm breeding program. In, *Proceedings of the Dutch elm disease symposium and workshop*, Winnipeg, Manitoba; October 5–9, 1981, eds. E.S. Kondo, Y. Hiratsuka and W.B.G. Denyer, 78–90. Department of Natural Resources, Manitoba, Canada.

HEYBROEK, H.M. (1983). Resistant elms for Europe. In, *Research on Dutch elm disease in Europe*, ed. D.A. Burdekin. Forestry Commission Bulletin 60, 108–113. HMSO, London.

KONDO, E.S., HIRATSUKA Y. and DENYER, W.B.G. (Eds.) (1982). *Proceedings of the Dutch elm disease symposium and workshop*, Winnipeg, Manitoba, October 5–9, 1981. Department of Natural Resources, Manitoba, Canada.

LESTER, D.T. and SMALLEY, E.B. (1972). Response of *Ulmus pumila* and *U. pumila x U. rubra* hybrids to inoculation with *Ceratocystis ulmi*. *Phytopathology* **62**, 848–852.

MASLOV, A.D. (1970). *Insects harmful to elm species and their control*. Forest Industries Publications, Moscow.

RICARD, J.L. (1983). Field observations on the biocontrol of Dutch elm disease with *Trichoderma viride* pellets. *European Journal of Forest Pathology* **13**, 60–62.

ROGERS, H.J., BUCK, K.W. and BRASIER, C.M. (1986). The molecular nature of the d-factor in *Ceratocystis ulmi*. In *Fungal virology*, ed. K.W. Buck, 221–236. C.R.C. Press, Florida.

SCHEFFER, R.J. (1983). Biological control of Dutch elm disease by *Pseudomonas* species. *Annals of Applied Biology* **103**, 21–30.

SCHEFFER, R.J. (1984). *Dutch elm disease. Aspects of pathogenesis and control*. PhD thesis, Willie Commelin Scholten Phytopathological Laboratory, Baarn, The Netherlands.

SHI, J.L. and BRASIER, C.M. (1986). Experiments on the control of Dutch elm disease by injection of *Pseudomonas* species. *European Journal of Forest Pathology* **16**, 280–292.

SMALLEY, E.B. and LESTER, D.T. (1983). 'Regal' elm. *Hortscience* **18** (6), 960–961.

STENNES, M. (1981). The efficacy of thiabendazole hypophosphate in Dutch elm disease control. In, *Dutch elm disease symposium and workshop, Abstracts*, Winnipeg, Manitoba, October 5–9, 1981. Department of Natural Resources, Manitoba, Canada.

STROBLE, G.A. and LANIER, G.N. (1981). Dutch elm disease. *Scientific American* **245** (2), 40–50.

WEBBER, J.F. (1981). A natural biological control of Dutch elm disease. *Nature* **292**, 449–451.

WEBBER, J.F. and BRASIER, C.M. (1984). Transmission of Dutch elm disease: a study of the processes involved. In, *Invertebrate-microbial interactions*, eds. J. Anderson, A.D.M. Rayner and D. Walton, 271–306. Cambridge University Press.

WEBBER. J.F. and GIBBS, J.N. (1984). Colonisation of elm bark by *Phomopsis oblonga*. *Transactions of the British Mycological Society* **82**, 348–352.

WEBBER, J.F. and KIRBY, S.G. (1983). Host feeding preference of *Scolytus scloytus*. In, *Research on Dutch elm disease in Europe*, ed. D.A. Burdekin. Forestry Commission Bulletin 60, 47–49. HMSO, London.

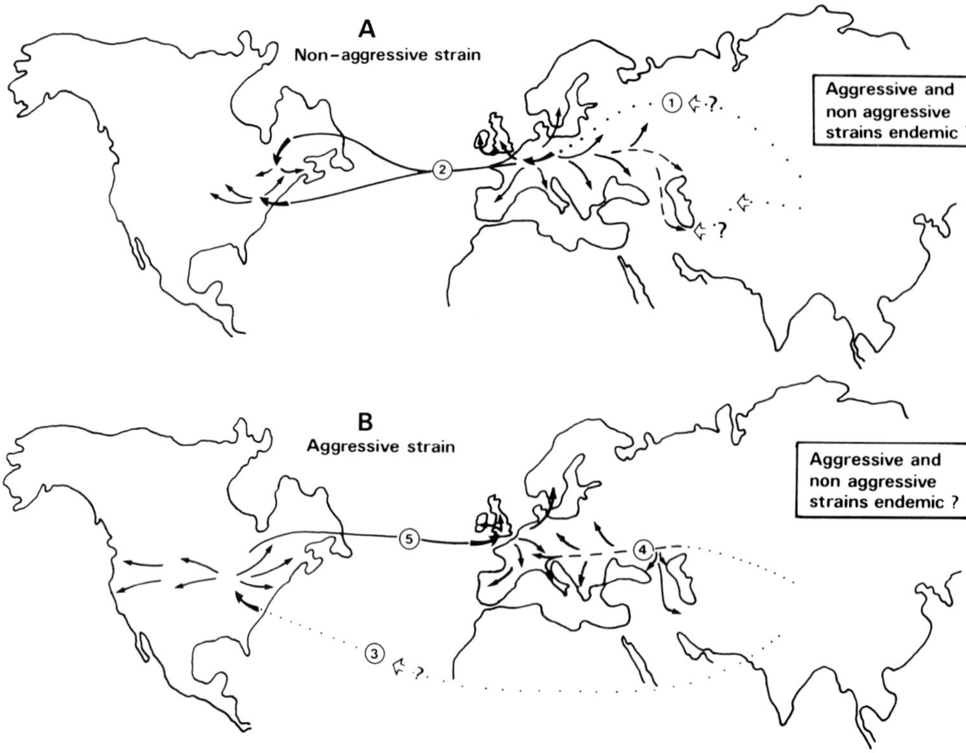

Discussion

A.D. BRADSHAW (Department of Botany, Liverpool University)

Will you please clarify the origins of the new aggressive strains of Dutch elm disease.

C.M. BRASIER This is an important but rather vexing question. Dutch elm disease as we know it was probably not present in Europe and North America prior to about 1900. From circumstantial evidence of higher levels of disease resistance among many Asiatic elm species, and on the evidence of rapid spread of the disease both eastward and westward from France in the 1920s, an origin in an area outside Europe and North America such as central or eastern Asia seems likely. A possible course of events consistent with recent historical evidence of the spread of the non-aggressive, EAN aggressive and NAN aggressive strains is shown in Figure 7.

Figure 7. Possible course of events in the spread of Dutch elm disease from a putative geographical origin in Asia (the Himalayas/China region).
a, *First epidemics; 1920s–1940s.* 1, Importation of the non-aggressive strain into north-west Europe from the 'origin' around 1900. 2, The subsequent importation of the non-aggressive strain into the eastern seaboard of the U.S.A. and into the Great Lakes area (*c.*1927); and its spread eastwards into central and southern Europe, the Black Sea and into south-east and central Asia from *c.*1940.
b, *Second epidemics; 1940s onwards.* 3, Introduction of a form of the aggressive strain (possibly close to the EAN form) into the mid-western United States (Illinois) in the 1940s–1950s; and its subsequent evolution into the NAN aggressive form coupled with its spread across the United States and Canada. 4, Migration of the EAN aggressive strain across Europe from the 1940s onwards following its introduction into the Black Sea area (Romania). 5, Importation of the NAN aggressive strain from the Great Lakes area into Britain in the 1950s–60s and its subsequent spread into neighbouring parts of Europe from the 1970s onwards.
C.M. Brasier, previously unpublished.

Honey Fungus

J. Rishbeth
Botany School, University of Cambridge

Summary

Three species of *Armillaria* attack trees in southern England. *A. mellea* kills a wide variety of broadleaved species and some conifers, *A. ostoyae* mainly kills resinous conifers such as pines, whilst *A. bulbosa* chiefly invades seriously weakened trees. Trees may be pre-disposed to attack by many agents, which include suppression, moisture stress and mechanical damage. Outbreaks generally originate from stumps, which become infected by rhizomorphs or less commonly by spores. At high-risk sites losses can be reduced by planting relatively resistant species, but not so closely that mutual suppression results. Stumps of unwanted trees should be removed or treated with ammonium sulphamate. Methods for controlling outbreaks are discussed.

Introduction

There have been so many misconceptions about Honey fungus that it is important to start by mentioning some of the main features of its biology and behaviour. This will involve consideration of the infection cycle, the species of *Armillaria* involved and their main characteristics, the types of tree attacked and the influence of external factors. It will then be easier to appreciate the possibilities and difficulties of control.

Infection Cycle

In outline the infection cycle is as follows. *Armillaria* produces fruit bodies in the form of toadstools mainly during the autumn. Vast numbers of spores are released and freshly cut stumps occasionally become infected. After many years the fungus may have occupied much of the stump and its root system, and is then in a position to colonize further woody tissues. These may be in other stumps, trees weakened in various ways, or apparently vigorous trees, depending partly upon the pathogenicity of the *Armillaria* involved. Such colonization may occur by means of rhizomorphs, up to several metres from the original stump if the system is extensive, or less commonly by direct transfer from root to root. The cycle is continued by exploitation of substrates and in many cases by the production of further fruit bodies and rhizomorphs. In old woodlands or other places where *Armillaria* is well established, colonization of stumps by rhizomorphs is much commoner than by spores.

Direct evidence that *Armillaria* can infect stumps by means of spores has been provided by inoculation experiments. Indirect evidence that this occurs naturally was obtained by surveying first-rotation stands of broadleaved trees where no infection sources existed at the time of planting: scattered foci were often present, usually associated with stumps created by thinning (Rishbeth, 1978). In four plantations of oak, (*Quercus robur* L.), the mean frequency of such foci was 4 per ha, and they had developed over a period of 18–30 years. Since types of *Armillaria* not producing large rhizomorph systems were undetectable by the method used, the frequency of foci is probably greater than that quoted. On the time-scale of plantation forestry, therefore, the origin of new *Armillaria* foci in broadleaved stands is not a particularly rare event. Similarly, stump infection by spores is probably common enough in parks and gardens to be borne in mind when control is being considered. Experiments have shown that *Armillaria* can colonize stumps of a wide variety of trees, including conifers, by this means, but very few naturally occurring foci have been detected in conifers so far.

A serious difficulty arises from the long period elapsing before it becomes obvious that a focus has been created.

For example, after the stump of a plum tree had been inoculated with spores of *Armillaria*, fruit bodies were only produced after 15 years and a small apple tree nearby only showed symptoms of infection one year after that. Thus by the time characteristic signs appear, the fungus will often have grown some distance below ground. Another unfortunate aspect is the potentially long survival of *Armillaria*: in one case it was present in an oak stump 50 years after the tree had been felled. Observations suggest that the risk of infection to surrounding trees must often exist for 10 years or more after the period of initial colonization.

Species of *Armillaria* Involved

Until comparatively recently all attacks on trees were attributed to *Armillaria mellea* (Vahl ex Fr.) Kummer, but it is now known that several species may be involved. Those most commonly encountered in southern England are *A. mellea,* in the strict sense, *A. ostoyae* (Romagn.) Herink, now usually known as *A. obscura* (Secrétan) Romagn. in much of Europe, and *A. bulbosa* (Barla) Kile and Watling (Rishbeth, 1982). The fruit bodies of these species have been briefly described by Rishbeth (1983).

The rhizomorphs of *A. mellea* may be abundant around the base of recently killed trees but seldom extend more than 10–20 cm into the soil. Fruit bodies commonly occur in dense clumps and grow out directly from stumps or shallow roots, whose position may be indicated by radial rows of toadstools. *A. ostoyae* varies considerably in its ability to produce rhizomorphs, but sometimes fairly extensive branching systems are formed. Fruit bodies are often aggregated but may also appear individually between stumps or trees. Rhizomorphs of *A. bulbosa* are very common in woodlands, may be up to 5 mm in diameter and often form extensive networks in the soil. They are often present on the surface of living roots and, misleadingly, on roots killed by other fungi, including *A. mellea* and *A. ostoyae*. Fruit bodies are sometimes found in clusters, as on stumps, but on the ground they are often widely separated. A fourth species, *A. tabescens* (Scop. ex Fr.) Eml., is occasionally seen in woodlands on heavy-textured soil. The types of *Amillaria* present in northern England and Scotland are being investigated by Dr. S.C. Gregory at the Forestry Commission's Northern Research Station; they include species less common or absent in the south.

Host Specialization

Some information about attacks by *Armillaria* comes from sampling trees which were apparently growing well prior to infection. *A. mellea* is found to kill a wide variety of broadleaved trees and also less resinous conifers such as *Sequoiadendron, Cupressus* and *Thuja*. This wide host range may be illustrated by analysis of casualties in 24 Cambridge gardens where trees of 20 genera had been killed. Deaths were commonest in genera in the following list, which also records the number of occurrences out of 68: *Prunus*, 12; *Betula, Salix* and *Malus*, 7; *Acer*, 5; *Aesculus* and *Fraxinus*, 4; *Pyrus* and *Cupressocyparis*, 3.

By contrast, *A. ostoyae* is more specialized, chiefly attacking very resinous conifers such as *Pinus, Picea* and *Larix*. In all outbreaks so far discovered, trees killed by this species were growing on acidic soil. This may account for the failure to record *A. ostoyae* in Cambridge gardens, which have *circum*-neutral or alkaline soil. *A. ostoyae* also causes butt-rot in Norway spruce (*Picea abies*) and possibly other conifers, whereas in broadleaved trees this is caused by *A. bulbosa*. Although not very pathogenic, *A. bulbosa* sometimes causes alarm in gardens because a well established but unseen rhizomorph system can invade a wide variety of organic material, such as sawdust, pulverized bark, composted leaves, or even wooden posts, and may give rise to large numbers of toadstools. *A. tabescens* has not been found to kill standing trees in England, where it is probably at the northern end of its range, but in warmer regions, such as southern Europe and the southern U.S.A., it causes many deaths. Since roots may be extensively decayed by this species, as well as by the other species of *Armillaria*, the possibility of wind-throw should be borne in mind.

Effects of Debilitating Factors

More can be learnt about the behaviour of *Armillaria* by sampling trees known to have been weakened in some way before they were infected. Species of *Armillaria* are often found that would not be expected in comparable unstressed trees. In particular, *A. bulbosa* frequently occurs in trees weakened by suppression, moisture stress, mechanical damage or infection by other fungi. The ability of this species to exploit such weakness is partly due to the opportunities created by its wide-ranging rhizomorphs. An important consequence of this debilitation, regardless of the species involved, is that more trees may die than would be the case with stress alone. This probably accounts for the greatly increased tree killing by *Armillaria* after the hot dry summers of 1975 and 1976. Examples of artificially induced moisture stress are also known. Large numbers of willows, (*Salix alba* L.), were killed by *A. mellea* at a site in Suffolk 2 years after the soil water level had been lowered by drainage. On a smaller scale, adverse factors may re-activate *Armillaria* in gardens. Thus in a situation where *A. mellea* had not killed a tree for over 20

Figure 1. Diagram illustrating spread of *Armillaria* in two gardens.
Above: *A. mellea* colonized the stump of a *Prunus* cut down in 1946. The type of tree attacked and the year in which it died are shown.
Below: spread of *A. bulbosa* rhizomorphs from the stump of an ash felled in 1955. ● stump from which the fungus originated; ○ position of trees killed; —— extent of rhizomorph spread.

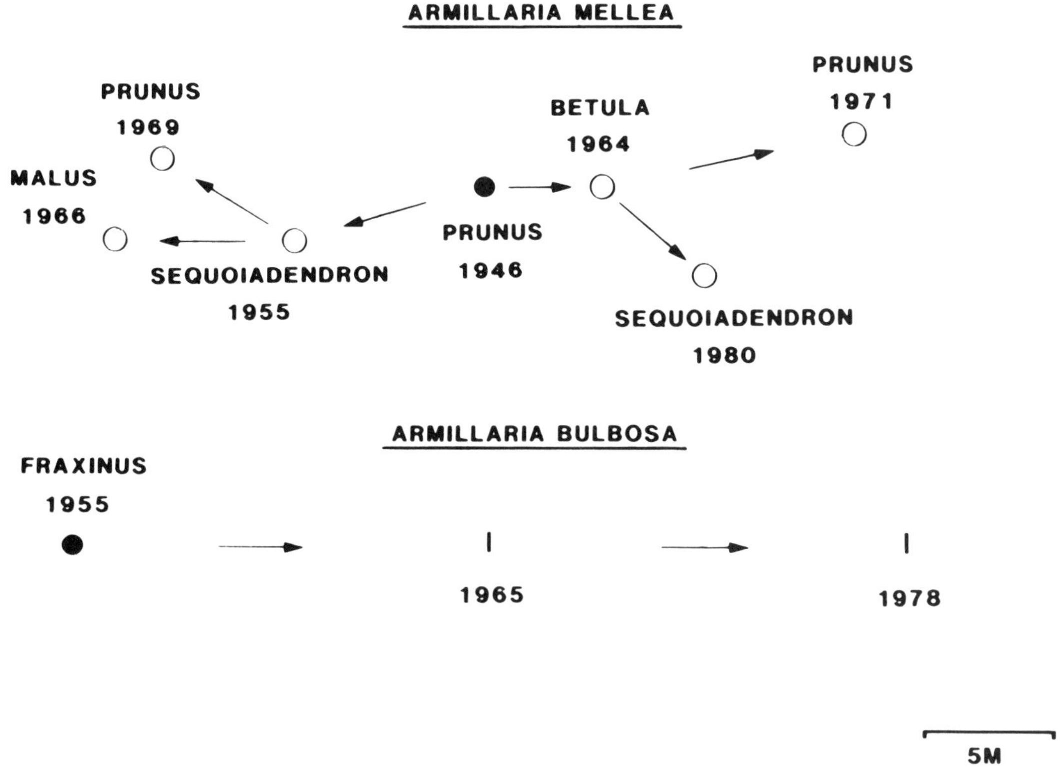

years, a Leyland cypress (x *Cupressocyparis leylandii*) was invaded by the fungus after it had become constricted at the base by a wire intended to support a neighbouring tree. It seems reasonable to conclude that weakening by a variety of agents plays an important role in the incidence of *Armillaria* attacks, and this conclusion is also supported by experimental evidence about the effects of shading and pruning, for example (Redfern, 1978).

Rate of Spread

It was mentioned above that *Armillaria* may take many years to colonize a stump and infect nearby roots; however, once it has become established it often spreads steadily. Around foci created by *A. bulbosa* in broadleaved stands, rhizomorphs grow at rates of 0.9–1.6 m per year; in such places only heavily suppressed trees are likely to be killed. In ancient woodland this species sometimes creates foci 200–300 m in diameter. The situation is rather different with *A. mellea*. In plantations of a susceptible tree such as birch, (*Betula pendula* Roth) the fungus may spread at a rate of about 0.6 m per year along infected roots and kill most of the trees. In gardens much depends upon the susceptibility of the species present and the distance between them, *A. mellea* having only a limited ability to spread through soil in the absence of roots. The maximum diameter of foci so far discovered is about 60 m. A typical example of the way *A. mellea* spreads in gardens and kills trees is shown in Figure 1, which also records the extent to which *A. bulbosa* rhizomorphs grew from the stump of a large ash, (*Fraxinus excelsior* L.) in a garden where no trees were killed. The mean rates of fungal advance were 0.7 and 1.4 m per year respectively.

Prevention and Control in Ornamental Plantings

Action to diminish the risk from *Armillaria* can be taken when first planting. The likelihood of early attack depends on the history of the site, and is virtually nil in the case of gardens established on arable land, unless a stump or former hedgerow was incorporated, or was nearby. Early attack is more probable if the site was formerly an orchard, and even more likely if it was broadleaved or mixed woodland, recently felled. It would be prudent in these situations to avoid planting trees particularly susceptible to *Armillaria* (Greig and Strouts, 1983). Unfortunately many of the debilitating conditions that favour attack by the fungus cannot readily be ameliorated, so at high-risk sites it might also be wise to avoid planting trees markedly intolerant of drought or unsuited to local soil conditions. Crowded planting is probably best avoided: on high-risk sites *Armillaria* might be activated early, following suppression of slower-growing trees, while on any site cutting out of unwanted trees may occasionally lead to stump infection by *Armillaria* spores. This may be illustrated by reference to Figure 1: the *Prunus*, from the stump of which *Armillaria* spread, was removed because it had been planted too close to the adjacent trees and was misshapen. At a later stage in parks, gardens or orchards, the potential weakening effect of very heavy pruning should be taken into account.

In situations where apparently healthy trees have to be cut down and it is hoped to reduce the chance of infection by *Armillaria*, whether from spores or infection sources below ground, several courses are open. It is always worth removing as much stump and root material as possible. In the case of larger trees, methods of dealing with stumps have been described by Wilson (1981). Where a chipper is used to remove most of the stump body, there is a risk that at sites where *Armillaria* is present the remaining large roots may be colonized. If removal is impracticable, prompt treatment of a hardwood stump surface with a 30 per cent aqueous solution of ammonium sulphamate should be considered. With certain types of stump, such as sycamore (*Acer pseudoplatanus*) two or more applications may be needed. This treatment generally kills the stump rapidly, preventing regrowth and promoting rapid colonization of harmless decay fungi, some of which compete well with *Armillaria* (Rishbeth, 1976). Inoculation with such fungi has been carried out experimentally, but no treatments of this sort are commercially available at present.

Once attack by *Armillaria* is diagnosed what is the best thing to do? Since the root system is usually much infected by the time any crown symptoms appear, it is probably seldom worth attempting to save the tree, despite occasional claims of successful treatment. It is important to remove or inactivate as much infected material below ground as possible. If the tree is small, the bulk of the root system should be removed. Larger trees should be felled, chiefly for reasons of safety. Removal of as much of the stump as possible by mechanical means can also be considered. It seems unlikely that introduction of any competitive fungus, such as *Trichoderma*, could lead to the replacement of *Armillaria* rapidly enough in stumps to prevent infection of adjacent roots.

If the infected stump has to be left in position, attention must be concentrated on limiting the spread of *Armillaria* from it. One potentially useful method is to create a barrier to prevent the fungus growing along living roots or through the soil by means of rhizomorphs. This must be sited beyond the limits of infection, as determined by digging. In a suitably located trench, at least 50 cm deep (more in porous soils), all roots should be severed. A vertical sheet of durable material, such as thick polythene or PVC (Greig and Strouts, 1983), should then be introduced and the soil replaced; open trenches are less satisfactory.

Although soil fumigation following stump removal is successful in some parts of the world, it is very doubtful whether it would be effective under most conditions in Britain. The fungicide Armillatox is recommended by the makers for use in this situation. In field trials, Pawsey and Rahman (1976) showed that although this preparation killed many rhizomorphs in light-textured soil, it was less effective in heavier soils and caused appreciable damage to roots of some tree species. The treatment will not kill *Armillaria* in roots and stumps and fresh rhizomorphs may eventually be produced.

Another possibility is to re-plant the area with species considered relatively resistant (Greig and Strouts, 1983). Probably no tree is immune, although yews, for instance, are rarely if ever killed. At sites, probably the great majority, where *A. mellea* rather than *A. ostoyae* is involved, more resinous conifers such as pine (*Pinus* spp.), spruce (*Picea* spp.), fir (*Abies* spp.) and Douglas fir (*Pseudotsuga menziesii*) might also be tried, although very young transplants are vulnerable and should not be used. Careful thought should be given to the position of planting. Since the aim is to avoid close proximity to infected roots, it is advisable to make wider and deeper planting holes than usual in order to extract as much infected material as possible. Places where fruit bodies of *A. mellea* have been seen should be avoided. Sometimes the problem is so intractable that the best solution is not to plant woody species in the vicinity for 10 years or more.

References

GREIG, B.J.W. and STROUTS, R.G. (1983). *Honey fungus*. Arboricultural Leaflet 2. HMSO, London.

PAWSEY, R.G. and RAHMAN, M.A. (1976). Field trials with Armillatox against *Armillariella mellea*. *PANS* **22**, 49–56.

REDFERN, D.B. (1978). Infection by *Armillaria mellea* and some factors affecting host resistance and the severity of disease. *Forestry* **51**, 121–135.

RISHBETH, J. (1976). Chemical treatment and inoculation of hardwood stumps for control of *Armillaria mellea*. *Annals of Applied Biology* **82**, 57–70.

RISHBETH, J. (1978). Infection foci of *Armillaria mellea* in first-rotation hardwoods. *Annals of Botany* **42**, 1131–1139.

RISHBETH, J. (1982). Species of *Armillaria* in southern England. *Plant Pathology* **31**, 9–17.

RISHBETH, J. (1983). The importance of honey fungus (*Armillaria*) in urban forestry. *Arboricultural Journal* **7**, 217–225.

WILSON, K.W. (1981). *Removal of tree stumps*. Arboricultural Leaflet 7. HMSO, London.

Discussion

C. CHAMBERS (London Borough of Hillingdon)

When removing stumps by chipper it is common practice to use some of the chipped material for backfill. Is this likely to create problems with Honey fungus?

J. RISHBETH If wood chips are mixed with the soil the fungus will quickly die.

C. CHAMBERS (London Borough of Hillingdon)

Will the remains of tree stakes left in the ground give rise to infection?

J. RISHBETH Stake points in the ground will be colonized by other organisms within a year or two and are unlikely to pose a threat to the adjacent tree.

R. FINCH (Roy Finch Tree Care Specialists)

Oak are occasionally found with tufts of rhizomorphs attacking bark just below ground level. Can this situation last for many years without damage to the tree?

J. RISHBETH Yes, although a major weakening of the tree could lead to the oak's resistance being weakening to the point where the fungus could gain a hold. Even so, ringbarked oak have been known to resist *A. bulbosa*.

R.P. DENTON (R.P. Denton & Co Limited)

Is a drainage ditch 50–60 cm deep a natural barrier to Honey fungus?

J. RISHBETH Heavy waterlogged clay at the bottom of the ditch might be an effective barrier, but rhizomorphs can penetrate in light soils or in dry summers.

R.P. DENTON (R.P. Denton & Co Limited)

When replanting diseased sites, will very healthy, well fertilised trees stand a better chance of resisting Honey fungus?

J. RISHBETH Correct choice of species is by far the most important factor affecting resistance.

Summary and Conclusions

R.C. Steele
Director General, Nature Conservancy Council

It is difficult to know where to start at the end of a group of papers such as this, but what better point than the last joint meeting of the Forestry Commission and the Arboricultural Association held 5 years ago in Preston on *Research for Practical Arboriculture*.

The primary objective of that meeting was to encourage communication with the industry by reporting research funded by the Department of the Environment and highlighting areas of related work (Brenan, 1981). The five sessions of the Preston meeting were very similar to those of this meeting. This helps us to assess advances but also emphasised the long-term nature of much arboricultural work.

I will run through the papers, not seeking to summarise so much detailed work, but rather to pick out points here and there and finish with some general points.

Plant Production

The first set of speakers was concerned with the production of plants both from seed and vegetatively. Vegetative propagation in its various forms is clearly of great significance in arboriculture because it is often more certain than the production of seedlings and it produces a more uniform and predictable plant which is important for ornamentals.

The references to mycorrhizal relationships of plants reminded me of my student days when the controversy over the merits of mycorrhiza and fertilisers in establishing trees on poor soils was still raging. We have come a good way since then but I wonder if our research effort in this field is commensurate with the very considerable benefits that a better understanding of the biology and ecology of mycorrhiza could bring.

Tree Establishment and Growth

The two papers on land reclamation hammered home an important message, namely the long-lasting and serious problems for tree establishment and growth resulting from compaction and the need to prepare the ground properly before planting. The right ground conditions, appropriate plants and good aftercare were recurring themes throughout the seminar and indeed are basic to arboriculture and forestry.

The differences between engineers and arboriculturists in the soil treatments they consider appropriate reminded me of similar problems the Nature Conservancy Council had with water engineers. These were eventually resolved to a large extent by the preparation of guidelines acceptable to both sides. The Arboricultural Association may wish to prepare similar guidelines in consultation with soil engineers.

One of several important points brought out in the paper on 'Tree improvement by selection' was the danger of characterising a species and even a variety on the much too narrow information base of limited seed samples. Another point was the possibility of developing predictive tests for tree performance at a very early stage to cut down the risk of expensive long term mistakes. The suggestion that disease-free clones of known performance of a range of species should be maintained as a central service is worth pursuing but who would bear the costs?

The development and use of tree shelters are a major aid to arboriculture and it was interesting to hear that they are also being tried on natural regeneration. I did a limited experiment on direct sowing of acorns in Monks Wood some years ago and the growth rates were spectacular. Direct sowing using Shelters could eliminate the expense and time of raising 1+1 oak seedlings only 20 cm high.

The papers on staking gave dramatic examples of a necessary technique carried to damaging extremes and also provide sound advice on how staking should be done. The harmful effects of weeds on tree growth particularly when establishing was frequently and rightly emphasised. However, the contradictory findings on the effects of nutrients on growth in the first year of establishment are a cause for concern. The findings may well reflect differences between sites and species but until they are reconciled and a more coherent view is obtained it will be difficult to provide consistent advice to the non-specialist.

Damage

The session dealt with damage caused to trees or damage caused by trees. Much has been learned about the treatment of wounds and cankers to minimise infection and the work on the relationship between trees and buildings is doing much to help tree planting and management near houses. As someone said – the association of trees with damage to houses gives arboriculture a bad name.

Epicormics have always been considered a bad thing in forestry but in many situations do they really matter? If you are growing veneer quality oak then yes they probably do matter; if they occur on trees in an amenity belt what is being lost? Should arboriculture always be governed by the requirements of timber production? In the course of our work it is salutory to ask ourselves from time to time the question 'need or habit?' Are we doing something because it is necessary or helpful or simply because we are in the habit of doing it?

Protection

Certainly some few species of insects can be a pest and we need to be on our guard but the great majority of insects are positively beneficial either as objects to admire, e.g. butterflies, or as important links in food webs. I would not like to think that the only tree for an arboriculturist is one free from all other forms of life.

Ash is visually dominant in some parts of Britain and its decline is a matter of great concern. We do need to understand better the associated factors so that we can seek to reverse the decline.

The authoritative paper on Honey fungus brought us up-to-date with the occurrence and treatment of what can be a very damaging pest.

Dutch Elm Disease

Knowledge of the beetle vector and the fungal pathogen have greatly increased although one vector, *Scolytus laevis*, has only just been confirmed as occurring in Britain and another, *S. pygmaeus*, which occurs just across the Channel, may also occur here. I wonder what influence we could have had on the spread of the disease if the significance of brood trees had been known in the early days?

This presentation emphasised the point that came through in other sessions namely that practical advice and information depends on a knowledge of the biology and ecology of the species concerned which in turn depends on thorough, broad-based and often long term, research.

Wildlife Research for Conservation

Few wildlife problems seem to be specific to arboriculture and many of the solutions used in forestry are applicable. The paper concentrated on problems of damage prevention but we must also emphasise the wildlife opportunities provided by trees. A vole may damage young trees but it is the food of owls, kestrels and foxes; grey squirrels may be a menace to tree growth but in urban situations they are a source of great pleasure to people; *Tortrix* caterpillars make holes in oak leaves (severe defoliation is not common) but they are a most important food for breeding birds. For whom and why do arboriculturists plant trees?

Conclusions

i. Progress

I believe the past 5 years have seen advances in arboricultural research and practice. Good practice depends on knowledge; knowledge depends on learning, experience and research. The Department of the Environment, in conjunction with the Forestry Commission, have performed a very useful public service in supporting research in arboriculture and all of us urge them to continue and indeed to extend this support.

ii. Research

Research in arboriculture is spread through Government Departments and Agencies, Universities and Polytechnics, Research Councils, Local Authorities and private organisations and individuals. This spread of work is a source of strength and initiative but it demands good communication. Communication to ensure that there is no wasteful duplication of effort or important gaps in research; and communication of results between research workers, with practitioners and with the general public. Meetings such as this help both aspects of communication. Department of the Environment provide a valuable service in helping both to plan research and to communicate its results but all of us can help with communication.

iii. Why plant trees?

What ends does arboriculture serve? I have the impression that the thinking on objectives has not developed as far or as quickly as arboricultural practice. If arboriculture is to be recognised as a profession in its own right then it has to have an identity related to, but distinct from, forestry and horticulture. There is more to arboriculture than growing trees faster and straighter and without a hole in a leaf or a snag on a stem. The hole in the leaf is the price of a butterfly; and snag is the price of a woodpecker. If I emphasise the nature conservation aspects of arboriculture it is because I believe this to be important and neglected with loss to both the general public and the arboriculturist. We have had, and continue to have, through the huge amount of planting in and around towns, a marvellous opportunity to bring a variety of pleasures and interests to many millions of people. I am not convinced that we are making the most of this opportunity. We shall only do so if we remember that a tree is also a habitat and patterns of planting and the species used should reflect the many properties of trees.

iv. The wider countryside

It is not only in and around towns that the arboriculturist will have the opportunity to develop and use his skills. Changes in agriculture both locally induced and through the mechanisms of the Common Agricultural Policy will, I believe, lead to a big increase in tree planting in the countryside. Whether this is called arboriculture or forestry there will be a big increase in requests for advice, help, plants and materials. The advice given and the work done over the coming years will determine the character and content of the countryside for decades to come.

Finally, may I thank you for inviting me to bring this most interesting and successful seminar to a conclusion and my good wishes for the future go to arboriculture and all who practise it.

Reference

BRENAN, J.P.M. (1981). Introduction. In, *Research for practical arboriculture*. Occasional Paper 10, 5–7. Forestry Commission, Edinburgh.

Advances in Practical Arboriculture

10–12 April 1985 University of York

List of Delegates

L. ADAMS	
A. AMOS	Nottinghamshire County Council
G. ANDREWS	Somerset County Council
P.M. ANNETT	The Royal Borough of Kensington and Chelsea
D.H. ARCHER	Arrow Tree Services
R.J.T. ASHWORTH	J.J. Harrison (Properties) Ltd
R. ASKQUITH-ELLIS	Bath City Council
D.G. AUSTIN	Merseyside Parks Training Centre
J.A. BAGNALL	S.W.O.A.(C) Ltd
M.J. BAGULEY	Forest Investment Services Ltd
C. BAKER	Harborough District Council
B.H. BALLARD	Property Services Agency
M. BARBER	
G. BARFIELD	Wimborne District Council
R. BEAL	Boothferry Borough Council
N. BEARDMORE	London Borough of Sutton
P. BENHAM	Derby City Council
D.N. BENSON	Economic Forestry Ltd
R.W.J. BERENDSEN	Firm Eijkelboom BV
D. BLACKBURN	Property Services Agency
S.M. BLUNT	Robinson Jones Partnership Ltd
T.C. BOOTH	Forestry Commission
R.F.L. BOTHAMLEY	Harrogate Borough Council
G.H. BRADLEY	Wakefield Metropolitan District Council
T. BRADNAM	Hampshire County Council
M. BRIGHTMAN	Lords Tree Specialists
J. BROCKIE	Hamilton District Council
G. BROOME	Kent County Council
M.F. BROWELL	A.E. Weddle
Mr BROWNING	Knowsley District Council
P. BROWNLEE	East Lothian District Council
M. BULFIN	Agricultural Research Institute Ireland
P. BULLIMORE	Cambridge County Council
A. BURR	London Borough of Southwark
A. BURTON	Devon Tree Services
A.J. BUTCHER	Forestry Services
D. BUTCHER	Cherwood Tree Nursery
J.W. CALLOW	Gillingham Borough Council
I.G. CAROLAN	National Coal Board

W. CATHCART	Dartford Borough Council
C. CHAMBERS	London Borough of Hillingdon
T. CHILDS	Thanet District Council
N.W. CLARK	
D.M. CLARKE	London Borough of Wandsworth
F.R. CLAXON	Epping Forest
J. CLAYTON	Ryedale District Council
J.D. COLLINS	West Derbyshire District Council
P. COLLINS	Bracknell District Council
S.A. COOMBES	Raven Tree Services
M. COOPER	Doncaster Metropolitan Borough Council
P. COOPER	Property Services Agency
C. COTTINGHAM	East Lothian District Council
T.J. CRAMB	Durham County Council
S. CRESSWELL	Essex County Council
M. CROOKES	Metropolitan Borough of Rochdale
J. CROUCH	Newark District Council
J. CROWTHER	Reading Borough Council
R.E. CROWTHER	Forest Research Station
W. CULBERT	Craigavon Borough Council
R. CURE	The Department of Transport
P. CURRAN	Dublin County Council
P.H. CUTHBERTSON	Bedfordshire County Council
H. DAVIES	Forestry Commission
C.L.A. DAVIES	London Borough of Hillingdon
H. DE HAAS	Advies Buro
B. DEAN	
R.P DENTON	R.P. Denton & Co. Ltd.
H. DICKINSON	Lincs County Council
J.A. DOLWIN	Dolwin & Gray
R. DRAKE	Westminster City Council
F. EDWARDS	Doncaster Metropolitan Borough Council
J. EDWARDS	Basildon District Council
K. EMERY	Stockton on Tees Borough Council
W. ENGELS	Department of Transport
C. EVANS	Men of the Trees
R.I. FAIRLEY	Countryside Commission for Scotland
G.W. FARNELL	Kirklees Metropolitan Council
R. FEARNEHOUGH	Property Services Agency
M. FEATHER	
R. FINCH	Roy Finch Tree Care Specialists
A. FINDLAY	Grampian Regional Council
D. FLEMING	Arboricultural and Forestry Contractor
J.G. FLEMING	West Yorkshire Metropolitan County Council
I. FLETCHER	The National Trust
P. FLORIS	Pius Floris Tree Care Ltd
K. FORBES	Motherwell District Council
D. FORD	Askham Bryan College
D. FORD	Central Electricity Generating Board
T.M. FORDER-STENT	Winchester City Council

C. FOSTER	Forestry Commission
M.E. FOSTER	Forestry Commission
D. FRANCIS	Springfield Park Group Training Centre
A.W. FROST	Trafford Borough Council
D. FROST	London Borough of Ealing
A. FURNESS-HUSON	Solihull Metropolitan Borough Council
M.D. GALLOWAY	
J.B.H. GARDINER	Forestry Commission
P.F. GARTHWAITE	
D. GASCOIGNE	Pendle Borough Council
G.M. GASKIN	Wyre Forest District Council
A. GEAR	Henry Doubleday Research Association
K. GIBB	Kirkcaldy District Council
S. GIBSON	Hillier Nurseries (Winchester) Ltd
I. GLADSTON	Tyne and Wear County Council
B. GOLDSTONE	Fountain Forestry
J.W. GOODY	The National Trust
A.G. GORDON	EFG Nurseries
J.G. GRACE	Royal Borough of Kingston
E.N. GREENSILL	Welsh Office Transport & Highways Group
D.A. GREGORY	Ryland Horticulture
P.A. GREGORY	Westonbirt Arboretum
D. HADLEY	Kilmarnock and Loudoun District Council
C.P. HAINES	Corruplast Limited
A.M. HALL	Mid-Suffolk District Council
T.H.R. HALL	Oxford University Parks
P. HAMBIDGE	North Bedfordshire Borough Council
E. HAMILTON	Dundee District Council
E.A. HAMILTON	The Woodland Trust
D.F. HAMMER	Borough of Poole
D.J.L. HARDING	Woodlands Research Group
T.W. HARFORD	Amey Roadstone Corporation Ltd
N. HARRIS	
M.J. HAXELTINE	Domestic Recreation Services
G. HAY	Economic Forestry Ltd.
R. HELLIWELL	
A. HEMINGWAY	Holderness Borough Council
M. HEMMING	Coventry City Council
P.A. HEMSLEY	Askham Bryan College
R. HEWLETT	Weymouth and Portland Borough Council
S. HEY	Humberside County Council
B.G. HIBBERD	Forestry Commission
J. HICKLING	Bishop Burton College of Agriculture
N.H.M. HOGG	Corruplast Limited
B. HOLDING	London Borough of Brent
J.Q. HOLMES	Urban Forest Group
D.R. HONOUR	
L. HOOGSTAD	Forestry and Arboricultural Practical College
E.M. HOPKINS	Amenity Tree Surgery
R.W. HOWARD	Royal Botanic Gardens, Kew
J.B. HOWELLS	Durham County Council

R.N. HUMPHRIES	Midland Research Project
W.B. HUNT	City of Newcastle Metropolitan Borough Council
S. HUNTER	Kilmarnock and Loudoun District Council
R.T. HURST	Forestry Commission
N. HVASS	Denmark
D. IVISON	Department of the Environment
D.A. JACKSON	Metropolitan Borough of Knowsley
H.A. JACKSON	Property Services Agency
G.J. JANSEN	Research Institute for Forestry and Landscape
D.C. JOHN	Hampshire County Council
M.A. JOHNSON	Writtle Agricultural College
P.D. KEEBLE	Department of Transport
D. KENNEDY	Aberdeen University
P. KERLEY	University Botanic Gardens, Cambridge
J.M. KERR	Durham County Council
L.A. KING	Tree and Landscape Consultant
G. KING	Tree Surgeon
F. KINMOUTH	Kinmouth Tree Surgery
J. KOPINGA	Research Institute for Forestry and Landscape
D.M. LAMB	A.E. Weddle
G. LAMB-SHINE	London Borough Greenwich
C. LAUGHRIDGE	Lords Tree Specialists
W. LAVERY	Glasgow Parks and Recreation Department
B. LEIGH	Property Services Agency
A.D. LONGLEY	Berkshire County Council
A.J. LOW	Forestry Commission
I. LUSCOMBE	Eastbourne Health Authority
S.E. MALONE	Forestry Commission
J.S. MANCHEE	Northumberland County Council
N. MANNING	Dales Forestry Co.
I. MANSFIELD	Wakefield Metropolitan District Council
G. MARCH	
P. MARSDEN	Countryside Commission
A. MARSH	
T.P. MARSH	Southampton City Council
B. MASKELL	Property Services Agency
S. MASLEN	Macclesfield Borough Council
C. MATTHEWS	Norwich City Council
R.C. MELVILLE	Forestry Commission
L. MITCHELL	Sheffield City Council
M.A. MITCHELL	Department of Transport
S. MITCHELL	
W. MOORHOUSE	Wakefield Metropolitan District Council
D. MORRIS	Hamilton District Council
S.J. MORRIS	Devon Tree Services
T.E.C. MOUSLEY	T. Mousley & Sons
J.M. MUNRO	Edinburgh City Council
G.S. MURGATROYD	West Dorset District Council

T.W. McDONALD	East Lindsey District Council
R. McDONNELL	Macclesfield Borough Council
P. McGOWAN	Peter McGowan Associates
D. McMILLAN	Property Services Agency
B. McNEILL	Estate and Forestry Services Ltd.
P.A. McNEILL	Belfast City Council
R. McNEILL	Estate and Forestry Services Ltd.
R. McPARLIN	Northumberland College of Agriculture
E. MACPHERSON	Property Services Agency
G.S. MACPHERSON	Tilhill
R. McSORLEY	Southern Tree Surgeons Ltd.
M.E. MOORE	Calderdale Metropolitan Borough Council
C.J. NEILAN	
R. NICHOLAS	Department of Transport
C. NICHOLAS	Property Services Agency
A. NICHOLSON	Property Services Agency
D. NOBLE	Park and Playing Fields Department
D.P. O'CALLAGHAN	Lancs College of Agriculture and Horticulture
A. OGDIN	City of London
A. OSBOURNE	Kent County Council
T. OTENIYA	Stirling District Council
N. PALFREY	London Borough of Hackney
A. PARKER	Metropolitan Borough of Rochdale
D. PARKIN	Treecare Arboricultural Services
A. PEET	Lancaster City Council
P. PLEASANCE	Congleton Borough Council
R. POLLARD	Nottinghamshire County Council
J. POTTER	Forestry Commission
M.J. POTTER	Forest Research Station
A.R.J. POWELL	Greater Manchester Council
J. PRESCOTT	Metropolitan Borough of Knowsley
M. PRIOR	Norfolk County Council
B. PULFORD	Forestry Commission
E. RADZIWILLOWICZ	Treecare Arboricultural Services
G. RAE	Forestry Commission
T.A. RAPER	Chorley Borough Council
D. REDFEARN	Sheffield City Council
T. RENDELL	Gwynedd County Council
A. RICHARDSON	Lichfield District Council
G. RIGGS	Hove Borough Council
A. RILEY	Planning Department, Bucks County Council
J. ROBINSON	
K. ROBINSON	University College Dublin
C.G. ROSE	Askham Bryan College
C.J. ROSE	Milton Keynes Development Corporation
L.D. ROUND	Trafford Borough Council
T.A. ROWELL	Midland Research Project
R.J. ROWLETT	London Borough of Havering
K. ROWTON	Tunbridge Wells Borough Council
D.J. RUSSELL	Atomic Energy Authority

A. SANGWINE	Departments of Environment & Transport
J. SAVAGE	Wakefield Metropolitan District Council
M.J. SAVAGE	Hampshire County Council
R. SCRUTON	
J.E. SEARLE	London Borough of Southwark
T. SEYMOUR	Cardiff City Council
J. SHACKLETON	Torridge District Council
R.M. SHANKS	East Kilbridge District Council
P.J. SHAW	Cleanaway Ltd
P. SHEARD	Manning Clamp & Partners
P.J. SHELDRAKE	Shropshire County Council
D. SHIPWRIGHT	London Borough of Enfield
R. SIBLEY	London Borough of Greenwich
A. SIMMONDS	Lichfield District Council
K.R. SIMONS	Warwickshire County Council
C. SKELTON	
R. SKERRATT	
A. SMITH	London Borough of Camden
C.A. SMITH	Rochester upon Medway City Council
M. SMITH	Clacton on Sea
G. SOUTER	Norwich City Council
M.J. SOUTH	M.A.F.F.
S.J. SPENCER	Wakefield Metropolitan District Council
M.C. SPICKETT	South East Thames Regional Health Authority
D. STAMP	Warrington Borough Council
D. STANGER	Oxford University Parks
B.J. STAPLETON	Cambridgeshire County Council
C. STARR	Hastings Borough Council
R.J. STENSON	Eastbourne Borough Council
B.A. STEPHENS	Rushmoor Borough Council
F.R.W. STEVENS	Forestry Commission
F.G. STEWART	Hyndburn Borough Council
R.G.L. STIRRAT	Planning Environment & Forestry Services Ltd.
J.A. STORER	Langbaurgh Borough Council
J. STONEHAM	Oxfordshire Social Services
N. SUTTERBY	Sasquatch Silviculture
C. SUTTON	Property Services Agency
K.B. SWINSCOE	Derbyshire County Council
A. SYMONDS	Wakefield Metropolitan District Council
J.R. TALBOT	National Coal Board
C. TAYLOR	Economic Forestry Ltd
R. TAYLOR	Bolton Metropolitan Borough Council
R.G. TAYLOR	Wealden Woodlands Ltd
D TERRY	David Terry Tree Surgeons
P.R. THODAY	University of Bath
C.G. THOMAS	East Malling Research Station
J. THOMAS	Janie Thomas Associates
P.R. THOMAS	P.R. Thomas (Tree Work Contractors)
D. THORMAN	
C. TOBIN	West Yorkshire Metropolitan County Council
J.F. TOOTILL	North West Tree Services
M.N. TOWLER	Norfolk FWAG

J. TUCK	Department of Transport
J. TUCKER	Stirling District Council
M. WALLER	
D. WALSHAW	Basingstoke and Deane Borough Council
A.E. WARD	Department of the Environment
N.J. WARD	Robinson Jones Partnership Ltd.
G.D. WATSON	West of Scotland Agricultural College
G.P. WATSON	Northamptonshire County Council
R. WEBB	An Foras Forbartha
J. WEBBER	Forestry Commission
M. WEBBER	Parks Department, City of Bristol
A.E. WEDDLE	
J. WETHERELL	North Bedforshire Borough Council
C. WHITE	Colin White Tree Surgery & Forestry Contractor
J. WHITE	Cheshire County Planning Department
P.M. WHITE	London Borough of Harrow
K. WIGGINTON	Reading Borough Council
T.A. WIGNALL	East Malling Research Station
G. WIJCHMAN	Tuin & Landschap, Holland
R. WILES	Luton Borough Council
R.V. WILLIS	Forestry Commission
R.C. WOOLNOUGH	London Borough of Bexley
M. WORTLEY	Essex County Council
T. WORTLEY	
C. YARROW	Chris Yarrow & Associates

Delegates booked after 29 March 1985 not included in the above list.

Speakers

DR. D. ATKINSON	East Malling Research Station
DR. P.G. BIDDLE	Tree Conservation Ltd
PROF. A.D. BRADSHAW	Dept of Botany, University of Liverpool
DR. C.M. BRASIER	Forestry Commission
DR. I.R. BROWN	University of Aberdeen
DR. D.R. CLIFFORD	Long Ashton Research Station, University of Bristol
R.J. DAVIES	Forestry Commission
DR. H.F. EVANS	Forestry Commission
DR. J. EVANS	Forestry Commission
DR. C.P. FAIRHURST	University of Salford
D. GILBERT	A.D.A.S., Cambridge
DR. J.E. GOOD	Institute of Terrestrial Ecology, Bangor
DR. P.G. GOSLING	Forestry Commission
DR. J. GRACE	University of Edinburgh
DR. B.H. HOWARD	East Malling Research Station
J. JOBLING	Forestry Commission
DR. D. LONSDALE	Forestry Commission
D. PATCH	Forestry Commission
DR. R.G. PAWSEY	University of Oxford
H.W. PEPPER	Forestry Commission
J.C. PETERS	Department of the Environment
DR. J. RISHBETH	University of Cambridge
DR. R.C. STEELE	Nature Conservancy Council
K. WILSON	Forestry Commission

Chairmen

C.G. BASHFORD	Department of the Environment
D.A. BURDEKIN	Forestry Commission
J. CHAPLIN	Arboricultural Association
A.J. GRAYSON	Forestry Commission
I.J. KEEN	Arboricultural Association
PROF. H.G. MILLER	University of Aberdeen
A.L. WINNING	Arboricultural Association

Organisers

MRS. J. BERRY	Arboricultural Association
DR. P.G. BIDDLE	Arboricultural Association
MRS. M.E. FOSTER	Forestry Commission
D. PATCH	Forestry Commission